Modelling Soil–Biosphere Interactions

Dedicated to my teachers

Declan Barraclough

and

Lester Simmonds

Then God looked over all
that he had made, and it was
excellent in every way.
(Genesis 1:31)

Modelling Soil–Biosphere Interactions

Christoph Müller
Department of Applied Microbiology
Justus-Liebig University, Giessen
Germany

CABI *Publishing*

CABI *Publishing* – a division of CAB *International*

CABI *Publishing*
CAB INTERNATIONAL
Wallingford
Oxon OX10 8DE
UK

Tel: +44 (0)1491 832111
Fax: +44 (0)1491 833508
Email: cabi@cabi.org

CABI *Publishing*
10 E 40th Street
Suite 3203
New York, NY 10016
USA

Tel: +1 212 481 7018
Fax: +1 212 686 7993
Email: cabi-nao@cabi.org

A catalogue record for this book is available from the British Library, London.

Library of Congress Cataloging-in-Publication Data

Müller, Christoph, 1963–
Modelling soil : biosphere interactions : version 1.2:6 May, 1999/ Christoph Müller.
p. cm.
Includes bibliographical references.
ISBN 0-85199-353-2 (alk. paper)
1. Soils--Environmental aspects--Computer simulation.
I. Title.
S596.M85 1999
631.4′0113--dc21 99-35159
CIP

ISBN 0 85199 353 2

Typeset in 10/12pt Times by Columns Design Ltd, Reading
Printed and bound in the UK at the University Press, Cambridge

Contents

Preface **ix**

How to Use the Book **xi**

1 Introduction **1**
- 1.1 Processes in the Biosphere 1
- 1.2 Some Mathematical Essentials 2
  - 1.2.1 Mathematics – the universal language 2
  - 1.2.2 Translation into the language of mathematics 3
  - 1.2.3 Some mathematical tools 6
  - 1.2.4 Numerical and analytical solutions 21
- 1.3 The Transport Equation and the Finite Difference Notation 23
- 1.4 Mathematical Description of Kinetics 27
  - 1.4.1 Zero-order kinetics 27
  - 1.4.2 First-order kinetics 28
  - 1.4.3 Second-order kinetics 30
  - 1.4.4 Temperature dependency of the rate constant 31
  - 1.4.5 Chain reactions 32
  - 1.4.6 Michaelis–Menten kinetics 33
  - 1.4.7 Units 36
- 1.5 Introduction to ModelMaker 37
- 1.6 The First Model 39

2 Nitrogen Transformations in Soil **51**
- 2.1 Introduction 51
- 2.2 Modelling Kinetics 53

		2.2.1 Zero-, first- and second-order kinetics	53
		2.2.2 Temperature dependency of the rate constant	55
		2.2.3 Chain reactions	56
		2.2.4 Michaelis–Menten kinetics	57
	2.3	Nitrification	58
		2.3.1 Conceptual nitrification model	58
		2.3.2 Basic nitrification model	60
		2.3.3 Nitrification and gaseous N production	69
		2.3.4 Influence of soil moisture and soil temperature on nitrification	79
		2.3.5 Further additions to the nitrification model	85
	2.4	Denitrification	86
		2.4.1 Conceptual denitrification model	86
		2.4.2 Basic denitrification model	87
		2.4.3 Denitrification enzyme dynamics	93
		2.4.4 Denitrification and gaseous N emission	96
		2.4.5 Influence of soil moisture and soil temperature	100
	2.5	C/N Transformations in Soil Organic Matter	104
		2.5.1 Conceptual model	104
		2.5.2 C transformation concept	107
		2.5.3 N transformation concept	110
		2.5.4 Basic C/N transformation model	116
		2.5.5 Substrate-specific microbial growth and decay rates	129
		2.5.6 Protection capacity of microbial biomass	134
		2.5.7 Physiological state of microorganisms	136
		2.5.8 Influence of soil moisture and soil temperature	151
		2.5.9 Future development	155
3	**Soil Temperature**		**156**
	3.1	Introduction	156
	3.2	Soil Temperature Dynamics in Space and Time	156
	3.3	Volumetric Heat Capacity and Thermal Conductivity	159
	3.4	Heat Flow Model with Constant Thermal Properties	165
	3.5	Heat Flow Model with Dynamically Changing Thermal Properties	172
4	**Soil Water**		**179**
	4.1	Introduction	179
	4.2	Potential Concept	180
	4.3	Soil Moisture Characteristic	183
	4.4	Hydraulic Conductivity	184
	4.5	Basic Water Flow Model	188
	4.6	Effect of Other Boundary Conditions	194
	4.7	Infiltrability	198
	4.8	Future Development	202

5 Soil Energy Balance **203**
5.1 Introduction 203
5.2 Combined Soil Temperature–Moisture Model 204
5.3 Radiation Balance of Bare Soil 205
5.4 Calculation of the Various Terms of the Radiation Balance 211
5.5 Exchange Between Soil and Atmosphere 213
5.6 Calculation of the Various Terms of the Energy Balance 219
5.7 Calculation Procedure for Soil Surface Temperature and Evaporation 223
5.8 Contribution of Water Vapour Movement to Soil Thermal Conductivity 229
5.9 Contribution of Water Vapour Movement to Soil Water Flow 238
5.10 Non-isothermal Water and Water Vapour Flow in Soil 241
5.11 Future Development 247

6 Plant Growth **248**
6.1 Introduction 248
6.2 Conceptual Plant Growth Model 249
6.3 Photosynthesis 250
6.3.1 Simple model 251
6.3.2 Advanced model 257
6.4 Plant Growth–Substrate Relationships 266
6.4.1 Introduction 266
6.4.2 Growth above ground 266
6.4.3 Root growth and partitioning 275
6.5 Plant Growth – Relationships with Environmental Factors 292
6.5.1 Introduction 292
6.5.2 Temperature 292
6.5.3 Plant water status 294
6.6 Future Development 311

7 Leaching **315**
7.1 Introduction 315
7.2 Mathematical Notation of Transport Processes 316
7.2.1 Convection 316
7.2.2 Diffusion 317
7.2.3 Hydrodynamic dispersion 318
7.2.4 Combined notation for solute transport 318
7.3 Leaching Model with Constant Soil Parameters 320
7.4 Leaching Model with Dynamic Soil Parameters 324
7.5 Future Development 331

8 Final Comments **332**

Appendix 334
References 338
Index 343

Preface

Countless times I have read publications dealing with mathematical descriptions of processes in the soil or biosphere and not grasped the main idea. It is most annoying when authors assume that all the mathematical steps they have just developed are quite obvious and easy to understand. Very often these authors are juggling with differential or even partial differential equations and implying that all concepts are basic. In most cases, I have just given up and put those publications aside with feelings of incompetence and inadequacy. However, I never lost the feeling that the concepts and ideas the authors were trying to get across would help me to understand better the processes in question.

If you have also had similar experiences, you may be interested in this publication. The main aim is to revisit some of the concepts and processes in soils and their interaction with the biosphere and explain them in an easy-to-understand way. I hope this publication is helpful to students in natural science who want a basic understanding of various processes in the soil and the biosphere and also to scientists who desire to understand and apply mathematical descriptions to specific problems in their own field of research.

The reader is expected to have only basic mathematical skills to follow this text. To solve the mathematical models, an advanced modelling package is used which requires no formal programming skills. It is my hope that the examples given and the strategies presented may enable users to increase their understanding of the soil system and its interactions with the biosphere, and to translate original ideas into working mathematical expressions.

At this stage I would like to extend a big 'thank you' to all the people who have made this publication possible. Professor Ottow from the Department of Applied Microbiology (Giessen, Germany) has been very supportive of the idea of describing various microbial processes mathematically and therefore enabled this entire modelling project. Christine Thompson, Claudia Kammann, Jim Stevens, Rob Sherlock and Tim Clough have proofread the manuscript. They provided me with invaluable ideas on how to improve the text. Many thanks to Simone, my wife, who has supported and encouraged me throughout this time.

I would be very interested to receive any comments from the users of this book because this will finally determine how helpful this publication is and if there are aspects of the text which need further attention.

Christoph Müller
Giessen, May 1999

How to Use the Book

The book is aimed at readers with a whole range of modelling backgrounds. The primary function, of course, is to introduce basic modelling skills and to show how processes in the soil–biosphere can be described with the help of mathematics. The book starts at a very basic level (Chapter 1) by revisiting the mathematics needed for successful model development. The following chapters, which are each concerned with specific processes in the soil–biosphere, always begin with an introduction to the particular topic and then proceed to model development. The first stage of model development is aimed at the beginner's level, where basic concepts are introduced. Readers who are only interested in the mechanism of a certain process might stop at this point. Others who wish to go more into the details of a particular process may continue with the subsequent sections of the chapter. The culmination of each section is always a working model. There is therefore no need to work through the entire chapter.

Some of the chapters include many sections (e.g. Chapters 2 and 6). In writing this book, it was decided to include these additions and extensions so that readers who are interested in advanced applications in the area of interest can see how these concepts may be introduced into an already existing model. This approach is especially useful for scientists who are working in a particular area and want to create models which include advanced process-based considerations. From my own experience, once I found that I understood the basic principles involved in a process, I soon wanted to move rapidly to more sophisticated considerations. The book can therefore be used to gain a basic understanding of the concepts of the various processes described and the modelling of these processes and as a reference for the development of models for more advanced applications.

The book is designed to be used together with the supplied models (set up in the modelling package 'ModelMaker', Cherwell Scientific*). It is therefore advisable to use both the book and the models in parallel. How the various notations are implemented becomes obvious once the actual models are investigated. Every model is presented in the section 'Application in ModelMaker' with a diagram as it appears in ModelMaker which is hopefully an additional help for your own model developments.

* Cherwell Scientific Publishing, The Magdalene Centre, Oxford Science Park, Oxford OX4 4GA, UK.
Tel: +44 (0) 1865 784800, Fax: +44 (0) 1865 784801.
email: modelmaker@cherwell.com.
Internet: http://www.cherwell com.
A free run-time version can be downloaded from this website and on request a demo CD is available which contains the run-time version and all models described in this book. To request the CD contact sales@cherwell.com or use the order form at the back of this book.

Introduction 1

1.1 Processes in the Biosphere

Life on earth is dependent on processes in the biosphere. These include processes in the soil underneath our feet as well as the processes near the ground, such as photosynthesis in plants. Soil is an essential substrate sustaining life on earth. The processes in the biosphere can be divided roughly according to their nature into chemical, physical and biological. Examples of processes in soil include various chemical transformations, the physical processes of water and heat transport, and microbial transformations of organic nitrogen into plant-available N components. Above ground, photosynthesis, the process by which light energy is captured by plants to produce oxygen, is the prerequisite for most of the metabolic processes on earth.

However, it is not the individual process but the seemingly perfect interaction among the numerous processes which characterizes the biosphere. For example, microbial transformations in soil are greatly influenced by soil water and soil temperature. Research over the last century has provided insight into the complicated network of interactions in the biosphere; however, only a tiny fraction of the whole picture is well understood.

Observations in the last two decades have shown that interference by humankind may be having a devastating influence on the interaction of the biosphere processes. For example, toxic waste introduced into the soil or ocean may spread through the food-chain into the whole system. Similarly, the occurrence of global warming and ozone depletion through the accumulation of certain anthropogenic gases may affect the rates of processes in the biosphere. Even the life of human beings is knitted tightly into this system and may therefore be irreversibly influenced by our own mismanagement.

It is therefore important that we gain a better understanding of the processes in the soil and their interaction with other processes. The aim of this book is to provide a better understanding of some selected processes in which soil and atmosphere interact. If we regard the whole biosphere system as a huge multidimensional jigsaw puzzle, this book aims to put some pieces into place. It should be realized, however, that most of the pieces necessary to complete the whole picture are still missing.

1.2 Some Mathematical Essentials

1.2.1 Mathematics – the universal language

In order to develop a logical understanding of processes in the soil–atmosphere continuum, it is necessary to translate our knowledge into a universally understood language, capable of solving the problems in question. This 'universal language' is the language of mathematics. Therefore, we have to understand enough mathematical 'vocabulary' to be able to solve the problems in question. The complexity of the mathematical background needed is, of course, dependent on the complexity of the problems to be solved. In everyday life we are constantly translating into mathematical language, often without noticing. For instance, if we go into a supermarket in order to buy 1 kg of butter but the shop has only portions of 250 g or 500 g, we will have to buy several packages. If the available butter packages differ in price and our aim is to get the cheapest deal, we are faced with a problem which can only be solved via a translation into the language of mathematics. The translation and the solution often happen without us noticing that we used mathematical reasoning.

With more complex problems, we still perform the translation into the mathematical language in our mind but may find the required result only with the help of a hand-held calculator. If problems become very complex, the solution may not be possible by just using a hand-held calculator and more sophisticated ways of finding the required solutions are needed, such as specialized software on a personal computer. However, even if we use the most advanced computer technology available today, the solution of the problem is only the second step. As already seen in the butter-buying problem, the first step we perform is to realize that we are dealing with a problem which can be solved via translation into the language of mathematics. This first essential step is to actually find an appropriate translation. It cannot be stressed enough that this translation actually takes place in our mind and cannot be provided by computers or other means.

This text, therefore, is primarily concerned with the development of the first step, that is, to actually train our analytical thinking in the context of biosphere problems and initiate the important task of translation into the language of mathematics. The solution, utilizing sometimes complex processes, will be calculated with the help of advanced computer software. The software we use

(ModelMaker, Cherwell Scientific) belongs to a group of modelling packages which can be learned within a few days and do not require formal programming skills.

1.2.2 Translation into the language of mathematics

The 'mathematical vocabulary' needed to translate certain problems is dependent on the complexity of the problems. In this section, some basics are presented which can be seen as prerequisites to a successful understanding of the language of mathematics.

The conceptual model

The basic notation in mathematics is the equation form. In this section, it will be shown that this format is used very often to solve problems we are facing in everyday life. Suppose we want to buy 1 kg of butter at the supermarket. However the market only has packages of 250 g available each at a price of 1 Euro. How much do I have to pay for 1 kg of butter? For such a simple problem the translation into a mathematical expression and the solution of the problem are usually done very quickly in our mind, such as: 'In order to buy 1 kg of butter, I would have to buy four packages which means that I have to pay four times the price, i.e. 1 Euro multiplied by 4 equals 4 Euro.' If we translate this problem into a more general mathematical form by exchanging symbols for actual numbers, we can write the following equation:

$$P_{total} = \frac{P_{per\_unit}}{W_{per\_unit}} \cdot W_{required}$$ [1]

where:

P_{per_unit} = price per unit available (in our case 1 Euro)
W_{per_unit} = the weight per unit available (in our case 250 g)
$W_{required}$ = the total weight which we would like to buy (in our case 1000 g)
P_{total} = the total price which we would have to pay (our result = 4 Euro)

With this equation we have translated our problem (find the total price of butter) into the language of mathematics. Using this equation and exchanging actual numbers for the symbols, we are able to apply this generalized equation to a whole range of problems. However, it is still necessary to check that the available 'translation' is suitable for the problem in question. For instance, this equation may apply to problems similar to the one above. However, the equation is not adequate if we want to decide between two packages of butter in order to find the cheapest deal.

[1] The dot in the equation refers to the multiplication sign; this notation will be used throughout the text.

An example of that kind would be: we want to buy 1840 g butter and the supermarket has two butter packages available: one with 230 g butter at 0.99 Euro each and a 368 g package at 1.79 Euro each. Which butter package should we buy to get the least cost per unit?

Sometimes suitable mathematical translations of the problem in question are not as straightforward as in the above example. Hence, one of the main purposes in writing this text is to suggest methods for finding possible solutions in biosphere research. If we are faced with a problem which we do not know how to translate into a mathematical expression, it is always a good idea to begin simply. Usually there are some items we know that we can start with. It is also very useful to present ideas as a graphical solution. This leads us to the idea of a so-called 'conceptual model', which is always a precursor to the actual translated version. In a conceptual model, we can present our ideas without the need for finding suitable translations into a mathematical form. Even if we are not able to proceed to the second step, we should still make an effort to conceptualize our ideas in a way which is as realistic as possible. A good example of a conceptual model in biosphere research is the nitrogen cycle, which is presented in numerous textbooks. Usually a diagram is shown which brings together all (or the important) processes. In general, a conceptual model is a visualization of the ideas and processes which have been manifested in the mind of the author. A good conceptual model is a model which presents the ideas in an easy-to-understand way.

Example of a conceptual model

To demonstrate the value of a conceptual model, we look at an important group of processes in the soil which are concerned with transport. Processes in the soil very often deal with concentrations or quantities which are not steady but are changing in space and time. For instance, soil temperature is not constant throughout the soil profile but may change in different parts of the soil profile in response to different processes. The soil surface temperature may show a very dynamic response to absorbed radiation from the sun, which changes rapidly over the course of a day, whereas the soil temperature deeper in the soil profile may stay almost constant over a whole year. Soil temperature differences cause flows of energy from high to low temperature (i.e. high to low energy) and are referred to as heat flows. To illustrate heat flow, let us imagine a room in a house set at exactly the same temperature throughout the year. Outside the house, the temperature changes according to the season, with lower than room temperatures during the winter and higher than room temperatures during the summer (Fig. 1.1).

If we open a window either in summer or in winter, the room temperature will change in response to the outside temperature (assuming that the effect of the inside temperature on the outside temperature would be negligible; see also p. 6 on 'The importance of making assumptions'), i.e. opening the window in winter would cool the room temperature down (heat flow from the room to the outside) and opening the window in summer would increase the room temperature (flow of heat from outside into the room).

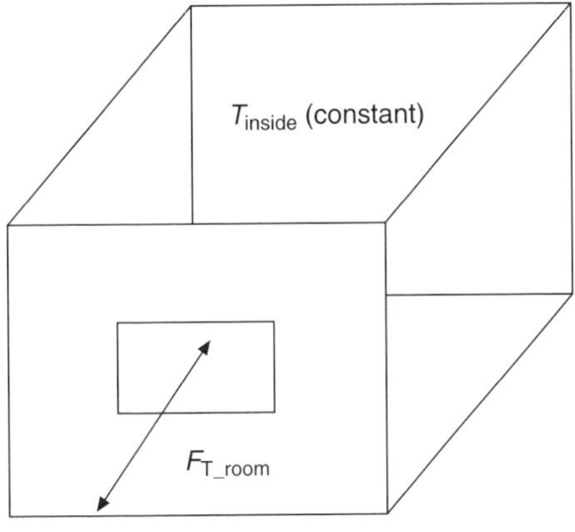

Fig. 1.1. Flow of heat (F_{T_room}) through a window in response to a temperature gradient.

The rate of change, i.e. how fast the room temperature changes, is related to the temperature difference between inside and outside the room (Fig. 1.1). If we do not have any further information, we can only say that the change in room temperature is proportional to the temperature difference or, expressed in a mathematical form:

$$F_{T\_Room} \propto (T_{inside} - T_{outside})$$

The sign $\propto$ means 'proportional to' and F_{T_Room} is the flow of heat between outside and inside. If the temperature difference is large, the room temperature will change quickly; if it is low, the temperature change will be slow. In order to transform this proportional expression into a mathematical equation, we have to introduce a proportionality factor, C, so that we can write an equation (see also Section 1.2.3):

$$F_{T\_Room} = C \cdot (T_{inside} - T_{outside})$$

Such a factor (as we shall see later in Chapters 3 and 4) has a physical meaning and can usually be determined quite readily for the process in question. Perhaps you can guess at this stage that C might be related to the physical size of the window: the larger the window, the greater the heat flow.

In many areas of biosphere research, we cannot write explicit equations for the process in question; however, we may still have an idea about how the process behaves. In biosphere studies, we are often faced with a situation where flows (e.g.

flow of heat) are proportional to a difference or gradient of a certain measurable quantity (e.g. difference in temperature over a certain distance).

The importance of making assumptions

In the previous example, we made an assumption that the effect of room temperature on the outside temperature would be negligible (i.e. the room is not able to heat up the outside temperature during the winter). It can easily be verified that this assumption is correct (we know this already by common sense). However, in other problems we cannot be so sure. As we shall see later, many assumptions are usually needed to tackle the problems in biosphere research. In many cases, we do not even recognize some assumptions as such because they are so obvious (such as the one above). However, we should always be very careful in examining our assumptions and state them clearly. Wrong assumptions can invalidate the whole modelling approach.

1.2.3 Some mathematical tools

In this section, we shall repeat and summarize some important mathematical rules and expressions which will be needed throughout the text. Rather than presenting a comprehensive collection of mathematical vocabulary (which is the task of mathematical textbooks), the focus is on presenting the most essential mathematical tools needed in natural science. Examples in this section are taken from Goldstein *et al.* (1990), a book which provides an easy-to-follow approach to calculus.

Independent and dependent variables

Usually we have a set of 'knowns' (values we actually know) and a set of 'unknowns' (values which we do not know). In the literature, we are often faced with the terms inputs and outputs. These terms originate from jargon used in the computer world. However, the term inputs refers to values we actually know and have available to carry out our calculation and the term outputs is an expression referring to a set of numbers which we do not know but want to calculate.

In the butter-buying problem, the 'knowns' or inputs we have available are:

- price per unit;
- weight per unit;
- weight required;

and the 'unknown' or output we want to calculate is:

- price to pay.

UNITS

As we have seen in the above example, the actual numbers have certain units. It is important to always check the units and their proper conversion because many difficulties (especially in more complex situations) can arise because the units are not chosen properly or conversion from one unit to another was not done correctly.

In our example above the units are:

$P_{\text{per\_unit}}$ in Euro
$W_{\text{per\_unit}}$ in g
W_{required} in g

Therefore, rewriting the equation for our butter-buying problem using these units gives:

$$[\text{Euro}] = \frac{[\text{Euro}]}{[\text{g}]} \cdot [\text{g}]$$

Later in the text, when biosphere processes and their solutions are presented, they will always be referred to by the appropriate units (generally the SI units).

The more formal mathematical terms for inputs and outputs are the terms independent and dependent variables, respectively. For example, the input or independent variable may be the price per unit of butter (e.g. Euro per unit) and the output or dependent variable may be the total price required to buy four units. Independent and dependent variables may be represented graphically by the x-values and the y-values (or f(x), 'function of x'), respectively. Examples of some mathematical functions are presented graphically in the next section. Pairs of x–y values are often expressed with the notation: (x, y) (e.g. Fig. 1.2).

Mathematical laws and basic transformations

A good background knowledge of basic mathematical laws is needed to transform equations and perform certain mathematical analyses. Some of the more important ones are summarized below.

LAWS OF EXPONENT

$$b^r \cdot b^s = b^{r+s} \qquad b^{-r} = \frac{1}{b^r}$$

$$\frac{b^r}{b^s} = b^r \cdot b^{-s} = b^{r-s} \qquad \left(b^r\right)^s = b^{r \cdot s}$$

$$(a \cdot b)^r = a^r \cdot b^r \qquad \left(\frac{a}{b}\right)^r = \frac{a^r}{b^r}$$

ROOTS

$$a^{\frac{1}{n}} \cdot b^{\frac{1}{n}} = (a \cdot b)^{\frac{1}{n}} \qquad \Rightarrow \qquad \sqrt[n]{a} \cdot \sqrt[n]{b} = \sqrt[n]{a \cdot b}$$

$$\frac{a^{\frac{1}{n}}}{b^{\frac{1}{n}}} \cdot = \left(\frac{a}{b}\right)^{\frac{1}{n}} \qquad \Rightarrow \qquad \frac{\sqrt[n]{a}}{\sqrt[n]{b}} = \sqrt[n]{\frac{a}{b}}$$

$$\left(a^{\frac{1}{n}}\right)^{\frac{1}{q}} = a^{\frac{1}{n \cdot q}} \qquad \Rightarrow \qquad \sqrt[q]{\sqrt[n]{a}} = \sqrt[n \cdot q]{a}$$

LAWS OF LOGARITHM

$$\ln(1) = 0 \qquad \ln(e) = 1$$

$$\ln(x \cdot y) = \ln(x) + \ln(y) \qquad \ln(x^a) = a \cdot \ln(x)$$

$$\ln\left(\frac{x}{y}\right) = \ln(x) - \ln(y)$$

$$e = 2.718281... \qquad (1.2/1)$$

The exponential and natural logarithm functions are very important in natural studies where growth rates or biological substrate transformations are simulated (kinetics, Section 1.4). The natural exponential function and the natural logarithm are presented in Fig. 1.2 (note: $\ln(x)$ is not defined for x-values equal to or below zero).

DETERMINING ZEROS OF FUNCTIONS

It is often necessary to determine the x-values so that the y-values are zero. This is necessary in applications where we want to determine which influencing factors yield a zero response (e.g. in plant–water relations where we want to determine the shoot water content at zero plant water potential; see Section 6.5.3). For linear functions such as $f(x) = 3 + 4 \cdot x$, it is very easy to find the zeros. In this case, the zero of the function can be calculated by setting $3 + 4 \cdot x = 0$ and solving for x (result: $x = -\frac{3}{4}$). However, very often we are dealing with functions of higher order, such as quadratic functions in the form:

$$f(x) = a \cdot x^2 + b \cdot x + c \qquad (1.2/2)$$

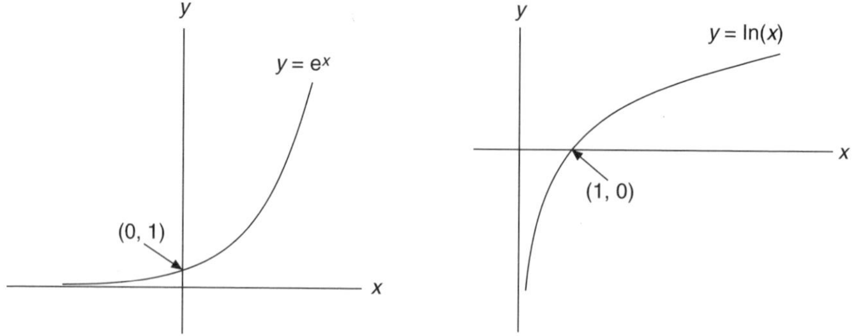

Fig. 1.2. Graph of the function $y = e^x$ and $y = \ln(x)$.

The zeros of a quadratic function: $a \cdot x^2 + b \cdot x + c = 0$ are given by the quadratic formula:

$$x = \frac{-b \pm \sqrt{b^2 - 4 \cdot a \cdot c}}{2 \cdot a} \tag{1.2/3}$$

Example. Find zeros of the equation: $y = 3 \cdot x^2 - 6 \cdot x + 2$ (Fig. 1.3). Here:

$a = 3;\ b = -6;\ c = 2$

Below, a detailed solution is provided for the above example, which includes some transformations according to eqn 1.2/1 and therefore some practice in basic mathematical transformations.

$$x = \frac{-(-6) \pm \sqrt{(-6)^2 - 4 \cdot 3 \cdot 2}}{2 \cdot 3}$$

$$\left(x = \frac{6 \pm \sqrt{12}}{6} \rightarrow x = \frac{6 \pm \sqrt{3 \cdot 4}}{6} \rightarrow x = \frac{6 \pm \sqrt{3} \cdot \sqrt{4}}{6} \rightarrow x = \frac{6 \pm 2 \cdot \sqrt{3}}{6} \rightarrow x = \frac{6}{6} \pm \frac{2 \cdot \sqrt{3}}{6} \right)$$

$$x = 1 \pm \frac{\sqrt{3}}{3}$$

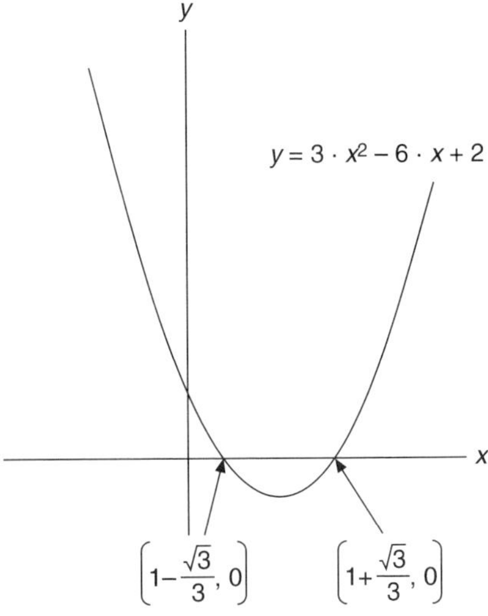

Fig. 1.3. Zeros of the function $y = 3 \cdot x^2 - 6 \cdot x + 2$.

An advanced method for determining zeros of complex functions is the Newton–Raphson algorithm, which is often used in water flow studies (e.g. Campbell, 1985; Anlauf *et al.*, 1988). This method will not be applied throughout this text, but for the interested reader an example is presented in the Appendix.

Differential equations

In the previous section, we presented methods to find zeros of functions, which are an important characteristic of a mathematical function. In this section, we shall present another important property of a function: namely the slope or gradient of a function at each point. As mentioned above, in biosphere studies we are very often dealing with gradients, e.g. concentrations or other variables which are changing in time and/or space. Therefore the slope or gradient of a function describing a process is a very important property and is needed to understand its overall behaviour.

To develop the ideas of the slope (also called the 'derivative') we consider a linear function: $y = 3 + 4 \cdot x$ (Fig. 1.4). As described in the previous section, the zero of the function $y = 3 + 4 \cdot x$ is $(-\frac{3}{4}, 0)$. The slope or gradient of a linear function can be approximated with any two x–y data pairs by:

$$\text{slope } (m) = \frac{y_2 - y_1}{x_2 - x_1} \quad \text{(Fig. 1.5)} \tag{1.2/4}$$

Thus the slope of the function is the change in y (i.e. $y_2 - y_1$) divided by the change in x (i.e. $x_2 - x_1$) (Fig. 1.5). Usually the symbol Δ is used to depict a change of a variable. Thus, the slope (m) may also be written as: $\frac{\Delta y}{\Delta x}$

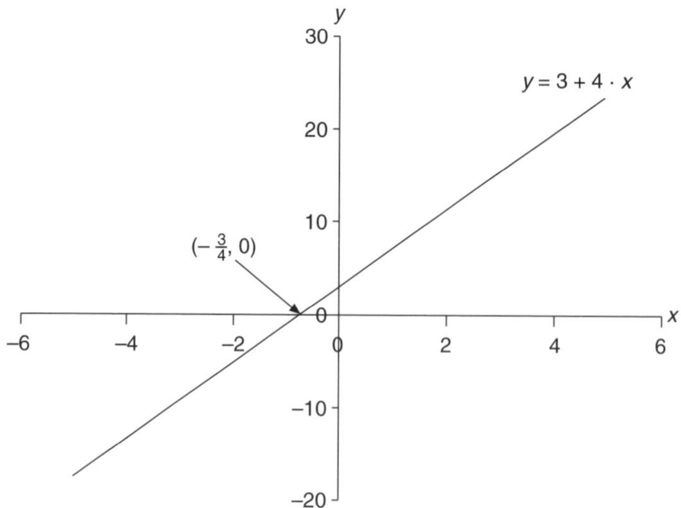

Fig. 1.4. Graph of the function $y = 3 + 4 \cdot x$.

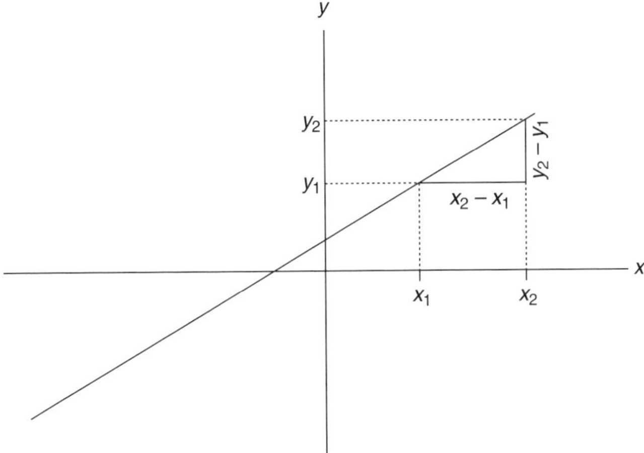

Fig. 1.5. General description of calculating the slope of a linear function.

(change in y over the change in x). This notation is referred to as the difference form (as opposed to the differential form!). Taking two x–y pairs of the equation $y = 3 + 4 \cdot x$ (Fig. 1.4), e.g. (1, 7) and (2, 11), we can determine m:

$$m = \frac{11-7}{2-1} = 4 \quad \text{or by using the difference form:} \quad m = \frac{4}{1} = 4$$

THE DERIVATIVE

To calculate the slope for any point on a straight line we are free to choose any two x–y pairs and will always get the correct result because the slope of a straight line does not change. However, this approach does not work with a non-linear equation because the slope will change (e.g. Fig. 1.6). To overcome this problem we might choose two x–y pairs which are close together and therefore can approximate the slope at the point in question more accurately. The approximation of the slope at the x–y pairs is also known as the 'derivative'. Moving closer and closer together the solution will become better till we are determining the slope just at the point in question. Let us take the example of the earlier quadratic equation: $y = 3 \cdot x^2 - 6 \cdot x + 2$. The slope at the point (1.5, −0.25) can be approximated by a straight line (a so-called tangent line, which is the derivative of the function). This straight line touches the function line just at the point (1.5, −0.25) (Fig. 1.6) which in turn provides the most accurate estimate of the slope at that point.

If we determine the slope at exactly one point in question, we are no longer using a difference form but a 'differential equation'. This equation is able to calculate the slope of every point of the original function. Such a differential equation can usually be calculated by following certain rules. Which rule to apply depends on the complexity of a function or whether it is made up of other

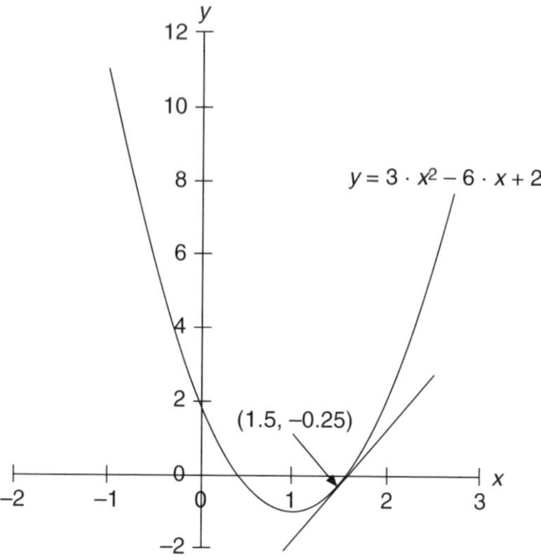

Fig. 1.6. Tangent line of the equation $y = 3 \cdot x^2 - 6 \cdot x + 2$ at the point (1.5, −0.25).

functions (composite function). If we want to obtain the derivative of the function $y = 3 \cdot x^2 - 6 \cdot x + 2$ (Fig. 1.6), we have to apply the sum in combination with the power rule. We obtain the derivatives of the three parts of the above equation, i.e. $3 \cdot x^2$, $-6 \cdot x$ and 2, separately and then add them together (see later for further details). The derivatives of each part can be calculated by the power rule $\left(\frac{\mathrm{d}}{\mathrm{d}x} x^r = r \cdot x^{r-1}\right)$. Applying this rule to the sections of the above equation we obtain $\frac{\mathrm{d}}{\mathrm{d}x} 3 \cdot x^2 = 6 \cdot x$, $\frac{\mathrm{d}}{\mathrm{d}x} 6 \cdot x = 6$ and $\frac{\mathrm{d}}{\mathrm{d}x} 2 = 0$ and all added together. $\frac{\mathrm{d}y}{\mathrm{d}x} = 6 \cdot x - 6$. Therefore the slope at the point (1.5, −0.25) equals 3 ($6 \cdot 1.5 - 6 = 3$).

The notation we use in the text to define a derivative is $\frac{\mathrm{d}y}{\mathrm{d}x}$, which expressed in words would be 'derivative of y with respect to x'. Other notations for derivatives are:

- $\frac{\mathrm{d}y}{\mathrm{d}x}$ (e.g. for differentiating a function f(x) the notation would be: $\frac{\mathrm{d}}{\mathrm{d}x} \mathrm{f}(x)$ ('differentiate function f(x)')
- y' ('derivative of y with respect to x').

RULES FOR DIFFERENTIATION

In this section, the most important rules for differentiating functions will be presented and illustrated by typical examples. The rules are presented in the form in which they are generally found in textbooks (i.e. $\frac{\mathrm{d}}{\mathrm{d}x}\mathrm{f}(x)$) (e.g. Goldstein *et al.*, 1990). However, the examples are expressed in the form in which they appear in the text (i.e. $\frac{\mathrm{d}y}{\mathrm{d}x}$).

$$\text{Power rule: } \frac{\mathrm{d}}{\mathrm{d}x}x^r = r \cdot x^{r-1} \qquad (1.2/5)$$

Example. (Fig. 1.4): $y = 3 + 4 \cdot x$;

$$\frac{dy}{dx} = 1 \cdot 4 \cdot x^{1-1} = 4 \text{ (note: constants fall away, i.e.} = 0\text{)}.$$

$$\text{General power rule: } \frac{\mathrm{d}}{\mathrm{d}x}[\mathrm{f}(x)]^r = r \cdot [\mathrm{f}(x)]^{r-1} \cdot \frac{\mathrm{d}}{\mathrm{d}x}\mathrm{f}(x) \qquad (1.2/6)$$

(a special form of the chain rule below)

Example. $y = (3 \cdot x^2 - 1)^8$ with $r = 8$ and $f(x) = 3 \cdot x^2 - 1$.

Following the general power rule we obtain for the derivative:

$$\frac{dy}{dx} = 8 \cdot (3 \cdot x^2 - 1)^7 \cdot 6 \cdot x = 48 \cdot x \cdot (3 \cdot x^2 - 1)^7.$$

$$\text{Sum rule: } \frac{\mathrm{d}}{\mathrm{d}x}[\mathrm{f}(x) + \mathrm{g}(x)] = \frac{\mathrm{d}}{\mathrm{d}x}[\mathrm{f}(x)] + \frac{\mathrm{d}}{\mathrm{d}x}[\mathrm{g}(x)] \qquad (1.2/7)$$

Example. (Fig. 1.6): $y = 3 \cdot x^2 - 6 \cdot x + 2$ with $f(x) = 3 \cdot x^2$ and $g(x) = 6 \cdot x$. We can derive the overall derivative as:

$$\frac{dy}{dx} = 6x - 6;$$

the slope at the point (1.5, – 0.25) would be: $m = 6 \cdot 1.5 - 6 = 3$.

$$\text{Constant multiple rule: } \frac{\mathrm{d}}{\mathrm{d}x}[\mathrm{k} \cdot \mathrm{f}(x)] = \mathrm{k} \cdot \frac{\mathrm{d}}{\mathrm{d}x}[\mathrm{f}(x)] \qquad (1.2/8)$$

Example. $y = 5 \cdot (3 \cdot x + 13)$

$$5 \cdot \frac{dy}{dx} = 5 \cdot 3 = 15.$$

Product rule:

$$\frac{d}{dx}[f(x)\cdot g(x)] = f(x)\cdot\frac{d}{dx}\cdot g(x) + g(x)\cdot\frac{d}{dx}f(x) \qquad (1.2/9)$$

Example: $y = x^2 \cdot x^3$ with $f(x) = x^2$ and $g(x) = x^3$.

We can derive the derivative:

$$\frac{dy}{dx} = x^2\frac{d}{dx}x^3 + x^3\frac{d}{dx}x^2$$

$$\frac{dy}{dx} = x^2\cdot(3\cdot x^2) + x^3\cdot(2\cdot x)$$

$$\frac{dy}{dx} = 3\cdot x^4 + 2\cdot x^4$$

$$\frac{dy}{dx} = 5\cdot x^4$$

The same result is obtained when first transforming the function by a power rule ($x^a \cdot x^b = x^{a+b}$) (eqn 1.2/1) and then using eqn 1.2/5:

$$y = x^2\cdot x^3 = x^5;\ \frac{dy}{dx} = 5\cdot x^4$$

This example shows that it is often a good idea to simplify the equation first, before differentiating it.

$$\text{Quotient rule: } \frac{d}{dx}\left[\frac{f(x)}{g(x)}\right] = \frac{g(x)\cdot\frac{d}{dx}f(x) - f(x)\cdot\frac{d}{dx}g(x)}{[g(x)]^2} \qquad (1.2/10)$$

Example. $y = \frac{x}{2\cdot x + 3}$ with $f(x) = x$ and $g(x) = 2 \cdot x + 3$ and and the derivatives of the two functions: $\frac{d}{dx}f(x) = 1$ and $\frac{d}{dx}g(x) = 2$.

We can derive the following expression:

$$\frac{dy}{dx} = \frac{(2\cdot x+3)\cdot\frac{d}{dx}(x) - (x)\cdot\frac{d}{dx}(2\cdot x+3)}{(2\cdot x+3)^2}$$

$$\frac{dy}{dx} = \frac{(2\cdot x+3)\cdot(1) - (x)\cdot(2)}{(2\cdot x+3)^2}$$

$$\frac{dy}{dx} = \frac{3}{(2\cdot x+3)^2}$$

$$\text{Chain rule: } \frac{d}{dx} f(g(x)) = \frac{d}{dg(x)} f(g(x)) \cdot \frac{d}{dx} g(x)$$

$$or \frac{dy}{dx} = \frac{dy}{du} \cdot \frac{du}{dx} \text{ with } y = f(u) \qquad (1.2/11)$$

$$u = g(x)$$

Example. $y = (6 \cdot x - 1)^5$ with $g(x) = 6 \cdot x - 1$, $u = g(x)$, $f(u) = u^5$ and the derivatives:

$$\frac{d}{dx} g(x) = 6, \quad \frac{d}{du} f(u) = 5 \cdot u^4 \text{ or } \frac{d}{dg(x)} f(g(x)) = 5 \cdot g(x)^4.$$

We can write the derivative as:

$$\frac{dy}{dx} = 5 \cdot (g(x))^4 \cdot 6 \quad \text{or} \quad \frac{dy}{dx} = 5 \cdot (u)^4 \cdot 6$$

After combination with $g(x) = 6 \cdot x - 1$ or $u = 6 \cdot x - 1$ we obtain:

$$\frac{dy}{dx} = 5 \cdot (6 \cdot x - 1)^4 \cdot 6$$

$$\frac{dy}{dx} = 30 \cdot (6 \cdot x - 1)^4$$

DERIVATIVES OF COMMON FUNCTIONS

$$\frac{d}{dx}(x^n) = n \cdot x^{n-1}$$

$$\frac{d}{dx}(e^{k \cdot x}) = k \cdot e^{k \cdot x}$$

$$\frac{d}{dx}(\ln x) = \frac{1}{x}$$

INTEGRATION

In many applications, particularly in studies of natural systems, the rates of certain processes are often monitored. In mathematical terms, these rates represent a derivative of a function. Although we may be interested in the rate of change of a quantity at any one time (i.e. the derivative), we are also interested in the amount of this quantity at any one time. The total amount can be calculated by reversing the procedure described in the last section (differentiation). This process is known as antidifferentiation or integration. An illustration is the emission of gas from soil. If the rate is given in units of [amount $\cdot$ time^{-1}] then the entire amount of gas emitted over a certain time period can be calculated by summing all individual rates or performing an integration. The resulting unit is in [amount].

The notation when integrating a function is given by:

$\int$ f(x) dx = F(x) + C ('integrate the function f(x) with respect to x')[2] (1.2/12)

Note: it is important that an integration constant (C) is always added to the final equation. This arises from the fact that functions which just differ by a constant produce the same derivative.

INTEGRATION RULES

$$\int x^r \mathrm{d}x = \frac{1}{r+1} x^{r+1} + \mathrm{C}, \quad r \neq -1 \qquad (1.2/13)$$

Example.

$$\int \mathrm{x}^2 dx = \frac{1}{2+1} \cdot \mathrm{x}^{2+1} + \mathrm{C} = \frac{1}{3} \cdot \mathrm{x}^3 + \mathrm{C}$$

or $\int 4d\mathrm{x} = 4 \cdot \mathrm{x} + \mathrm{C}$. One solution is the equation: $\mathrm{y} = 4 \cdot \mathrm{x} + 3$ (Fig. 1.4).

$$\int \frac{1}{x} \mathrm{d}x = \ln|x| + \mathrm{C} \qquad (1.2/14)$$ [3]

Example.

$$\int \frac{2}{\mathrm{x}} \, d\mathrm{x} = 2 \cdot \ln|\mathrm{x}| + \mathrm{C}$$

$$\int \mathrm{e}^{\mathrm{k} \cdot x} \mathrm{d}x = \frac{1}{\mathrm{k}} \mathrm{e}^{\mathrm{k} \cdot x} + \mathrm{C} \qquad (1.2/15)$$

Example.

$$\int \mathrm{e}^{4 \cdot \mathrm{x}} d\mathrm{x} = \frac{1}{4} \mathrm{e}^{4 \cdot \mathrm{x}} + \mathrm{C}$$

INTEGRATION BY SUBSTITUTION

This method is often used to transform a complicated integral into a simpler one.

[2] The sign '$\int$' is just an indicator for an integration.
[3] The notation $|x|$ refers to the absolute value of 'x', e.g. $|-5| = 5$.

To determine	$\int f(g(x)) \cdot \frac{d}{dx} g(x)\ dx = \int f(u)\ du$
1. Set	$u = g(x), du = \frac{d}{dx} g(x) dx$
2. Determine	$\int f(u)\ du = F(u)$
3. Substitute the value of u:	$\int f(g(x)) \cdot \frac{d}{dx} g(x)\ dx = F(g(x)) + C$

(1.2/16)

Example. Determine: $\int (x^2 + 1)^3 \cdot x\ dx$ with $f(g(x)) = f(u) = u^3$ and $g(x) = u = x^2 + 1$ with the derivative $du = 2 \cdot x\ dx$.
We can compute the overall integral as:

$$\int u^3 du = \frac{1}{4} u^4 + C$$

After substitution of $u = x^2 + 1$ and $du = 2x\ dx$ (or $\frac{du}{2} = x\ dx$) we obtain:

$$\int (x^2 + 1)^3 \cdot x\ dx = \int u^3 \frac{du}{2} = \frac{u^4}{8} + C = \frac{1}{8} \cdot (x^2 + 1) + C$$

Note: sometimes it is necessary to expand the equation first to obtain a suitable notation so that this rule can be applied.

INTEGRATION BY PARTS

$$\int f(x) \cdot g(x) dx = f(x) \cdot G(x) - \int \frac{d}{dx} f(x) \cdot G(x) dx$$

where G (x) is an antiderivative or integral of $g(x)$.
It only makes sense to apply this rule when it is easy to compute $\frac{d}{dx} f(x)$ and $G(x)$ and when the integral $\int \frac{d}{dx} f(x) \cdot G(x)\ dx$ can be calculated. (1.2/17)

Example. Evaluate $\int x \cdot e^x dx$ with $f(x) = x$ and $g(x) = e^x$ and the derivative $\frac{d}{dx} f(x) = 1$ and the integral: $G(x) = e^x$.

We can follow the above rule and calculate the integral by:

$$\int f(x) \cdot g(x) dx = x \cdot e^x - \int 1 \cdot e^x\ dx$$

$$\int f(x)\cdot g(x)dx = x\cdot e^x - e^x + C$$ (note: do not forget to add the constant 'C')

EVALUATION OF DEFINITE INTEGRALS

If actual lower and upper bounds of the integral are known (values between which the integration should be performed, e.g. time period), the integral can be calculated by:

$$\int_a^b f(x)dx = F(x)\Big|_a^b = F(b) - F(a) \qquad (1.2/18)$$

Due to the subtraction, the constant cancels out.

(The notation $\Big|_b^a$ is just another representation of the limits which are inserted into the integrated equation $F(x)$ to obtain the final result.)

Evaluate the example $\int x \cdot e^x dx$ between the limits 0 and 1:

$$\int_0^1 x\cdot e^x\, dx$$

First the integration is carried out (see example above for eqn 1.2/17):

$$\int x\cdot e^x\, dx = x\cdot e^x - e^x + C$$

The evaluation between the limits is:

$$\int_0^1 x\cdot e^x\, dx = \left(x\cdot e^x - e^x + C\right)\Big|_0^1 = \left(1\cdot e^1 - e^1\right) - \left(0\cdot e^0 - e^0\right) = 1$$

INTEGRAL TABLES

To ease the integration of complicated integrals, general solutions of numerous function types are published in tables (e.g. Gradshteyn and Ryzhik, 1995). Once the general form of a given function has been determined, the solution can often be obtained by reference to such tables. An example of a solution using a 'standard form' is used in Chapter 6 to find an analytical solution for instantaneous canopy photosynthesis (eqns 6.3/40 and 6.3/41).

SOLUTION OF DIFFERENTIAL EQUATIONS

This section outlines the process of determining solutions (i.e. integrations) to a differential equation. If the integration constant C is not specified, we can have numerous result functions differing only by this constant. In natural studies, the independent variable is often time (t) and we may know the initial value of a quantity by measurement, i.e. the value when $t = 0$. This value is sometimes written as y_0 and could, for example, be fertilizer applied to the soil at the beginning of the experiment.

The solution of the differential equation:

$$\frac{dy}{dt} = 3t^2 - 4$$

is:

$$y = t^3 - 4 \cdot t + C$$

However if the initial value of y, $y_0 = 4$, is given, then the specific solution of this equation is:

$$y = t^3 - 4 \cdot t + 4.$$

This can be verified by solving the equation for $t = 0$, i.e. $y = 0^3 - 4 \cdot 0 + 4 = 4$.

SEPARATION OF VARIABLES

This technique can be used to solve an important class of differential equations in the form: $\frac{dy}{dt} = p(t) \cdot q(y)$ where $p(t)$ is a function of t (usually time) and $q(y)$ is a function of y. If we have a differential equation such as:

$$\frac{dy}{dt} = \frac{3 \cdot t^2}{y^2}$$

we can rewrite the equation so that the variables are separated and appear on each side of the equation:

$$y^2\, dy = 3 \cdot t^2\, dt$$

Integration of both sides gives:

$$\int y^2\, dy = \int 3 \cdot t^2\, dt$$

The integrals are now solved on both sides and rewritten as an equation of y in terms of t, obtaining:

$$\frac{1}{3} y^3 + C_1 = t^3 + C_2$$

or

$$y^3 = 3 \cdot \left(t^3 + C_2 - C_1\right)$$

and finally:

$$y = \sqrt[3]{3 \cdot t^3 + C}$$

In the soil–atmosphere continuum, we very often have to deal with equations in the form:

$$\frac{dy}{dt} = -k \cdot y$$

which basically means that the rate of change of y with respect to t ($\frac{dy}{dt}$) is proportional to y. To obtain the rate equation, we multiply y by a proportionality factor (k). An example of such a process would be the nitrogen transformation rate in soil, which is proportional to the amount of nitrogen present (see Section 1.4 and Chapter 2):

$$\frac{1}{y} \cdot \frac{dy}{dt} = -k$$

integration with respect to t:

$$\int \frac{1}{y} \frac{dy}{dt} dt = -\int k dt$$

$$\int \frac{1}{y} dy = -\int k dt$$

and after integrating both sides we obtain:

$$\ln y = -k \cdot t + C$$

and solving for y:

$$y = e^{-k \cdot t + C}$$

or:

$$y = e^{C} \cdot e^{-k \cdot t}$$

determining y at $t = 0$ we solve:

$$y = e^{C} \cdot e^{-k \cdot 0}$$

or:

$$y = e^{C} \cdot 1$$

and obtain:

$$y_0 = e^{C}$$

The final equation is:

$$y = y_0 \cdot e^{-k \cdot t}$$

We shall revisit the last equation in Section 1.4. in the context of the description of transformation rates, where the importance of this equation in natural studies will become more obvious.

For readers who are not entirely confident with techniques of differentiation or integration, or who need further practice, it is recommended they follow a textbook such as Goldstein *et al.* (1990). That text is easy to follow and provides a step-by-step approach to calculus.

1.2.4 Numerical and analytical solutions

In Section 1.2.3, mathematical tools were presented to solve differential equations. In many cases solutions to differential equations can be found by integration. Such a solution is also called an analytical solution or exact mathematical solution for the given differential equation. For instance, $y = y_0 \cdot e^{-k \cdot t}$ is called an analytical solution of the differential equation $\frac{dy}{dt} = -k \cdot y$ (see previous section).

Once differential equations become more complex, it may be very cumbersome to find analytical solutions. Moreover, there are differential equations which cannot be solved analytically. However, to solve such differential equations mathematicians have developed techniques which can calculate the solutions numerically. These techniques use the actual differential equation and calculate the sum of the differentials between the limits of integration, i.e. find the 'area under the curve'. Compared with the analytical solution, these techniques usually require many more calculations; however, this additional work can easily be carried out by computers and therefore presents no problem.

The general idea of a numerical solution in comparison with an analytical solution is illustrated by solving the differential equation $\frac{dy}{dt} = 3 \cdot t^2$ (Fig. 1.7).

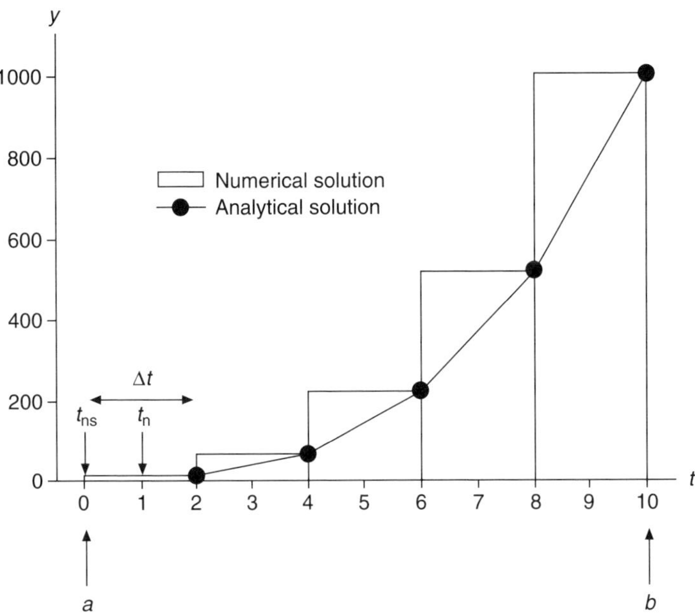

Fig. 1.7. Numerical and analytical solution of $y = \int_0^{10} 3 \cdot t^2 dt = t^3 \Big|_0^{10}$.

The analytical solution to this equation is: $\int_a^b 3 \cdot t^2 \mathrm{d}t = b^3 - a^3$. (Note: due to the subtraction, the constant cancels out (see eqn 1.2/18).)

The solution of the differential equation can also be approximated by a simple numerical method: the mid-point rule. The integration range (between the lower limit (a) and the upper limit (b)) is divided into n equal subintervals, each of length: $\Delta t = \frac{(b-a)}{n}$. The area of each subinterval (each row in Table 1.1) is calculated and the value of the total integral (the total area under the curve between a and b) is estimated by summing up all the segments (sum of all rows in Table 1.1). As the name of the rule indicates (see below), the area of each segment is evaluated at the mid-point (t_n) of each segment, defined by: $t_n = t_{ns} + \frac{\Delta t}{2}$, where t_{ns} indicates the start of the interval (Fig. 1.7). The general numerical solution via the mid-point rule is given by :

$$\int_a^b f(t)dt \approx f(t_1) \cdot \Delta t + f(t_2) \cdot \Delta t + \ldots + f(t_n) \cdot \Delta t = \left[f(t_1) + f(t_2) + \ldots + f(t_n)\right] \cdot \Delta t$$

(1.2/19)

For the above example, using: $a = 0$, $b = 10$, $n = 5$ the interval is: $\Delta t = \frac{(10-0)}{5} = 2$.

The values t_n and t_{ns}, the calculations for f(t_n) using the mid-point rule and the analytical solution are presented in Table 1.1 and Fig. 1.7. The numerical solution

Table 1.1. Analytical and numerical solution of $y = \int_0^{10} 3 \cdot t^2 dt$.

Label	t_{end}*	t_n	$f(ti) \cdot \Delta t$	Cumulative results: Numerical	Cumulative results: Analytical
t0	0			0	0
t1	2	1	$3 \cdot 1^2 \cdot 2 = 6$	6	8
t2	4	3	$3 \cdot 3^2 \cdot 2 = 54$	60	64
t3	6	5	$3 \cdot 5^2 \cdot 2 = 150$	210	216
t4	8	7	$3 \cdot 7^2 \cdot 2 = 294$	504	512
t5	10	9	$3 \cdot 9^2 \cdot 2 = 486$	**990**	**1000**

* End of interval.

(Table 1.1) is calculated by: $[3 \cdot 1^2 + 3 \cdot 3^2 + 3 \cdot 5^2 + 3 \cdot 7^2 + 3 \cdot 9^2] \cdot 2 = 990$. The numerical solution (990 (Table 1.1)) is very close to the analytically correct solution (1000). One reason for the slight discrepancy between the two solutions is the fact that relatively large intervals were chosen. By reducing the interval size (increasing the number of intervals), the numerical solution would approximate the analytical solution more accurately.

Other numerical methods may be used but they all work on similar principles. The most common numerical methods are: Euler's method, the Runge–Kutta method and the Bulirsch–Stoer method. The Runge–Kutta method, especially the fourth-order Runge–Kutta, is often used as a standard method for numerical integration because it is robust and precise. This is also the method which is used to solve the differential equations in this book. For an overview of the methods, see Walker (1997).

There are other methods of calculating approximations of mathematical functions. The most important one is the calculation of Taylor polynomials. Taylor polynomials are used, for example, in iteration methods, such as the Newton–Raphson method, which is often used to solve water flow problems in soil (e.g. Campbell, 1985; Anlauf *et al.*, 1988). This method is not used in this text, but for the interested reader, Taylor polynomials, including the Newton–Raphson algorithm, are outlined in the Appendix.

It should be pointed out that the reader does not actually have to implement numerical solutions of differential equations him/herself because this work is carried out by the modelling package (ModelMaker).

1.3 The Transport Equation and the Finite Difference Notation

Processes in soil are very often concerned with transport phenomena (e.g. flow of heat, water, material, etc.). In this section, the basic concepts of transport processes are presented, which are similar for all transport phenomena. Details and examples of transport processes, such as flow of heat or water in soil, are considered in the respective chapters. The basic idea of a transport process was introduced in Section 1.2.2, where we showed that the flow of heat in or out of a room was proportional to the temperature difference (Fig. 1.1). Here we want to extend the concept further to develop a transport concept as it generally occurs in natural processes. The flow or flux density may be presented by F. The 'concentration' in transport processes is usually represented by C. In transport processes, F, is generally proportional to the 'concentration gradient'[4] $\frac{dC}{dx}$, i.e. $F \propto \frac{dC}{dx}$ (read: F is propor-

[4] Note the term 'concentration gradient' refers to the difference in quantity appropriate for the flow in question, e.g. in heat flow it is the temperature or energy change, in water flow the change in water potential, in solute leaching the change in salt concentration, etc.

tional to the change in C over the change in *x*') or after introduction of a proportionality constant:

$$F = -k\frac{dC}{dx} \tag{1.3/1}$$

where *x* is the distance over which this flow occurs (the minus sign in front of k is needed to indicate that flow always occurs from high to low 'concentration' – an illustration will be given later on in this section).

Assuming that *F* and k are constant this differential equation can be integrated in order to find solutions (see previous section).

$$\int_{x_1}^{x_2} F dx = \int_{C_1}^{C_2} -k\ dC$$

$$(F \cdot x)\Big|_{x1}^{x2} = (-k)\Big|_{C1}^{C2}$$

$$F \cdot x_2 - F \cdot x_1 = -k \cdot C_2 - (-k \cdot C_1)$$

$$F \cdot (x_2 - x_1) = -k \cdot (C_2 - C_1)$$

$$F = -k\frac{C_2 - C_1}{x_2 - x_1} \tag{1.3/2}$$

To illustrate the transport equation, we imagine a hypothetical soil profile which is divided into two layers and the flow *F* occurs according to a 'concentration gradient'. An appropriate graphical representation is given in Fig. 1.8. The quantities C_1 and C_2 are taken to be representative for the respective layer. The distances x_1 and x_2 are measured from the soil surface to the middle of each soil layer. Therefore the flow *F* can be calculated according to eqn 1.3/2. Replacing the variables and constants with dimensionless values ($C_1 = 10$, $C_2 = 20$, $x_1 = 5$, $x_2 = 15$ and $k = 0.01$), we can calculate the flux (*F*):

$$F = -0.01 \cdot \frac{20 - 10}{15 - 5} = -0.01.$$

In this example, the flux has a negative value, which means that within the soil profile the flux is directed towards the soil surface. This makes sense since the 'concentration' in the lower soil layer ($C_2 = 20$) is actually higher than in the top soil layer ($C_1 = 10$), resulting in movement from x_2 to x_1. So, in this particular example, the flow would therefore be in the opposite direction to that shown in Fig. 1.8.

The assumption that the quantities *F* and k are constant is only correct if the

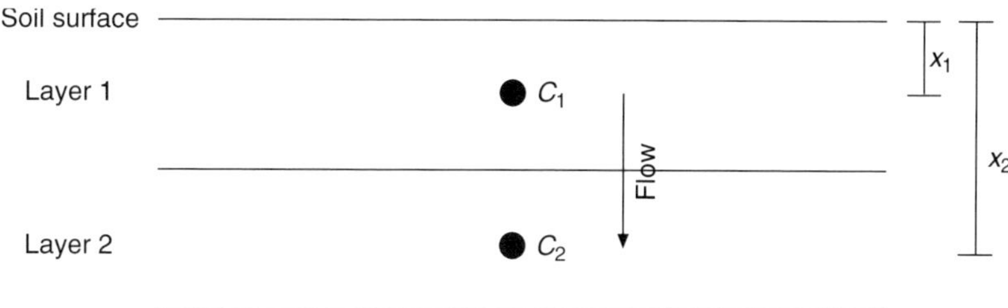

Fig. 1.8. Representation of flow in a two-layered soil profile.

distance over which the flow occurs is very small. Therefore, in calculating the flow within a deep profile the two-layered approach is not sufficient and should be replaced with a multilayered system by dividing the soil profile into a lot of small layers. Flows can then be calculated between adjacent layers. A graphical representation of such a multilayered soil profile (provided all soil layers have the same thickness) is given in Fig. 1.9; or, in general terms, exchanging the layer numbers with indices,[5] a soil profile containing *n* layers can be represented as in Fig. 1.10.

The flow into layer *i* (Fig. 1.10) is generally given by:

$$F_i = -k \cdot \frac{C_i - C_{i-1}}{x_i - x_{i-1}} \tag{1.3/3}$$

or

$$F_i = -k \cdot \frac{C_i - C_{i-1}}{\Delta x} \tag{1.3/4}$$

BOUNDARY CONDITIONS

We are now able to calculate the various flows *F* within the whole profile except for the flow into the topmost compartment (F_1, Fig. 1.10) and the flow out of the bottom compartment (F_{n+1}, Fig. 1.10). To determine the flow into the top layer (layer 1) we have to calculate:

$$F_1 = -k \cdot \frac{C_1 - C_0}{x_1 - x_0} \tag{1.3/5}$$

and to determine the flow out of the bottom layer (F_{n+1}) we calculate:

$$F_{n+1} = -k \cdot \frac{C_{n+1} - C_n}{x_{n+1} - x_n} \tag{1.3/6}$$

[5] The term 'index' (pl. indices) refers to the subscript numbers, which indicate the layer number.

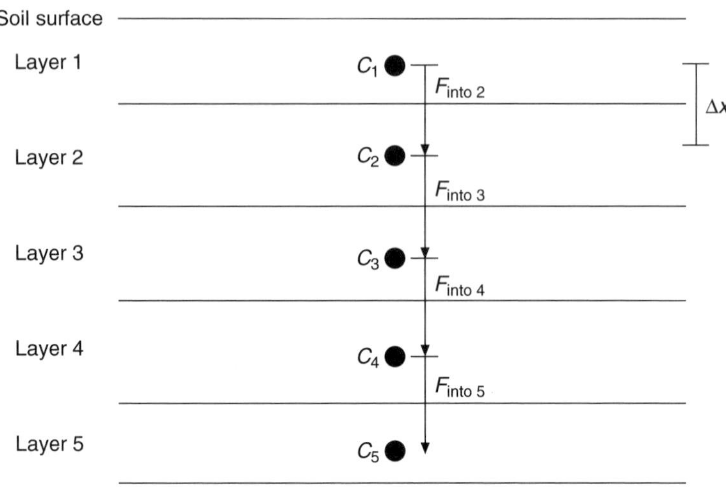

Fig. 1.9. Representation of flows in a multilayered soil profile.

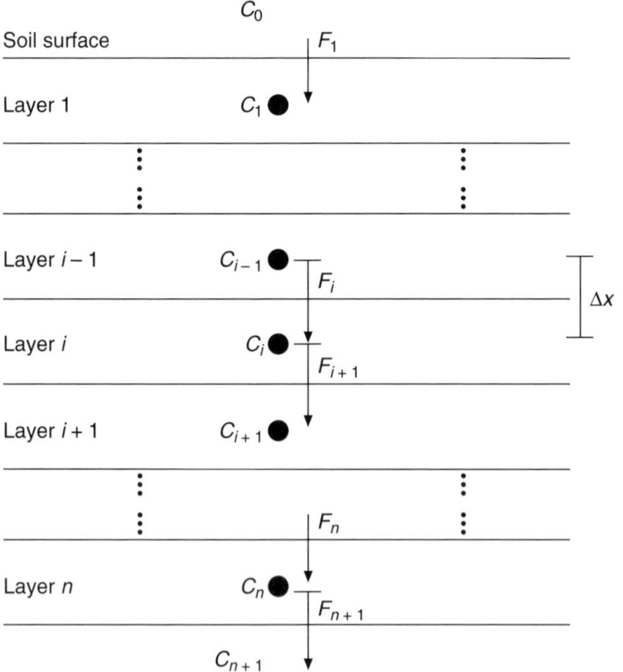

Fig. 1.10. General representation of flows in a multilayered soil profile.

However, neither C_0 nor C_{n+1} exists within the profile. Due to the lack of information at the boundaries of the profile, these problems are called *boundary problems*. In order to solve the flow equations for the whole profile, including the flows which occur at the boundaries, we have to supply additional information about C_0 and C_{n+1}. Usually these are some measured values or conditions which realistically represent the system outside the soil profile and are referred to as *boundary conditions*. For instance, in a problem involving heat flow in soil, we might choose measured air temperature as an approximation for C_0 and soil temperature C_{n+1} which has been measured just outside the bottom layer, as suitable boundary conditions. In water flow problems, the gain or loss of water in the soil surface layer due to rainfall and evaporation and a water-table or impermeable soil layer just outside the bottom layer might represent realistic boundary conditions. These few examples already show that it is important to supply realistic boundary conditions. For instance, the dynamics of soil moisture in a soil profile would be affected quite differently depending on whether we have a water-table or a rock layer as the bottom boundary. Therefore the boundary conditions are crucial for a realistic solution of the flow problem in question. This important aspect of choosing appropriate boundary conditions will become clearer when the effect of different boundary conditions on the solution of the heat and water flow patterns in soil are examined (see Chapters 3 and 4).

So far we have looked at the instantaneous flow characteristics and developed ideas on how a flow (F) can realistically be described in relation to depth. However, the flow will also change with time in response to various conditions having an impact on the system, such as climatic conditions. The analysis of the change in flow with time will be described in detail in Chapter 3, where we shall develop the 'continuity equation' for heat flow in soil, which defines the change both in space and in time. The prediction of flow characteristics in response to space and time are crucial for an understanding of flow processes.

1.4 Mathematical Description of Kinetics

Another important group of processes are transformation reactions. They are generally referred to as kinetics. A very good overview of various reaction types, including complex reactions, is given in Richter (1986). The most important ones which will be used during model developments in subsequent chapters are presented below.

1.4.1 Zero-order kinetics

A transformation

$$A \xrightarrow{k} B$$

which is independent of the initial concentration *A* is called a zero-order reaction. Irrespective of the concentrations of *A* or *B*, the transformation rate k stays constant. However, it should be noted that a zero-order reaction is an approximation, because it is not possible for the reaction to proceed at a constant rate if there is no substrate present to sustain it. If the sum of *A* and *B* is constant (i.e. no loss or gain of *A* or *B* due to other processes), zero-order kinetics is defined by:

$$\frac{dB}{dt} = k \text{ and } -\frac{dA}{dt} = k \qquad (1.4/1)$$

Integration yields:

$$\begin{aligned} \int dB &= \int k \cdot dt \quad \rightarrow \quad B = k \cdot t + B_0 \\ \int -dA &= \int k \cdot dt \quad \rightarrow \quad A = -k \cdot t + A_0 \end{aligned} \qquad (1.4/2)$$

The half-life of the substrate (the time when half of the initial substrate, A_0, is transformed to B) is reached at:

$$\frac{A_0}{2} = -k \cdot t_{half} + A_0$$

and after rearranging for t_{half}:

$$t_{half} = \frac{A_0}{2 \cdot k} \qquad (1.4/3)$$

1.4.2 First-order kinetics

If the transformation

$$A \xrightarrow{k} B$$

is proportional to the concentration of substrate *A* with a proportionality factor k (rate constant), we are describing a first-order reaction, which is defined by:

$$\frac{dA}{dt} = -k \cdot A \qquad (1.4/4)$$

or in the integrated form (where A_0 = initial substrate concentration):

$$\left(\int \frac{1}{A} \, dA = \int -k \, dt \rightarrow \ln(A) = -k \cdot t + C \rightarrow A = e^{-k \cdot t + C}; \; A_0 = e^C \right)$$

$$A = A_0 \cdot e^{-k \cdot t} \qquad (1.4/5)$$

For the product *B*, the differential equation is:

$$\frac{dB}{dt} = k \cdot A \qquad (1.4/6)$$

(Note: the quantity: k • *A* is subtracted from the substrate (*A*) and added to the product (*B*). For numerical solutions it is convenient to define a flow (*F*): F = k • *A* and perform mathematical operations with this flow.)

Provided that the sum of *A* and *B* is constant, i.e.:

$$A + B = A_0$$

we can rewrite the equation for the product *B* as:

$$\frac{dB}{dt} = k \cdot (A_0 - B) \qquad (1.4/7)$$

Integration of the last equation yields:

$$\int_0^B \frac{1}{A_0 - B}\, dB = \int_0^t k\ dt$$

$$\begin{pmatrix} \ln(A_0 - B) - \ln(A_0 - 0) = -k \cdot t - (-k \cdot 0) \rightarrow \ln\left(\frac{A_0 - B}{A_0}\right) = -k \cdot t \\ \rightarrow A_0 - B = A_0 e^{-k \cdot t} \end{pmatrix}$$

$$B = A_0 \cdot (1 - e^{-k \cdot t}) \qquad (1.4/8)$$

The half-life (t_{half}) of the substrate *A* for a first-order reaction can be calculated by solving eqn 1.4/5 when $A = \frac{A_0}{2}$:

$$\frac{A_0}{2} = A_0 \cdot e^{-k \cdot t_{half}}$$

Rearranging and solving for t_{half} yields:

$$\frac{A_0}{A_0 \cdot 2} = e^{-k \cdot t_{half}}$$

$$\left(\frac{1}{2} = e^{-k \cdot t_{half}} \rightarrow \ln(\tfrac{1}{2}) = -k \cdot t_{half} \rightarrow t_{half} = \frac{\ln(\tfrac{1}{2})}{-k} \rightarrow t_{half} = \frac{\ln(1) - \ln(2)}{-k}\right)$$

$$t_{half} = \frac{\ln 2}{k} \text{ or } t_{half} \approx \frac{0.693}{k} \qquad (1.4/9)$$

1.4.3 Second-order kinetics

If a transformation rate is proportional to the product of two substrates:

$$A_1 + A_2 \xrightarrow{k} B$$

we are describing a second-order reaction. In the simplest case, when the two substrates are the same ($A_1 \cdot A_2 = A^2$), the second-order reaction is defined by:

$$\frac{dA}{dt} = -k \cdot A^2 \text{ and } \frac{dB}{dt} = k \cdot A^2 \tag{1.4/10}$$

Integration of this equation for substrate A yields:

$$\int_{A_0}^{A} \frac{1}{A^2} dA = \int_0^t -k dt$$

$$\frac{1}{A} - \frac{1}{A_0} = k \cdot t$$

$$\left(\frac{1}{A} = k \cdot t + \frac{1}{A_0} \rightarrow \frac{1}{A} = \frac{k \cdot t \cdot A_0 + 1}{A_0} \right)$$

$$A = \frac{A_0}{A_0 \cdot k \cdot t + 1} \tag{1.4/11}$$

The time when half of the initial substrate is transformed to B is defined by:

$$\frac{A_0}{2} = \frac{A_0}{A_0 \cdot k \cdot t_{half} + 1}$$

and solving for t_{half} yields:

$$t_{half} = \frac{1}{A_0 \cdot k} \tag{1.4/12}$$

Provided that the sum of A and B is constant, i.e. $A_0 = A + B$ or $A = A_0 - B$, integration of eqn 1.4/10 for the product B yields:

$$\int_0^B \frac{1}{(A_0 - B)^2} \; dB = \int_0^t k \; dt$$

$$\left.\frac{1}{(A_0 - B)}\right|_0^B = k \cdot t\Big|_0^t$$

$$\left(\frac{1}{A_0 - B} - \frac{1}{A_0} = k \cdot t \rightarrow \frac{1}{A_0 - B} = k \cdot t + \frac{1}{A_0} \rightarrow A_0 - B = \frac{1}{k \cdot t + \frac{1}{A_0}}\right)$$

$$B = A_0 - \frac{1}{k \cdot t + \frac{1}{A_0}} \qquad (1.4/13)$$

1.4.4 Temperature dependency of the rate constant

Biological reactions exhibit a temperature response because enzymes denature or microorganisms die. The rate constant is therefore not static but will change in response to environmental conditions. The temperature dependency of *k*, for instance, is often modelled with the Arrhenius equation:

$$k = \alpha \cdot e^{-E/_{R \cdot T}} \qquad (1.4/14)$$

or in a linearized form:

$$\ln k = -\frac{E}{R} \cdot \frac{1}{T} + \ln(\alpha)$$

where:

α = Arrhenius constant [same as units as k] {e.g. 3}
E = activation energy [$J \cdot mol^{-1}$] {e.g. 10,000}
R = gas constant [$J \cdot mol^{-1} \cdot K^{-1}$] {8.314}
T = temperature [K]

Via a linear regression of ln k versus $1/_T$, the slope, $-\frac{E}{R}$, and the intercept, ln (α), are obtained (Fig. 1.11).

Note that, if temperature values are expressed in dimensions of °C, then they have to be converted to K first by adding a value of 273.16.

The Arrhenius equation applies accurately to only a few inorganic reactions. In many situations, it only works satisfactorily if applied over a very limited temperature range. Therefore, in the N transformation model developed later (Chapter 2, Sections 2.3–2.5) equations are applied which model the dynamics of biological reactions more realistically. Reduction factors for temperature, pH and moisture

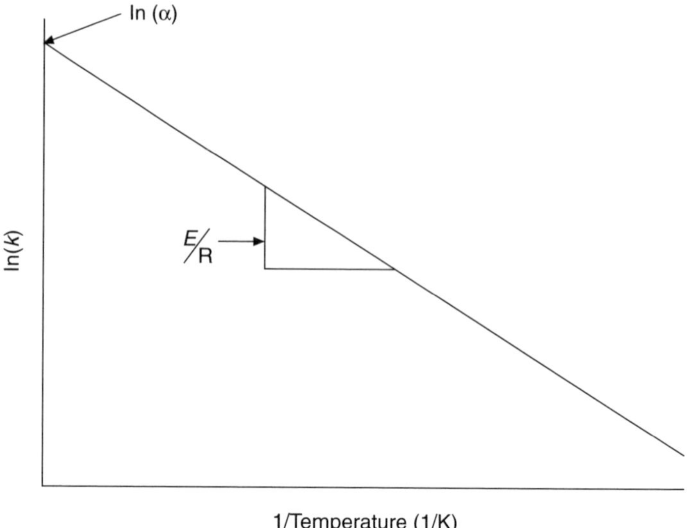

Fig. 1.11. Linearized Arrhenius relationship.

are introduced to scale the rate constant down in response to suboptimal conditions. Dependent on the conditions, such a reduction factor takes on a value between 0 and 1, with 1 = optimum conditions (highest activity) and 0 = no activity.

1.4.5 Chain reactions

Rarely do we have a system where we are dealing with only one kinetic transformation. Much more often, several reactions run in parallel or are linked to each other in a chain reaction. Often one or more intermediates occur within a reaction sequence, such as in the nitrification or denitrification processes in soil.

A first-order kinetic reaction system with three components *A*, *B* and *C* and two kinetic constants k_1 and k_2:

$$A \xrightarrow{k_1} B \xrightarrow{k_2} C$$

can mathematically be described as:

$$\frac{dA}{dt} = -k_1 \cdot A$$

$$\frac{dB}{dt} = +k_1 \cdot A - k_2 \cdot B \qquad (1.4/15)$$

$$\frac{dC}{dt} = +k_2 \cdot B$$

1.4.6 Michaelis–Menten kinetics

A very important group of reactions in microbial ecology are enzyme kinetics, which proceed as:

$$A + E \underset{k_2}{\overset{k_1}{\rightleftarrows}} X \xrightarrow{k_3} E + B \qquad (1.4/16)$$

where *A* and *B* are the substrate and product concentration, respectively, *X* is the enzyme–substrate complex which then reacts to give the product (*B*) and free enzymes (*E*). The rate constants k_1 and k_3 refer to the forward reactions, and the rate constant k_2 to the reverse reaction.

In a steady state, the rate of production of *X* (influenced by k_1) equals the rate of degradation of *X* (influenced by k_2 and k_3), i.e.:

$$k_1 \cdot A \cdot E = (k_2 + k_3) \cdot X \qquad (1.4/17)$$

The total amount of enzymes (E_{tot}) can be calculated as:

$$E_{tot} = E + X$$

(free enzymes plus enzymes within the enzyme–substrate complex) or:

$$E = E_{tot} - X$$

Assuming that the total amount of enzymes is constant, we can write the following equation for *X*:

$$k_1 \cdot A \cdot (E_{tot} - X) = (k_2 + k_3) \cdot X \qquad (1.4/18)$$

Rearranging yields:

$$X = \frac{E_{tot} \cdot k_1 \cdot A}{k_1 \cdot A + k_2 + k_3} \qquad (1.4/19)$$

The overall reaction speed will be controlled by the speed of the degradation process, k_3, to yield *B*. From eqn 1.4/16 this rate is given by: $k_3 \cdot X$. Therefore the overall reaction speed is proportional to the concentration of the enzyme–substrate complex (*X*) and the overall production equation of *B* can be written as:

$$k_3 \cdot X = \frac{E_{tot} \cdot k_1 \cdot k_3 \cdot A}{k_1 \cdot A + k_2 + k_3} \qquad (1.4/20)$$

After eliminating k_1 from the numerator of the right-hand side we obtain:

$$k_3 \cdot X = \frac{E_{tot} \cdot k_3 \cdot A}{A + \frac{k_2 + k_3}{k_1}}$$

This equation can be simplified to:

$$V = \frac{V_{max} \cdot A}{A + K_m} \quad (1.4/21)$$

with:

$$V_{max} = E_{tot} \cdot k_3$$

$$K_m = \frac{k_2 + k_3}{k_1}$$

The reaction speed with which the product (*B*) is generated is given by the symbol *V*. However, the reaction rate (*V*) can also be replaced by its differential form, $\frac{dB}{dt}$ (the rate at which B is produced, if B is not degraded), which yields:

$$V = \frac{dB}{dt} = \frac{V_{max} \cdot A}{A + K_m} \quad (1.4/22)$$

V_{max} refers to the maximum speed of the reaction when all sites on the enzymes are occupied by the substrate (*A*) (i.e. when the maximum concentration of the enzyme–substrate complex is obtained). The Michaelis–Menten constant K_m refers to the substrate concentration when half the maximum reaction speed is reached. Equation 1.4/21 describes formally a rectangular hyperbola which approaches a maximum set by V_{max} (Fig. 1.12).

The Michaelis–Menten equation describes the reaction rate relative to the substrate concentration, combining both zero- and first-order kinetics. At low substrate concentrations ($A << K_m$), the Michaelis–Menten equation can be approximated by:

$$V \approx \frac{V_{max}}{K_m} \cdot A \quad (1.4/23)$$

which describes first-order kinetics. At high substrate concentrations ($A >> K_m$) the Michaelis–Menten equation can be approximated by:

$$V \approx \frac{V_{max} \cdot A}{A} = V_{max} \quad (1.4/24)$$

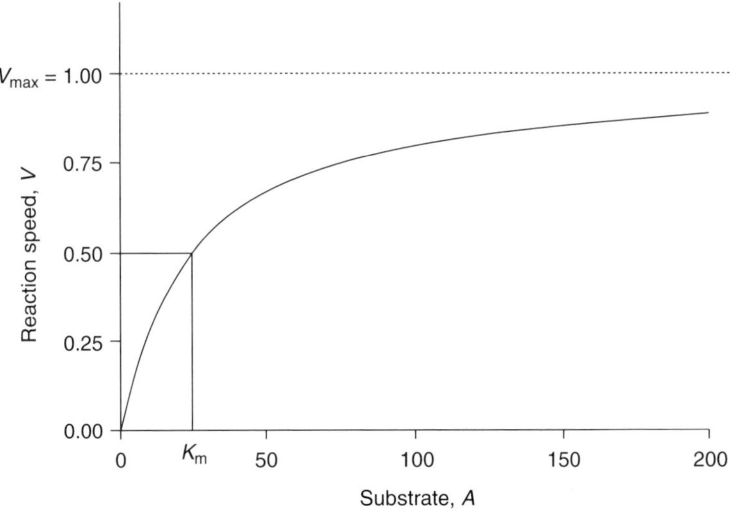

Fig. 1.12. Michaelis–Menten equation ($V_{max} = 1$, $K_m = 25$).

and therefore approaches zero-order kinetics. These properties of the Michaelis–Menten kinetics are graphically displayed in Fig. 2.4 where typical outputs of model calculations are presented. Comparing the slopes depicted in Fig. 2.4 with the graphical representation of the Michaelis–Menten equation (Fig. 1.12) shows that initially a first-order reaction and later on a zero-order reaction dominates.

The Michaelis–Menten parameters V_{max} and K_m can be determined from measurements (e.g. V = rate of transformation from ammonium ($NH_4^+{-}N$) to nitrite ($NO_2^-{-}N$); $A = NH_4^+{-}N$) either by non-linear curve fitting to equation 1.4/21 or by converting eqn 1.4/21 into a linear form and determining the slope and intercept by linear regression (e.g. Tate, 1995, p. 137). Common conversions are given below.

LINEWEAVER BURK TRANSFORMATION

$$\frac{1}{V} = \frac{1}{V_{max}} + \frac{K_m}{V_{max}} \cdot \frac{1}{A}$$

$$\text{Plot } \frac{1}{V} \text{ versus } \frac{1}{A},$$

$$\text{slope} = \frac{K_m}{V_{max}}, \text{ intercept} = \frac{1}{V_{max}}$$

HANES–WOLF TRANSFORMATION

$$\frac{A}{V} = \frac{K_m}{V_{max}} + \frac{1}{V_{max}} \cdot A$$

Plot $\frac{A}{V}$ versus A,

$$\text{slope} = \frac{1}{V_{max}}, \ \text{intercept} = \frac{K_m}{V_{max}}$$

EADIE–HOFSTEE TRANSFORMATION

$$V = V_{max} - K_m \cdot \frac{V}{A}$$

Plot V versus $\frac{V}{A}$,

$$\text{slope} = -K_m, \ \text{intercept} = V_{max}$$

In soil it is often difficult to obtain parameters for a particular microbiological transformation process because a whole range of microorganisms may perform the same transformation. Therefore, the overall Michaelis–Menten parameters obtained represent a kind of average parameter set. Since the Michaelis–Menten parameters usually refer to purified enzymes, it is better to use the term 'apparent' Michaelis–Menten parameters when we are dealing with soil.

1.4.7 Units

In the models so far, no units or dimensions have been used. If in the system: $A \xrightarrow{k} B$ the substrate (A) and the product (B) have units of concentration (e.g. $\mu g\ g^{-1}$ soil) the flow rate from A to B will have units of concentration per time (e.g. $\mu g\ g^{-1}$ soil day^{-1}). Depending on the type of kinetics, the rate constant, k, will have the following units:

$$\text{Zero-order kinetics} = \mu g \cdot g^{-1} \cdot day^{-1} \left(\text{rate} = k;\ \left[\mu g \cdot g^{-1} \cdot day^{-1}\right] \right)$$

$$\text{First-order kinetics} = day^{-1} \left(\text{rate} = k \cdot A;\ \left[day^{-1}\right] \cdot \left[\mu g \cdot g^{-1}\right] \right)$$

$$\text{Michaelis–Menten kinetics} = V_{max}: \left[\mu g \cdot g^{-1}\ day^{-1}\right],\ k_m: \left[\mu g \cdot g^{-1}\right]$$

$$[V] = \frac{\left[\mu g \cdot g^{-1} \cdot day^{-1}\right] \cdot \left[\mu g \cdot g^{-1}\right]}{\left[\mu g \cdot g^{-1}\right] + \left[\mu g \cdot g^{-1}\right]} = \left[\mu g \cdot g^{-1} \cdot day^{-1}\right]$$

In Chapter 2, models will be developed for nitrification (Section 2.3), denitrification (Section 2.4) and C/N transformation (Section 2.5) in soils which are based on kinetics presented in this section. The goal is for the models to present real situations as closely as possible. The model development will move from simple introductory steps up to a quite advanced level, thereby catering for a wide range of model development. In this process, the step-by-step approach of model development is extremely important, useful for both beginners and advanced users.

1.5 Introduction to ModelMaker

The main idea behind this text is to provide techniques and strategies on how to translate processes in the soil–atmosphere continuum into mathematical models. This analysis would be incomplete without performing actual model calculations. In recent years, computer software programs have been developed which fall into the category 'modelling packages'. These programs are developed with users in mind who have only limited time available to transform their ideas into mathematical models. This time factor is particularly important when programming skills are required. To learn programming languages, such as FORTRAN, C++ or any other language, takes a long time. Even if programming skills are available, the time factor in actually writing specialized code is very important. Therefore the modelling packages developed in recent years are a major advantage for people who know the mathematical vocabulary and want to transform their ideas into user-friendly mathematical models. We are therefore free to concentrate on the development of the necessary conceptual ideas within biosphere research because the computational side is taken care of by the use of modelling packages.

The ideas presented here can be implemented in many currently available modelling packages. However the examples are specifically presented in ModelMaker (Version 3 or higher; Cherwell Scientific – see footnote in Preface), because currently it is a widely used modelling package among scientists working in the field of natural sciences.

The most important components available for model development within ModelMaker are outlined below. An excellent step-by-step tutorial to learn ModelMaker is provided in the manual (Walker, 1997). It only takes a few hours to become sufficiently familiar with the program that simple models can be developed.

Some of the components available in ModelMaker used in this book are as follows.

DEFINED VALUES <DEFINE>[6]

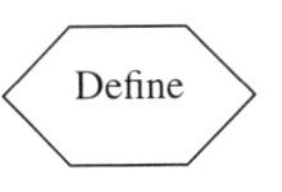

Defined values are components which are only calculated at the start of a model run or in response to an event action (see below). Therefore they are ideal for holding values which do not change very often.

[6] The term in brackets (< >) will be used throughout the text to indicate which component is used.

VARIABLES <VARIABLE>

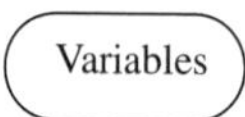

Variables are recalculated at every calculation step and are therefore ideal for holding values which change very often.

COMPARTMENTS <COMPARTMENT>

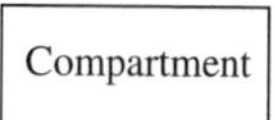

Differential equations are solved with compartments. In Model/Run options, the integration method can be chosen as well as the range of the independent variable over which the integration should be performed.

It is possible to link compartments via the following:

FLOWS <FLOW>

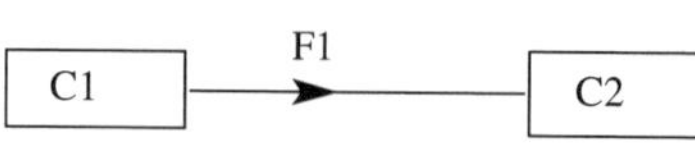

Flows can take actual mathematical expressions. They are useful for holding quantities which are moving from one compartment to another.

INFLUENCES <INFLUENCE>

If a ModelMaker component is referred to in another calculation (e.g. C1 is used in the calculation of component C2), it is necessary to introduce links (in ModelMaker called <influences>, indicated by a dotted line with an arrow). It is also possible to make components globally available (indicated by green hatching or grey shading), so that influences are not needed. However, to make it easier to understand and visualize the mathematical links among model components it is advisable to keep as many influences as possible.

DELAYS <DELAY>

Delays can hold the value of another model component. Delays operate for a defined period, with the previous value available in the current model calculation.

PARAMETERS <PARAMETER>

Parameters are constants or values which do not change during a model run (e.g. the number pi (π) = 3.14…).

LOOKUP TABLES <LOOKUP TABLE> AND LOOKUP FILES <LOOKUP FILE>

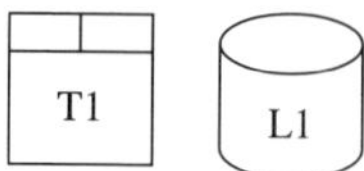

Lookup tables and lookup files are used to hold input parameters such as climate data.

INDEPENDENT EVENTS <INDEPENDENT EVENT>

Independent events are implemented to cater for discontinuous models and for actions which should be carried out while the model is running. Calculations in independent events are carried out under defined conditions. Calling actions, redefinition of defined values or advanced mathematical calculations such as calculations in iteration loops (e.g. Newton–Raphson method in the Appendix), can be performed in independent events.

COMPONENT EVENTS <COMPONENT EVENTS>

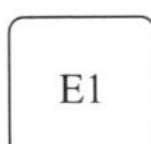

Component events work in a similar way to <independent events> with the only difference that they are carried out according to a conditional statement rather than an exact defined value. This provides additional flexibility in the execution of these statements.

ARRAY NOTATION <ARRAY>

Arrays are another term for components which are written in the form of an index notation. For instance, soil layers may be indicated by the component 'layer'. However, with many soil layers, it is convenient to work with an array or index notation which allocates numbers to the various layers (e.g. Fig. 1.9: layer[1], layer[2], etc.). With such a notation, it is easy to create multilayered models (Fig. 1.10; see also Chapters 3 and 4).

Other components available in ModelMaker include submodels and DLL functions[7], but they are not used here.

A detailed description of ModelMaker, as well as instructions on how to develop models in ModelMaker, is given in the manual (Walker, 1997) and on the Cherwell website. However, the development of the first model, in the next section, will provide an insight into how to work with ModelMaker.

1.6 The First Model

In the first model, a simple one-compartment mixing problem is considered. This model allows the reader to apply some of the mathematical tools presented earlier. Throughout the model development, emphasis is placed on developing a clear definition of the research question and the development of a conceptual model. This will be translated into a mathematical model and later solved with the help of ModelMaker. It is a good idea always to follow a certain sequence during model development. The following steps are usually part of the process after the research question is posed.

[7] Dynamic Link Library: specialized code which is written outside of ModelMaker.

Step 1
Formulate a conceptual model so that all the ideas and concepts which are important are represented in the most realistic and appropriate way. The clearer the conceptual model is, the easier it will be to translate the ideas into a mathematical notation.

Step 2
This step includes the translation of the conceptual model into a mathematical notation. The most important advice which can be given is always to start at the simplest level and gradually increase the complexity of the concepts and ideas. If no mathematical equations can be given for the processes in question, it is often possible to state relationships which are proportional to each other. Very often we learn from ideas and concepts other people have developed. Often, the best way is to learn by example. Therefore a good knowledge of the related literature and how others have tackled similar problems may be invaluable. Of course, a knowledge of mathematical tools and confidence in applying them is extremely important. Usually, all the above factors will work together in finding a suitable translation for the problem in question. The confidence and ease in the translation of particular problems into mathematical notation will usually increase with practice. The step-by-step approach followed in this text helps the reader to become gradually familiar with the concepts and their translation into various aspects of soil processes and their interaction with the biosphere.

Step 3
This step includes the actual process of finding a solution to the mathematical notation developed in step 2. This can range from calculations on a hand-held calculator or using commercially available spreadsheet software (e.g. ‘Excel’, ‘Lotus’, ‘Quattro Pro’, ‘Supercalc’) up to specialized computer code written in a programming language (e.g. ‘Fortran’, ‘Pascal’, ‘C’, ‘Visual Basic’, ‘StarCalc’). We shall use ModelMaker to solve the equations.

Research question

Consider a flask containing 5 l of salt water with an initial salt content of 5 g. The flask is connected to an inlet and an outlet. Solution is steadily pumped in and out of the flask at a rate of 2 l h^{-1}. Suppose that the salt solution entering the flask contains 30 g salt l^{-1} and the mixture in the flask is well stirred. The research question we want to solve is:

How does the salt content change in the flask with time?

To answer this question we shall work through the following tasks:

1. Present a picture (a conceptual model) of the system (step 1).
2. Find a differential equation which calculates the amount of salt in the flask at time t and calculate the analytical solution of this equation (step 2).
3. Solve the differential equation both numerically and analytically with the help of ModelMaker and present the results in the form of a graph (step 3).

Note: for this simple introductory problem it is possible to find an analytical solution and it is therefore calculated alongside the numerical solution. However, in later model developments, the step of finding an analytical solution is omitted.

REPRESENTATION OF THE SYSTEM IN THE FORM OF A CONCEPTUAL MODEL (FIG. 1.13)

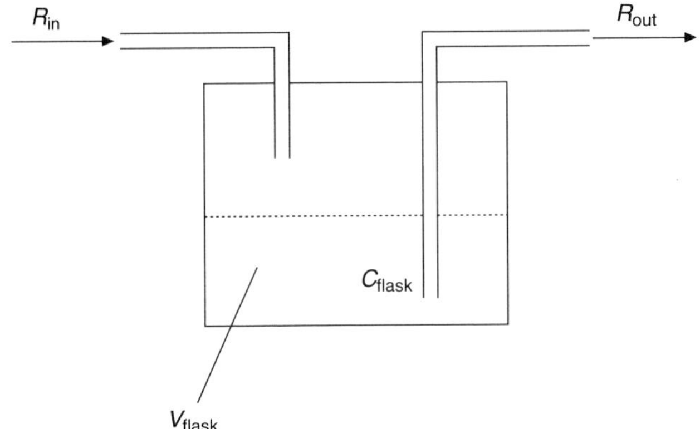

Fig. 1.13. Conceptual model of the one-compartment mixing process (first model, *Mod1-1a.mod*).

MATHEMATICAL DESCRIPTION OF THE CONCEPTUAL MODEL (FIG. 1.13)

The salt concentration in the flask (C_{flask}) is calculated by:

$$C_{flask} = \frac{W_{salt}}{V_{flask}} \quad [g \cdot l^{-1}] \qquad (1.6/1)$$

where:

W_{salt} = weight of salt in the flask [g]
V_{flask} = volume of flask [l]

The rate of salt entering the flask (R_{in}) is given by:

$$R_{in} = C_{in} \cdot F_{in} \quad [g \cdot h^{-1}] \qquad (1.6/2)$$

where:

C_{in} = concentration of salt in the salt solution entering the flask (eqn 1.6/1) $[g \cdot l^{-1}]$
F_{in} = flow rate of salt solution entering the flask $[l \cdot h^{-1}]$

The rate of salt leaving the flask (R_{out}) is given by:

$$R_{out} = C_{flask} \cdot F_{out} \quad [g \cdot h^{-1}] \qquad (1.6/3)$$

where:

C_{flask} = concentration of salt in the flask [$g \cdot l^{-1}$]
F_{out} = flow rate of salt solution leaving the flask [$l \cdot h^{-1}$]

The change in the weight of salt (W_{salt}) in the flask over time can be described by the differential equation:

$$\frac{dW_{salt}}{dt} = R_{in} - R_{out} \qquad (1.6/4)$$

with:

W_{salt_init} = initial amount of salt in the flask [g]

($W_{salt\_init} \geq 0$)

Equation 1.6/4 describes the rate of change (g h^{-1}) in the amount of salt in response to the salt moving in and out of the system.

The full equation of the entire system combining eqns 1.6/1–1.6/4 is:

$$\frac{dW_{salt}}{dt} = C_{in} \cdot F_{in} - \frac{W_{salt}}{V_{flask}} \cdot F_{out} \qquad (1.6/5)$$

Since the flowrates F_{in} and F_{out} are equal, we can define $F_{in\_out} = F_{in} = F_{out}$ and rewrite the eqn 1.6/5 as:

$$\frac{dW_{salt}}{dt} = F_{in\_out} \cdot C_{in} - F_{in\_out} \cdot \frac{W_{salt}}{V_{flask}}$$

or:

$$\frac{dW_{salt}}{dt} = F_{in\_out} \cdot \left(C_{in} - \frac{W_{salt}}{V_{flask}} \right) \qquad (1.6/6)$$

To obtain the total salt content in the flask (W_{salt}) we have to integrate eqn 1.6/6, i.e. sum up the change of salt content which has been accumulated over the time the system was allowed to run. This involves again bringing all variables belonging to each other on one side (e.g. all W_{salt} terms on one side). Integration of eqn 1.6/6 yields:

$$\int \frac{1}{C_{in} - \frac{W_{salt}}{V_{flask}}} \, dW_{salt} = \int F_{in\_out} \, dt \qquad (1.6/7)$$

To present the nomenclature more clearly we can also write the last equation as:

$$\int \frac{1}{C_{in} - \frac{1}{V_{flask}} W_{salt}} \, dW_{salt} = \int F_{in\_out} \, dt$$

which makes clear that the constant $-\frac{1}{V_{flask}}$ is in front of W_{salt} on the left-hand side. During integration the term on the left-hand side will be multiplied with 'one over this constant', i.e. $\frac{1}{-\frac{1}{V_{flask}}} = -V_{flask}$, and the resulting equation is (see also eqn 1.2/14):

$$-V_{flask} \cdot \ln\left(C_{in} - \frac{W_{salt}}{V_{flask}}\right) + C_1 = F_{in\_out} \cdot t + Const + C_2$$

After combining the two integration constants (C_1 and C_2) into one constant (Const), we obtain:

$$-V_{flask} \cdot \ln\left(C_{in} - \frac{W_{salt}}{V_{flask}}\right) = F_{in\_out} \cdot t + Const \qquad (1.6/8)$$

Rewriting the last equation, taking antilogs and finally rewriting the equation in terms of W_{salt}, we have:

$$\ln\left(C_{in} - \frac{W_{salt}}{V_{flask}}\right) = -\frac{F_{in\_out} \cdot t + Const}{V_{flask}}$$

$$C_{in} - \frac{W_{salt}}{V_{flask}} = e^{\left(-\frac{F_{in\_out} \cdot t + Const}{V_{flask}}\right)}$$

$$-\frac{W_{salt}}{V_{flask}} = e^{\left(-\frac{F_{in\_out} \cdot t + Const}{V_{flask}}\right)} - C_{in}$$

and finally:

$$W_{salt} = -V_{flask} \cdot \left[e^{\left(-\frac{F_{in\_out} \cdot t + Const}{V_{flask}}\right)} - C_{in} \right] \qquad (1.6/9)$$

The integration constant (Const) can be specified at $t = 0$ when the weight of W_{salt} has been initially set to W_{salt_init}. This leads to the following equation for Const:

$$W_{salt\_init} = -V_{flask} \cdot \left[e^{\left(-\frac{F_{in\_out} \cdot 0 + Const}{V_{flask}}\right)} - C_{in} \right]$$

$$W_{salt\_init} = -V_{flask} \cdot \left[e^{\left(-\frac{Const}{V_{flask}}\right)} - C_{in} \right]$$

$$-\frac{W_{salt\_init}}{V_{flask}} = e^{\left(-\frac{Const}{V_{flask}}\right)} - C_{in}$$

$$C_{in} - \frac{W_{salt\_init}}{V_{flask}} = e^{\left(-\frac{Const}{V_{flask}}\right)}$$

and finally after taking the antilog and multiplication with $-V_{flask}$:

$$Const = -\ln\left[C_{in} - \frac{W_{salt\_init}}{V_{flask}} \right] \cdot V_{flask} \qquad (1.6/10)$$

If W_{salt_init} is zero (i.e. the flask initially contained no salt), the integration constant reduces to:

$$Const = -\ln[C_{in}] \cdot V_{flask}$$

Instead of finding a general solution to eqn 1.6/6, it is sometimes preferred to integrate directly within the limits, which would lead to the same result. The nomenclature in this case would be:

$$\int_{W_0}^{W} \frac{1}{C_{in} - \frac{W_{salt}}{V_{flask}}} \, dW_{salt} = \int_{0}^{t} F_{in\_out} \, dt$$

$$-V_{flask} \cdot \ln\left(C_{in} - \frac{W_{salt}}{V_{flask}} \right) \Bigg|_{W_0}^{W} = F_{in\_out} \cdot t \Big|_0^t$$

$$\ln\left(\frac{C_{in} - \frac{W_{salt}}{V_{flask}}}{C_{in} - \frac{W_{salt\_init}}{V_{flask}}} \right) = -\frac{F_{in\_out} \cdot t}{V_{flask}}$$

After rearrangement, we obtain eqns 1.6/9 and 1.6/10.

The assumptions are:

1. An infinite supply of salt solution with the concentration C_{in} is available to enter the flask.
2. The solution in the flask is well mixed.

The parameters used in the model calculation are given in Table 1.2.

To view the implementation of the model in ModelMaker and 'run' the model open the file: *Mod1–1a.mod* in ModelMaker.

The graphical display of the model (*Mod1–1a.mod*) in ModelMaker is given in Fig. 1.14. The numerical solution of the problem (solving eqn 1.6/6) is set out in the upper part and the analytical solution (solving eqn 1.6/9) is presented in the lower part of the visual display of the ModelMaker model. For the numerical solution two variables, R_{in} (eqn 1.6/2) and R_{out} (eqn 1.6/3) as well as the compartment W_{salt} (eqn 1.6/6), were added and connected by influences (Fig. 1.14). Influences have to be inserted between components which are needed for the calculation of another component. For instance, R_{out} is used in the calculation performed in W_{salt} and W_{salt} is used in the calculations of R_{out} (Fig. 1.14). For the analytical solution, only two variables: 'Wsalt_analytical' (eqn 1.6/9) and Const, the integration constant (eqn 1.6/10), are needed. The mathematical descriptions in the various components are identical to notations in eqns 1.6/1–1.6/10.

The development of the salt content in the flask with the parameters given in Table 1.2 is presented in Fig. 1.15. Both numerical and analytical solutions provide almost identical results (Fig. 1.15). Wherever possible, such a comparison of the numerical solution with the analytically correct solution should be carried out to check the accuracy of the numerical solution. Furthermore, such a comparison can greatly increase the 'trust' we shall have in the numerical approximation.

Even with this very simple example we see that performing an integration to obtain a general analytical solution of the problem is not as straightforward as using a numerical solution. With increasing complexity of the differential equation, this becomes even more of a problem and very often we have to just 'trust' that the numerical solution provides correct results because either it is too difficult to integrate the equation or simply no analytical solution is available.

Extending the first model

Once we are happy with the performance of the basic model, we might want to extend the model to include other aspects. The first model represented a

Table 1.2. Parameters for the first model (*Mod1-1a.mod*).

F_{in_out}	2 l h^{-1}
C_{in}	30 g l^{-1}
V_{flask}	5 l
W_{salt_init}	5 g

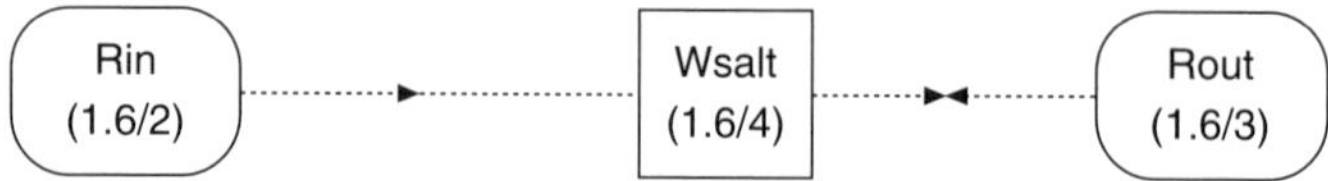

Fig. 1.14. Implementation of the first model in ModelMaker (*Mod1-1a.mod*),

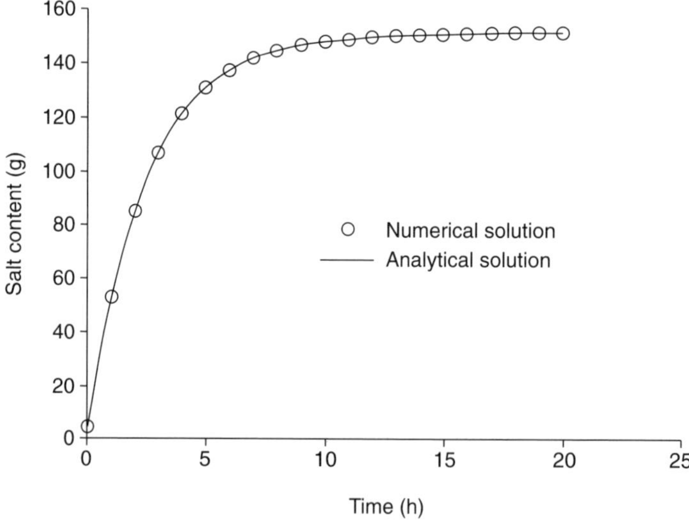

Fig. 1.15. Numerical and analytical solution of the one-compartment mixing model (*Mod1-1a.mod*).

one-compartment mixing process. However, we might want to investigate what happens if, instead of having an infinite supply of salt solution available (assumption 1), the supply of salt solution is limited to only 5 l and is then replaced by water which contains either no salt or only a very small salt content. The conceptual model is extended by the addition of a flask which contains the salt solution and where mixing with the incoming water takes place (Fig. 1.16).

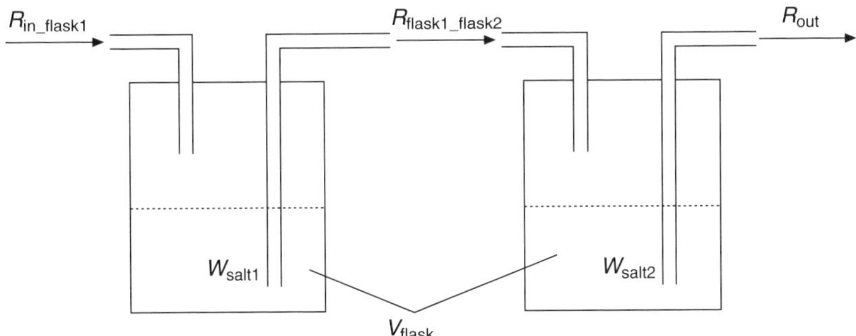

Fig. 1.16. Conceptual model for a two-compartment mixing model (extension to first model, *Mod1-1b.mod*).

The main change to the previous model is the addition of flask 1, with an initial amount of salt in flask 1 calculated by:

$$W_{salt1\_init} = C_{in} \cdot V_{flask} \quad [g] \tag{1.6/11}$$

The salt solution which is pumped from flask 1 into flask 2 is replaced by water with a given salt content (C_{in_water}). Therefore the salt concentration in the first flask will gradually be diluted in response to the salt content of the water. Salt will enter flask 1 at a rate of:

$$R_{in\_flask1} = F_{in\_out} \cdot C_{in\_water} \quad \left[g \cdot hour^{-1}\right] \tag{1.6/12}$$

The transfer of salt solution from flask 1 to flask 2 (R_{flask1_flask2}) is calculated according to eqn 1.6/2 as:

$$R_{flask1\_flask2} = \frac{W_{salt1}}{V_{flask}} \cdot F_{in\_out} \tag{1.6/13}$$

The rest of the system is similar to the one compartment mixing model with parameters presented in Table 1.2.

To view the implementation in ModelMaker and run the model open the file: *Mod1–1b.mod.*

The development of the salt content in flask 1 and flask 2 is presented in response to different salt concentrations in the water entering flask 1 (Fig. 1.17).

Results in Fig. 1.17 show that, after approximately 20 h, the salt concentration in both flasks has approached the salt concentration of the water entering flask 1.

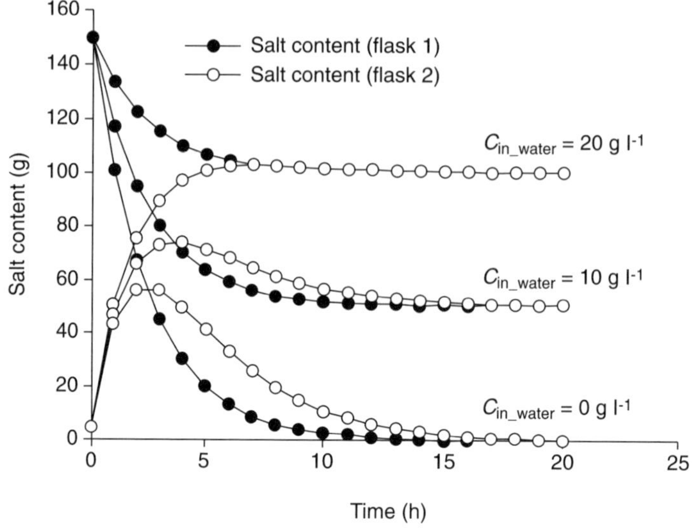

Fig. 1.17. Development of the salt content in flask 1 and flask 2 for different salt concentrations in the water entering flask 1 (*Mod1-1b.mod*).

Checking the results

In order to avoid mathematical error, we have already stressed the importance of checking, where possible, that the numerical approximation gives the same result as the analytical solution. Here we take a further step and check that the results obtained by the model calculations are in line with our conceptual ideas formulated at the beginning of the model development.

To do this, it is useful to sketch a possible solution for both flasks under the given conditions. This sketch should present the expected trend rather than an exact numerical solution. By sketching a possible solution, we also check that we have understood the system properly. We started the simulation with an initially high salt concentration in flask 1 and a very low salt concentration in flask 2. Water entering flask 1 gradually replaced the salt solution in the first flask. Using these conceptual ideas we would expect the following:

1. The salt content in flask 1 will decrease with time (there is no source which could possibly increase the salt content).
2. The dilution effect in flask 1 will be somehow governed by the salt content entering flask 1 and the flow rate of the system.
3. The salt content in flask 2 will, at least initially, increase with time because a solution with a high salt concentration enters the flask.
4. If the simulation runs long enough, all the salt solution in the system will be

replaced by the water entering flask 1 and therefore the salt concentration in both flasks should approach the salt concentration of the water.

With the parameter settings (Table 1.2) and the salt concentration in the water entering flask 1 (C_{in_water}), we are able to calculate the final salt content in the flasks (point 4). With a flask volume of 5 l and the salt concentrations entering flask 1 of 0, 10 and 20 g l^{-1} (Fig. 1.17), the final salt content in flask 1 and 2 would be: 0, 50 and 100 g (calculation: $C_{in\_water} \cdot V_{flask}$).

Comparing the conceptual ideas formulated in points 1–4, including the calculation of the expected final salt content, with results in Fig. 1.17, it seems that the system is behaving in a way we would expect it to behave.

This kind of 'conceptual check-up' should always be done. The best way, of course, is to check against measured values. However, very often measurements are not available. Comparing the output of the model with the initial conceptual ideas and our expectations, we can often get a 'feeling' as to whether the model is behaving in a realistic way or not. This whole procedure is therefore also a good check on whether our initial conceptual model is actually realistic enough to represent the system we wanted to model. Sometimes even an unexpected behaviour of the model may provide us with ideas on how to appropriately alter the conceptual model.

In complex systems, it is often not possible to perform a check once the whole model has been put together. The idea behind large systems (e.g. ecosystem models) is often to predict processes which cannot easily be measured. While developing a complex model, it is essential to check every single step. Only when we are happy with one stage of the model development should we proceed with the next. The additional advantage of following such a step-by-step approach is that our understanding of the model and its dynamic behaviour will grow alongside the model development. Such an approach of developing models step by step is followed in the next chapters when we are dealing with actual processes in the biosphere.

Before going ahead with the development of models which represent processes in the soil–biosphere, it might be a good idea to investigate the models developed so far in more detail. It might be interesting to observe the model behaviour if one of the parameters is changed to a different value. This can be done in ModelMaker by going to <View> <Parameters> and double-clicking on the parameters which should be altered. Furthermore, it might be of interest to have a look at some other outputs of the model, e.g. the flow rate 'Rflask1_flask2'. This can be done by creating a new graph or table (<View> <Graph> or <Table>) and choosing the appropriate component. As a new user, to become familiar with ModelMaker the best idea is to follow the 'learning-by-doing' approach. This can be done by creating the models again from the beginning and comparing the outputs of your own models with the supplied versions.

Exercise

In a more realistic situation (e.g. input of salt into a lake or stream from a factory), the salt input would not be constant but would vary with time. Suppose you want to allow for variable inputs of salt. Which are the alterations you have to consider to *Mod1–1b.mod* ?

SOLUTION

See *Mod1–1c.mod*.

Nitrogen Transformations in Soil 2

2.1 Introduction

Nitrogen is an important element in the soil and the biosphere. The inorganic forms, nitrate (NO_3^-−N) and ammonium (NH_4^+−N), are taken up by plants and are used by numerous microbiological processes catalysed by specialized enzymes. Two very important N transformation processes in soil are nitrification and denitrification. Nitrification refers to the transformation of ammonium into nitrate while denitrification transforms nitrate into gaseous N forms and thereby returns it to the atmosphere. A store of potentially available nitrogen is present in the soil organic matter and held tightly in the interlayer of clay minerals. Nitrogen in the soil organic matter becomes available through microbial decomposition (mineralization). The uptake of inorganic N by microorganisms and subsequent transfer into humic material is referred to as N immobilization. All the above-mentioned N transformations can occur simultaneously in soil, probably in niches which create favourable conditions for the various processes. How fast the N transformations proceed is primarily a combination of the amount of substrate available, the pH of the soil, the soil temperature and the oxygen conditions. The last of these is itself a function of the soil moisture content. Detailed descriptions of the N transformation processes are presented in numerous textbooks (e.g. Tate, 1995; Paul and Clark, 1996). However, some important aspects needed to understand the conceptual models developed later on in this chapter are presented below.

Mineralization and immobilization

In most soils, organic fractions are present as plant residues, fast and easily decomposable soil organic-matter fractions and dead microbial material.

Numerous groups of microbes, each specialized in feeding on a particular fraction of the soil organic matter, will use the carbon in these soil organic-matter fractions for maintenance and growth (van Veen and Paul, 1981; Paul and Clark, 1996). Depending upon the N content of the decomposing organic material, a flow of mineralized N enters the inorganic nitrogen pools of the soil (mainly in the form of ammonium). The transformation processes are catalysed by various groups of the soil microbial biomass. The processes occur at quite different speeds, dependent on the kind of organic source (e.g. lignin is more resistant to decomposition than other constituents) and whether they are physically or chemically protected from degradation (Hunt, 1977; van Veen and Paul, 1981; Parton *et al.*, 1987).

Whether the mineralized nitrogen is directly immobilized again is mainly dependent on the C:N ratio of the microbial biomass (i.e. the living part of the soil C–N pool which performs the transformations). In order to grow, microbes need carbon and nitrogen. If the N requirement for microbial growth is not met, the microorganism will immobilize the available nitrogen, mainly as ammonium but also as nitrate, especially under conditions when carbon-rich substrates are digested (e.g. straw with a high C:N ratio) (Jansson *et al.*, 1955; van Veen *et al.*, 1984; Tate, 1995).

The microbial biomass consists of numerous bacterial groups, each with a specific activity which is dependent upon the availability of substrates and suitable growth conditions (Hunt, 1977). Hence, whether the various microbial biomass fractions are active or dormant (awaiting suitable conditions) will determine the dynamics of the C/N transformations in soil (Hunt, 1977; Blagodatsky and Richter, 1998).

Nitrification

NH_4^+–N available in soil as a result of net mineralization can be taken up by plants, adsorbed by clay minerals and organic matter or utilized for microbial nitrification. Autotrophic nitrification proceeds in two steps, whereby carbon dioxide (CO_2) is generally used as a carbon source and the energy required is obtained by the oxidation of NH_4^+–N to nitrite (NO_2^-–N) by *nitroso*bacteria (e.g. *Nitrosomonas europaea*) and the oxidation of NO_2^-–N to NO_3^-–N by *nitro*bacteria (e.g. *Nitrobacter*). In most habitats, the two steps occur simultaneously and the intermediate NO_2^-–N almost never accumulates (Paul and Clark, 1996). Under O_2-limiting conditions in a process called 'nitrifier denitrification', ammonium oxidizers may use NO_2^-–N as an alternative electron acceptor and produce nitric oxide (NO) and nitrous oxide (N_2O) (Granli and Bøckman, 1994). However, it is unclear whether a reduction sequence $NO_2^- \rightarrow NO \rightarrow N_2O$ exists, as in denitrification (Conrad, 1996a). NO and N_2O may also be produced by chemical decomposition of NO_2^-–N together with hydroxylamine (NH_2OH), an intermediate between NH_4^+–N and NO_2^-–N. The NO/N_2O emission ratios vary to a large extent because the production rates are affected differently by environmental conditions. Usually only a small percentage of the key substance, nitrite, is transformed into gaseous nitrogen compounds (Conrad, 1996a; Paul and Clark, 1996).

Denitrification

Denitrification is generally referred to as the microbial reduction of NO_3^-–N to NO_2^-–N and further to gaseous forms: NO, N_2O and molecular nitrogen (N_2). Specialized enzymes catalyse each step. Denitrifiers use NO_3^-–N as their primary electron acceptor for obtaining energy from organic C, which proceeds under anaerobic conditions (Paul and Clark, 1996). A pH value of 6–8 and temperatures around 20°C provide optimum conditions. The diverse pattern of the relative emission rates of NO:N_2O:N_2 observed is species-dependent, as well as being affected by the environmental conditions (Conrad, 1996a). The N_2O:N_2 emission ratio, for instance, increases with low temperatures and low pH values (Granli and Bøckman, 1994). The high affinity of NO (i.e. only small NO concentrations are needed to activate the enzyme) may even lead to microbial NO consumption (Conrad, 1996a).

Various enzymes catalyse each of the above-mentioned transformations. The rates of the transformation processes (kinetics), along with the transport processes (described in Chapters 3, 4 and 7), are the key process groups to aid the understanding of microbiologically mediated transformations in natural environments. The various kinetics were described in Chapter 1 (Section 1.4). In the following section we focus on the calculation of the various kinetics and its implementation in ModelMaker. This provides practice in basic model development and leads to an understanding of the basic kinetics.

2.2 Modelling Kinetics

2.2.1 Zero-, first- and second-order kinetics

The numerical and analytical solutions for A (substrate) and B (product) for zero- (eqns 1.4/1, 1.4/2), first- (eqns 1.4/4, 1.4/5, 1.4/7, 1.4/8) and second-order kinetics (eqns 1.4/10, 1.4/11, 1.4/13) are calculated in the ModelMaker model *Mod2–1a.mod.* Figure 2.1 presents components and the set up for the calculations in ModelMaker.

Open *Mod2–1a.mod* in ModelMaker and note the following: the <compartments> A and B are linked via a <flow> 'Rate'. The transformations from A to B occur by subtracting 'Rate' from A and adding this quantity to B. Therefore, the <flow> 'Rate' has to be defined only once. The analytical solutions (for A and B) as well as the half-life of A are calculated in the <variables> 'A_anal', 'B_anal' and 't_half'. The advantage of defining a separate flow equation (i.e. <flow> 'Rate') is that the model set-up can remain the same while the notation in the <flow> is redefined. A user choice is added to the model via the <independent event> 'Choice' which prompts the user to choose the type of kinetic to be used.

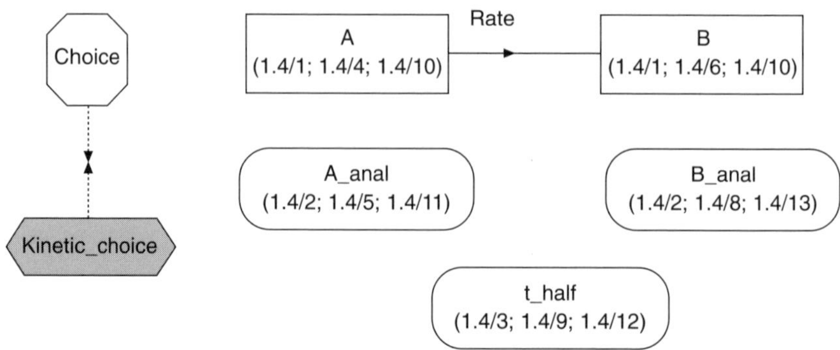

Fig. 2.1. Model to calculate numerical and analytical solutions of zero-, first- and second-order kinetics (*Mod2-1a.mod*).

The following <GetValue> statement is inserted under <actions> of the <independent event>:

```
GetValue("Which kind of Kinetics do you want to use: 0=zero
         order, 1=first order, 2=second order","Kinetic
         Choice",Kinetic_choice,Kinetic_choice=0 or
         Kinetic_choice=1 or Kinetic_choice=2);
```

The kinetic identifier (0 = zero order, 1 = first order or 2 = second order) is held in the <define> 'Kinetic_choice' and passed on to the <flow> 'Rate' and the <variable> 'A_anal' and 'B_anal' (where the analytical solutions are calculated). The components 'Rate', 'A_anal', 'B_anal' and 't_half' are made conditional so that the calculations are performed according to the kinetics chosen. This type of conditional calculation in response to user inputs is used very often in the forthcoming models. Event actions of this kind can simplify the model structure tremendously and should be used whenever a user choice is required. The implementation of a conditional statement can be seen by double clicking, for example, on the component <flow> 'Rate' i.e.

Conditions	**Equations**
`Kinetic_choice = 0`	`Kinetic_const`
`Kinetic_choice = 1`	`Kinetic_const*A`
`Kinetic_choice = 2`	`Kinetic_const*A^2`
`Default`	`0`

In addition to the model components (Fig. 2.1), the following parameters have to be inserted under the parameter section: 'A0' the initial concentration of substrate (set to 100), 'B0' the initial concentration of product (set to 0) and the rate constant 'Kinetic_const' (set to 0.05, dimensionless). Note: the rate constant 'Kinetic_const' has different dimensions according to the chosen kinetics (see Section 1.4.7).

Running the model

It is now up to the user to investigate the model. After starting the simulation with 'GO', the first user interaction with the program is the input of the desired type of kinetics. The model will perform the calculations according to user choice. The outputs of the concentration of substrate (A) and product (B) are presented in the graph 'Substrate_Product'. The numerical solutions for the various kinetics, using the current parameter settings (Kinetic_const = 0.05, A0 = 100, B0 = 0), are calculated for 100 time units and are presented in Fig. 2.2.

Other user actions may involve the change of some parameters to different values. Currently, the initial concentrations of 'A0', 'B0' and the rate constant 'Kinetic_const' are defined as parameters. It might be interesting to investigate the effect on the solution of the various components of the model if the value of one or more of the currently defined parameters is changed to a different value.

2.2.2 Temperature dependency of the rate constant

To model temperature dependency of the rate constant via the Arrhenius approach (Section 1.4.4), the basic model *Mod2–1a.mod* (Fig. 2.1) presented in the last section is extended by the notation described in Section 1.4.4. The parameter 'Kinetic_const' is replaced by a <variable> 'Kinetic_const' (eqn 1.4/14). Parameters alpha (α, called: 'alpha'), E and R (gas constant) have to be added. The

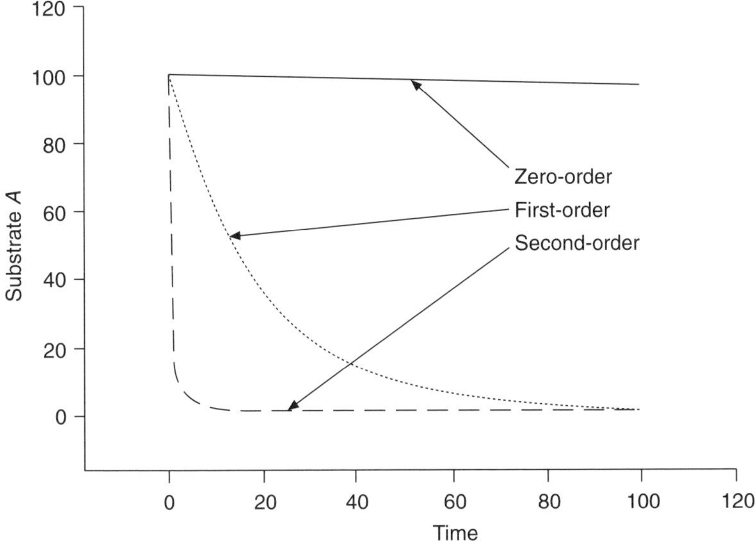

Fig. 2.2. Comparison between zero-, first- and second-order kinetics ($k = 0.05$, $A_0 = 100$) (*Mod2-1a.mod*).

<variable> 'Kinetic_const' is connected to 'Rate', 'A_anal' and 't_half', where it is required for the calculations. Temperatures are provided by a <Lookup table> 'Temp'. The adjusted model is presented in the ModelMaker file *Mod2–1b.mod.*

Running the model
Similar investigations to those already presented for *Mod2–1a.mod* may be suitable for this model (*Mod2–1b.mod*) too. This may involve the change of the parameters related to the Arrhenius equation and its effect on the rate constant (Kinetic_const).

2.2.3 Chain reactions

Chain reactions are characterized by three or more compartments (Section 1.4.5). The previous two-compartment models (Section 2.2.1) are extended by one or more compartments, depending on the 'chain length'. Figure 2.3 represents a three-compartment chain reaction.

The numerical solution of the chain reaction (Fig. 2.3) is calculated with the model *Mod2–1c.mod.* A difference from the previous kinetic models is that the kinetic constants k1 and k2 are now user-definable via a <GetValue> statement in the <independent event> 'Choice':

```
GetValue("Input kinetic constant k1 (from A to B) (0 - 1)",
        "Kinetic constant k1",k1,k1>0 and k1<1);
GetValue("Input kinetic constant k2 (from B to C) (0 - 1)",
        "Kinetic constant k2",k2,k2>0 and k2<1);
```

This statement interacts with the <defines> k1 and k2 which hold the kinetic rate constants, and therefore various scenarios of kinetic types in connection with different kinetic rate constants (k1 and k2) can be tested.

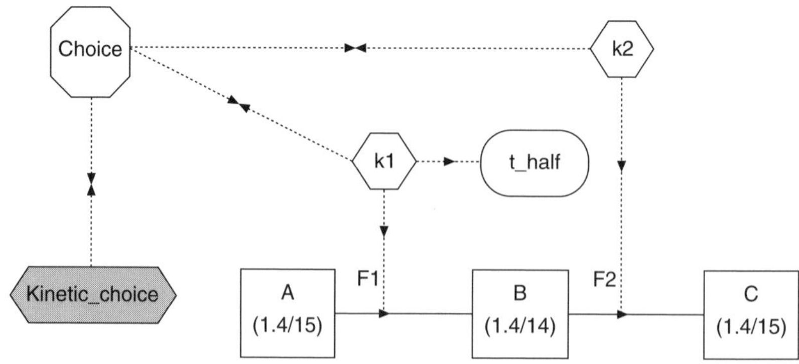

Fig. 2.3. Model for a three-compartment chain reaction (*Mod2-1c.mod*).

Running the model

With the user option to input rate constants (k1 and k2), it is easy to define different simulation scenarios. The model may be investigated by running the various kinetics with the same rate constants and then changing the rate constants and doing the same simulation exercise again. A large number of different combinations are available to customize the model output. Before going ahead, it is a good idea to try out as many user-defined scenarios as possible and investigate the effects on the solution of the chain reaction system.

2.2.4 Michaelis–Menten kinetics

An example of the Michaelis–Menten equation is presented in the file *Mod2–1d.mod* (parameters $K_m = 50$, $V_{max} = 10$). The notation for the Michaelis–Menten enzyme kinetics (eqn 1.4/21) is incorporated as an additional user choice into the notation of <flow> 'Rate' of model *Mod2–1a.mod* (Fig. 2.1) and can therefore be compared with zero-, first- and second-order kinetics (Fig. 2.4).

A comparison of the graphical description of a Michaelis–Menten reaction rate (Fig. 1.12) with the output of the model using the Michaelis–Menten option shows the gradual change from first- to zero-order kinetics with increasing substrate concentrations (Fig. 2.4).

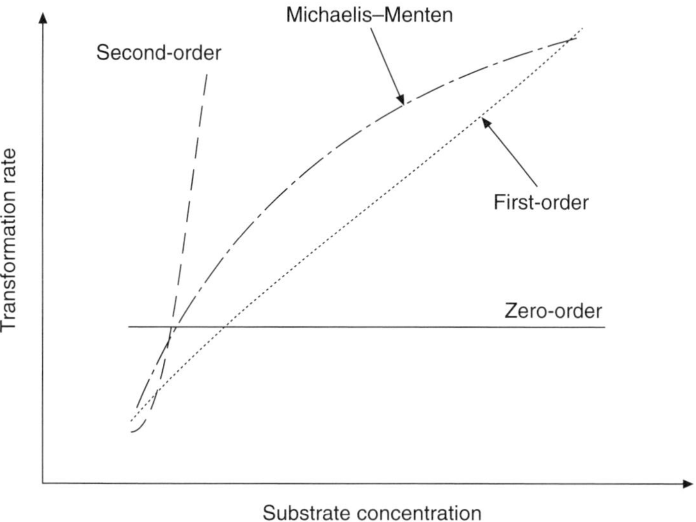

Fig. 2.4. Transformation rates of various kinetics in response to substrate concentrations (calculated with *Mod2-1d.mod*).

Running the model

The conditional notation for the calculation of all four kinetic reaction types is given in the <flow> 'Rate'. The implementation of the conditional statement can be viewed by double-clicking on the component 'Rate'. The theoretical considerations in Section 1.4 have already provided us with an idea of which output we can expect of the component 'Rate'. The default values for the Michaelis–Menten equation are $K_m = 50$ and $V_{max} = 10$ (see parameters in *Mod2–1d.mod*), which means that half of the reaction speed ($V_{max}/2 = 5$) should be reached at a substrate concentration of 50. This can be checked by running the model (Michaelis–Menten option, kinetic_choice = 3) and going to the graph 'Rate', where the output is plotted against substrate concentration (as opposed to time!). Clicking the left-hand mouse button over the graph and holding it down displays a cross with the corresponding data value displayed in the lower left-hand corner of the screen. For the Michaelis–Menten option, this value pair for a substrate concentration at $x = 50$ will display $y = 5$, which corresponds to the above considerations. Further investigations with different parameter settings may be useful in order to become more familiar with the dynamics of Michaelis–Menten kinetics.

As mentioned above, the various kinetics differ mainly in the 'shape' of the reaction rates in response to different substrate concentrations. To illustrate the effect of various kinetics, the output of the flow component 'Rate' from *Mod2–1d.mod* is plotted versus substrate concentrations in Fig. 2.4. The parameters given in *Mod2–1d.mod* are used and Fig. 2.4 is plotted in such a way that the shape of the rates and not the actual rates can be compared (scale of *y*-axis is different for the various kinetics). Typical 'shapes' of transformation rates with increasing substrate concentrations are obtained: zero-order = constant, first-order = linear change, Michaelis–Menten = gradual change from first- to zero-order, second-order = quadratic increase.

2.3 Nitrification

2.3.1 Conceptual nitrification model

Over the last century, the process of nitrification (oxidation of NH_4^+–N to NO_3^-–N) has been studied extensively. In the beginning, there was some debate as to whether this process of nitrate formation was based purely on a chemical reaction or was actually mediated by microorganisms (e.g. Warrington, 1879). However, as early as 1879 (Schloesing and Muntz, 1879), scientists were convinced that specialized microorganisms could perform this important step in the N cycle. Soon it became apparent that the process of nitrification in soil occurs in two steps (NH_4^+–N to NO_2^-–N and NO_2^-–N to NO_3^-–N) and that the entire process is regulated by factors such as soil pH, soil moisture and soil temperature conditions (Schloesing and Muntz, 1879; Warrington, 1884).

With the onset of the computer age in the 1960s, the first attempts were made to describe the process of nitrification mathematically and find numerical solutions for nitrification systems (e.g. Knowles *et al.*, 1965).

The nitrification model developed in this section is based on a description of the nitrification process made in the early 1970s by Paul and Domsch (1972). In this model it is assumed that microorganisms utilize for ammonium oxidation which is held on the surfaces of clay minerals and other soil particles. Therefore, first an exchange between the NH_4^+-N in the soil solution with the NH_4^+-N adsorbed has to be modelled. The oxidation occurs in two steps: from NH_4^+-N to NO_2^--N and from NO_2^--N to NO_3^--N. Each oxidation step is governed by specialized enzymes, which are produced by either *nitroso*bacteria or *nitro*bacteria. The production of oxidation enzymes is linked to the activity of the microorganisms, which is reflected in their generation time (the time taken for the population to double). The activities of the two microbial groups are governed by soil pH, which is itself affected by the oxidation of NH_4^+-N to NO_2^--N. In addition, some loss of N occurs through biosynthesis (uptake of N which is utilized for cell growth).

The basic nitrification model developed in the next section is extended in subsequent sections for NO and N_2O emission and the effects of soil moisture and soil temperature. The conceptual ideas of the basic model are presented in Fig. 2.5 and the implementation of this as a numerical model within ModelMaker is presented at the end of the next section in 'Application in ModelMaker' (see Fig. 2.7).

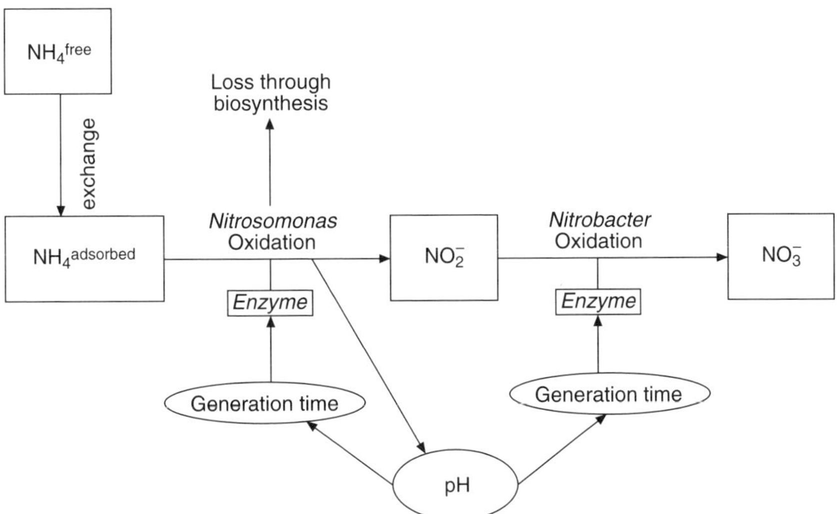

Fig. 2.5. Conceptual nitrification model (according to Paul and Domsch, 1972).

2.3.2 Basic nitrification model

For the adsorption of NH_4^+–N in the soil solution (NH4free[1]) on to soil particle surfaces, such as clay minerals (NH4adsorb), the following assumptions are made:

- The adsorption process is irreversible (i.e. occurs only from NH4free to NH4adsorb).
- The rate of adsorption is proportional to the concentration of NH4free and the adsorption capacity of the soil particles.

The capacity of the soil to adsorb NH_4^+–N is defined by the potential adsorption capacity (defined by parameter NH4adsorb_init) minus ammonium already adsorbed (NH4adsorb). The term 'NH4adsorb_init-NH4adsorb' is therefore a measure of the current adsorption capacity of the soil.

The rate of adsorption (F1) which is proportional to the actual adsorption capacity and the amount of NH_4^+–N in solution (NH4free) can be described by:

$$F1 = k\_free \cdot NH4free \cdot (NH4adsorb\_init - NH4adsorb) \quad (2.3/1)$$

with:

F1 = adsorption rate from NH4free to NH4adsorb [$\mu g\ N \cdot g^{-1}\ soil \cdot h^{-1}$][2]
k_free = adsorption constant from NH4free to NH4adsorb [$(\mu g\ N \cdot g^{-1}\ soil)^{-1} \cdot h^{-1}$] {0.004}
NH4free = NH_4^+–N concentration in soil solution [$\mu g\ N \cdot g^{-1}\ soil$]
NH4adsorb_init = NH_4^+–N concentration which is initially held on the soil particles; here it defines the potential adsorption capacity of the particle surfaces [$\mu g\ N \cdot g^{-1}\ soil$] {20}
NH4adsorb = actual NH_4^+–N concentration held on the soil particle surfaces [$\mu g\ N \cdot g^{-1}\ soil$]

The terms 'NH4free' and 'NH4adsorb' change with time and can be described by the differential equations:

$$\frac{dNH4free}{dt} = -F1 \quad (2.3/2)$$

with:

NH4free_init = initial value [$\mu g\ N \cdot g^{-1}\ soil$] {10}[3]

[1] Note: to facilitate the understanding of the mathematical models, the names in parentheses correspond to the notation used in the actual ModelMaker model presented later on.
[2] Note: dimensions or units are always given in square brackets and actual values of parameters used in the model are always given in curly brackets.
[3] Note: it is a good practice always to state the initial values of the differential equation.

and:

$$\frac{d\text{NH4adsorb}}{dt} = \text{F1} - \text{F2} \qquad (2.3/3)$$

with:

NH4adsorb_init = initial value [μg N $\cdot$ g^{-1} soil] {20, 100}

The flow F2 which is subtracted from the adsorption flux (F1) in eqn 2.3/3 is the rate of NH_4^+–N oxidation to NO_2^-–N (NO2nit). As illustrated by the conceptual model (Fig. 2.5) this rate depends on the activity of the enzyme which catalyses the first oxidation step in the nitrification sequence from NH_4^+–N to NO_2^-–N (called in the model: 'E1nit').

It is assumed that the rate of oxidation is proportional to the available 'NH4adsorb'. Therefore, the flow F2 can be described by:

$$\text{F2} = \text{E1nit} \cdot (\text{NH4adsorb} - \text{NH4adsorb\_end}) \qquad (2.3/4)$$

with:

F2 =	oxidation rate of NH4adsorb to NO2nit [μg N $\cdot$ g^{-1} soil $\cdot$ h^{-1}]
E1nit =	enzyme activity of *nitroso*bacteria [h^{-1}]
NH4adsorb =	total adsorbed NH_4^+–N on soil particles [μg N $\cdot$ g^{-1} soil]
NH4adsorb_end =	concentration of adsorbed NH_4^+–N which is so tightly held on the soil particles so that it is unavailable for oxidation [μg N $\cdot$ g^{-1} soil] {1}

The expression: (NH4adsorb − NH4adsorb_end) determines the NH_4^+–N concentration which is available for oxidation.

The change of NO2nit with time is determined by the differential equation:

$$\frac{d\text{NO2nit}}{dt} = (\text{F2} \cdot \text{k\_E1}) - \text{F3} \qquad (2.3/5)$$

with:

NO2nit_init =	initial soil nitrite concentration [μg N $\cdot$ g^{-1} soil] {0.5}
k_E1 =	fraction of the gross NH_4^+–N oxidation rate available for further oxidation to NO_3^-–N (NO3). The rest is being used for cell growth [–] {0.999}
F3 =	oxidation rate of NO2nit to NO3 [μg N $\cdot$ g^{-1} soil $\cdot$ h^{-1}]

The oxidation rate of NO_2^-–N is calculated in the same way as the NH_4^+–N oxidation rate by (no loss through biosynthesis is considered):

$$F3 = E2nit \cdot NO2nit \tag{2.3/6}$$

with:

E2nit = enzyme activity of *nitro*bacteria [h^{-1}]
NO2nit = nitrite concentration [μg N · g^{-1} soil]

Finally, the change in nitrate (NO3) is calculated as:

$$\frac{dNO3}{dt} = F3 \tag{2.3/7}$$

with:

NO3init = initial soil nitrate concentration [μg N · g^{-1} soil] {0}

Before we can find a solution to the basic nitrification model, it is necessary to define the enzyme system for both *nitroso*bacteria and *nitro*bacteria. In order to calculate the respective enzyme activities E1nit and E2nit, it is assumed that the enzyme activities for the two groups:

- are proportional to the enzyme activities already present in soil;
- are changing in response to the growth rates of the respective microorganisms.

The growth rates for the two groups are assumed to change according to Michaelis–Menten kinetics, where the maximum growth rates (a1 and a2) are calculated according to their optimal generation times (tgen1_opt and tgen2_opt), which are reduced by the pH value of the soil. The pH dependency has been included because H^+ ions are released during NH_4^+–N oxidation (thus decreasing the pH) according to the relationship:

$$NH_4^+ + 1.5O_2 \rightarrow NO_2^- + H_2O + 2H^+ \tag{2.3/8}$$

To calculate actual growth rates for the two nitrifier groups an additional degradation parameter (b1 and b2) is included, which allows for growth rate adjustments with respect to microbial death or transitions to dormancy.

The activities of the two nitrifier groups (E1nit and E2nit) change with time and can be described, according to the above considerations, for *nitroso*bacteria:

$$\frac{dE1nit}{dt} = \left(\frac{a1 \cdot (NH4adsorb - NH4adsorb\_end)}{K1nit + (NH4adsorb - NH4adsorb\_end)} - b1 \right) \cdot E1nit \tag{2.3/9}$$

with:

E1nit_init = initial enzyme activity[h^{-1}]{0.0035}
a1 = maximum growth rate [h^{-1}]
K1nit = Michaelis–Menten parameter [μg N · g^{-1} soil] {14}
b1 = degradation parameter [h^{-1}]{0.0071}

and for *nitro*bacteria:

$$\frac{d\text{E2nit}}{dt} = \left(\frac{\text{a2} \cdot \text{NO2nit}}{\text{K2nit} + \text{NO2nit}} - \text{b2}\right) \cdot \text{E2nit} \qquad (2.3/10)$$

with:

E2nit_init =	initial enzyme activity $[h^{-1}]\{0.1\}$
a2 =	maximum growth rate $[h^{-1}]$
K2nit =	Michaelis–Menten parameter $[\mu g\ N \cdot g^{-1}\ soil]\ \{1.4\}$
b2 =	degradation parameter $[h^{-1}]\{0.005\}$

The maximum growth rates (a1 and a2) occur under conditions of optimal substrate supply. Under these conditions the Michaelis–Menten notation (expression in brackets of eqn 2.3/10) reduces to a zero-order notation (see Section 1.4 eqn 1.4/24). Equation 2.3/9 therefore reduces to:

$$\frac{d\text{E1nit}}{dt} = (\text{a1} - \text{b1}) \cdot \text{E1nit}$$

or in the integrated form:

$$\text{E1nit} = \text{E1nit\_init} \cdot e^{(\text{a1}-\text{b1}) \cdot t}$$

The generation time is defined as the time which is needed to obtain twice the initial population. Therefore, following the procedure described in Section 1.4.2, we can calculate the maximum growth rates (a1 and a2) according to the generation times (tgen1 and tgen2). An example of the steps involved in finding a mathematical expression for maximum growth rates is carried out below for *nitroso*bacteria:

$$2 \cdot \text{E1nit\_init} = \text{E1nit\_init} \cdot e^{(\text{a1}-\text{b1}) \cdot \text{tgen1}} \rightarrow 2 = e^{(\text{a1}-\text{b1}) \cdot \text{tgen1}} \rightarrow$$
$$\ln(2) = (\text{a1} - \text{b1}) \cdot \text{tgen1} \rightarrow \ln(2) = \text{a1} \cdot \text{tgen1} - \text{b1} \cdot \text{tgen1}$$

Rearranging this equation yields for *nitroso*bacteria:

$$\text{a1} = \frac{\ln(2)}{\text{tgen1}} + \text{b1} \qquad (2.3/11)$$

with:

tgen1 =	generation time for *nitroso*bacteria under optimal conditions [h] {18}
b1 =	degradation parameter for *nitroso*bacteria $[h^{-1}]\ \{0.0071\}$

For *nitro*bacteria (performing the same transformations):

$$a2 = \frac{\ln(2)}{tgen2} + b2 \quad (2.3/12)$$

with:

tgen1 = generation time for *nitro*bacteria under optimal conditions [h] {25}
b2 = degradation parameter for *nitro*bacteria [h^{-1}] {0.005}

The pH dependency of maximum growth rates (a1 and a2) can be calculated by multiplying the maximum growth rates, as well as the degradation parameters, with pH reduction functions (fpHnit1 and fpHnit2) using the equation obtained by Parton *et al.* (1996, fig. 2c). The factors can be included into eqns 2.3/11 and 2.3/12 by multiplying a1 and a2 by fpHnit1 and fpHnit2, respectively.

The dimensionless pH reduction function has the general form:

$$fpH = fpH\_A + arctan\left(\frac{pi \cdot fpH\_B \cdot (fpH\_C + pH)}{pi}\right) \quad (2.3/13)$$

pi = 3.14159 … (usually written in Greek: π)

For *nitroso*bacteria the factor is called fpHnit1 and for *nitro*bacteria the factor is called fpHnit2.

Suitable parameters for these two functions were obtained by non-linear curve-fitting of data from Paul and Domsch (1972, p. 84), using the non-linear curve-fitting routine of the software package SigmaPlot (SPSS, 1997). The results for the two nitrifier groups were:

*Nitroso*bacteria		*Nitro*bacteria	
fpH1_A =	0.73	fpH2_A =	0.55
fpH1_B =	0.66	fpH2_B =	0.61
fpH1_C =	−7.52	fpH2_C =	−6.19

The output of eqn 2.3/13 with the parameters for *nitro*bacteria is presented in Fig. 2.6.

The change of pH is calculated in response to the actual NH_4^+–N oxidation rate (F2, eqn 2.3/4), according to the procedure described in Laudelout *et al.* (1976). For each mole NH_4^+ oxidized, 2 moles of H^+ will be produced. Therefore the H^+ ion increase (Hnit) can be calculated by:

$$Hnit = \frac{c\_pH\_nit \cdot F2}{molwt\_N \cdot wt\_conv} \quad (2.3/14)$$

with:

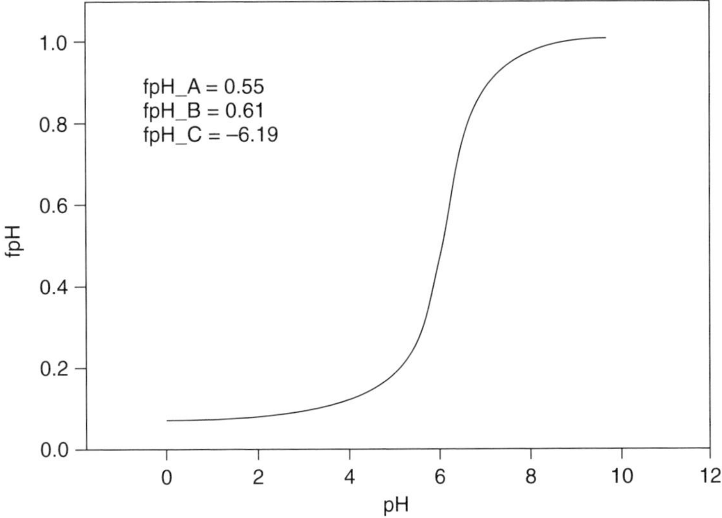

Fig. 2.6. Shape of the dimensionless pH reduction function (eqn 2.3/13).

Hnit =	H^+ ion concentration [mol]
c_pH_nit =	the moles of H^+ produced for each mole of N oxidized [–] {0.5}
F2 =	oxidation rate of NH4adsorb [μg N · g^{-1} soil · h^{-1}]
molwt_N =	molecular weight of nitrogen [g N · mol^{-1}]{14.0067}
wt_conv =	conversion factor [μg · g^{-1}]{1,000,000}

Note: to allow for buffering of soil organic matter, a value of 0.5 rather than the numerically correct value of 2 for the factor 'c_pH_nit' is used.

The total H^+ ion integrated over time is calculated by:

$$\frac{d\text{Htot}}{dt} = \text{Hnit} \tag{2.3/15}$$

The pH value is defined as the negative logarithm of the H^+ ion concentration, i.e.:

$$\text{pH} = -\log(\text{Htot}) \tag{2.3/16}$$

with:

pH_init =	initial pH value of the soil [–] {6.8}

Application in ModelMaker

Before we extend the nitrification model further, it is a good idea to compile the model developed so far as a first step in the model development. Once this first model works satisfactorily, it can then be extended to include other aspects.

The current definition of the nitrification model (eqns 2.3/1–2.3/16) contains seven differential equations or compartments: NH4free, NH4adsorb, NO2nit, NO3, E1nit, E2nit and Htot, three flows between compartments: F1, F2 and F3 and six variables: a1, a2, Hnit, ph, fpHnit1 and fpHnit2. The various model components are inserted into an empty ModelMaker sheet so that the entire model looks like the one presented in Fig. 2.7.

Next, the necessary model parameters listed below are inserted into the parameter definition. The model parameters are in alphabetical order. (For a description of the various parameters, see above.) In addition, an indication is given where time conversions would be needed if time steps other than per hour were used. At the moment the model is set up for hourly time steps. However, other time units could be used (e.g. minutes, days), provided that parameters with 'time' dimension are converted appropriately. The model parameters are:

b1 = 0.0071 $[h^{-1}]$ $\langle \bullet \frac{1}{\text{Time\_convert}} \rangle$

b2 = 0.005 $[h^{-1}]$ $\langle \bullet \frac{1}{\text{Time\_convert}} \rangle$

c_pH_nit = 0.5 [–]

E1nit_init = 0.0035 $[h^{-1}]$ $\langle \bullet \frac{1}{\text{Time\_convert}} \rangle$

E2nit_init = 0.1 $[h^{-1}]$ $\langle \bullet \frac{1}{\text{Time\_convert}} \rangle$

fpH1_A = 0.73 [–]

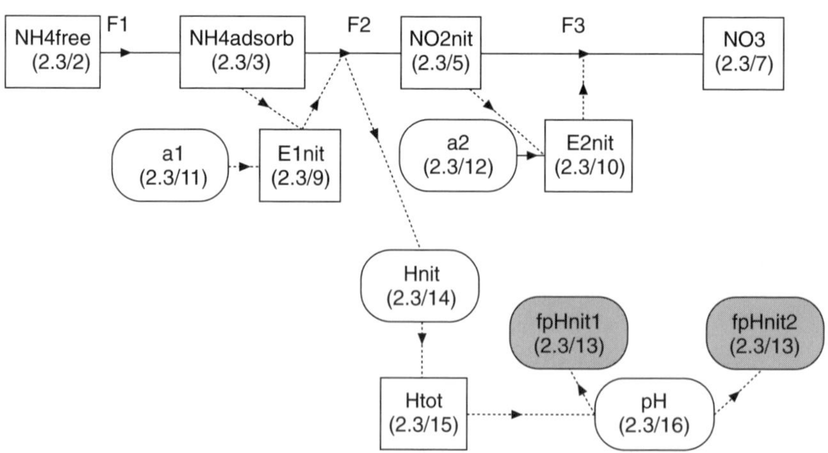

Fig. 2.7. Graphical display of the basic nitrification model (*Mod2-2a.mod*).

fpH1_B =	0.66 [–]
fpH1_C =	−7.52 [–]
fpH2_A =	0.55 [–]
fpH2_B =	0.61 [–]
fpH2_C =	−6.19 [–]
k_E1 =	0.999 [–]
k_free =	0.004 [μg N · g^{-1} soil · h^{-1}] $\langle \bullet 1/_{\text{Time\_convert}} \rangle$
K1nit =	14 [μg N · g^{-1} soil]
K2nit =	1.4 [μg N · g^{-1} soil]
molwt_N =	14.0067 [g N · mol^{-1}]
NH4adsorb_end =	1 [μg N · g^{-1} soil]
NH4adsorb_init =	20 [μg N · g^{-1} soil]
NH4free_init =	10 [μg N · g^{-1} soil]
NO2nit_init =	0.5 [μg N · g^{-1} soil]
NO3init =	0 [μg N · g^{-1} soil]
pH_init =	6.8 [–]
pi =	3.141592654 [–]
tgen1 =	18 [h] $\langle \bullet \text{ Time\_convert} \rangle$
tgen2 =	25 [h] $\langle \bullet \text{ Time\_convert} \rangle$
Time_convert =	$1[^{\text{Timeunits}}/_{\text{h}}]$
wt_conv =	1,000,000 [μg · g^{-1}]

The mathematical definitions as presented in eqns 2.3/1–2.3/16 have to be inserted into the corresponding model components. In addition to the above description, an <independent event> ‘Choice’ has been added to the model which prompts the user to decide whether the pH dependency on the maximum growth rates (a1 and a2) should be calculated or not. To carry out the calculation with or without the pH adjustment, the variables a1 and a2 have been made conditional and are calculated in response to the value of the <define> ‘function_adj’. The parameter ‘Time_convert’ allows for a variable time-step definition. At the moment, the model runs on an hourly basis. However, if we would like to run the model on a per minute basis, we would have to change a lot of parameters to ‘per-minute’ values (divide them by 60). To avoid this work, a parameter has been inserted in places where a time transformation would be needed (all the parameters with a time notation, e.g. tgen1). At the moment the parameter 'Time_convert' is set to 1.

Once the whole model is correctly defined, none of the model components should contain red hatching any more (red hatching in one of the components indicates that there is something wrong with the notation). Note that links must be inserted where mathematical connections are present. It is also possible to make the components globally available (indicated by green/black hatching), which means that they are available to all other components without the need of additional <influences>. This is done with the fpHnit1 and fpHnit2 functions and the <define> 'function_adj', so that additional links in these cases are not needed.

It cannot be stressed enough that all stages of model development should be carefully documented. This includes the exact definition of each model component with the correct dimensions.

Once the model is free from procedural errors, it can be tested and run for the first time (press the button <GO>). However, it is a good idea to check all mathematical definitions again to make sure that the notations in all the components correspond exactly to the definition in the mathematical description. Once the model is error-free, it should be similar to the model in the file: *Mod2–2a.mod*.

Before running the model, make sure that in the definition box: <Model><Run options> the integration method is set to Runge–Kutta and the 'Start' and 'Stop' value and the 'Output steps' are set to the desired values. For example, with 0, 240 and 24, respectively, it corresponds to an overall simulation period of 10 days (recall our time units in the model are hours) with model outputs every 10 h (240 divided by 24 output steps).

The development of the basic model was carried out in great detail in order to illustrate how to develop a model in ModelMaker. With more practice, it becomes easy to decide which model component must be added and which links must be inserted. As the model is extended, it might be a good idea to insert new components immediately into the first model. However, in order to be able to go back to the original model (to the one which worked satisfactorily), it is advisable to save model extensions under different file names (e.g. the first extension of *Mod2–2a.mod*, as *Mod2–2a_ext1.mod*).

The first model (*Mod2–2a.mod*) can now be used to simulate soil nitrification. To view the model output, it is advisable to create appropriate graphs. This can easily be done with the command: <View><Graph> and then selecting the desired components. The output of the component (y-axis) can be plotted against time (default) but can also be plotted against any of the other components. The graph already defined in *Mod2–2a.mod* is called 'Inorganic N' and shows the outputs of all inorganic N components with time.

A very common feature one might investigate with simulation models is the response of certain model components when a particular parameter is changed. This can easily be achieved with the tool: <Model><Sensitivity>. To illustrate the way this tool works, a set-up of NH4adsorb in response to changing NH4adsorb_init is already defined in *Mod2–2a.mod*. The original NH4adsorb_init was set to 20. However, if we want to investigate the effect of NH4adsorb_init values in the range from 3 to 20 and then run the model five times, the definition in the <Sensitivity Configuration> has to be changed to: Start value: 3, Stop value: 20, Steps: 4 (equals five runs). If you press 'Calculate' the model runs five times each time with the different NH4adsorb_init[4].

If you want to view the output of NH4adsorb with different NH4adsorb_init, you have to add the appropriate components into the graph 'Inorganic N' (click on graph 'Inorganic N' and then <view> <selection>, toggle down to the 'NH4adsorb:NH4adsorb_init...' components and add them to the current graph).

2.3.3 Nitrification and gaseous N production

The first extension to the basic nitrification model considers the process of gaseous N emission during nitrification. The basic mechanism of NO and N_2O production via nitrification has already been introduced in the introductory remarks (Section 2.1). For the model extensions in this section, the following are assumed:

- Gaseous N production during nitrification occurs through reduction from nitrite (NO2nit) to nitric oxide (NOnit) and further to nitrous oxide (N2Onit) (similar to the denitrification sequence, see Section 2.4).
- The production rates from NO2nit to NOnit (F4) and from NOnit to N2Onit (F5) follow first-order kinetics.
- The activity of the enzymes catalysing each step (E3nit and E4nit) follows Michaelis–Menten kinetics.
- The emission of gaseous N components out of soil can be characterized by a diffusion process.

First-order transformation rates F4 (NO2nit to NOnit) and F5 (NOnit to N2Onit) are defined by:

$$F4 = r1gas \cdot NO2nit \cdot E3nit \qquad (2.3/17)$$

with

F4 =	transformation rate from NO2nit to NOnit [μg N · g^{-1} soil · h^{-1}]
r1gas =	first-order transformation constant [h^{-1}] {0.004}
NO2nit =	nitrite concentration [μg N · g^{-1} soil]
E3nit =	relative nitrite reductase (nitrification) activity [–]

[4] Note: to avoid being prompted at the beginning of each model run for a decision on whether the model should calculate pH adjustment or not, you can input the value corresponding to the desired pH adjustment into 'function_adj' and then disable the <Trigger> in the <Independent event> 'Choice'.

and:

$$F5 = r2gas \cdot NOnit \cdot E4nit \tag{2.3/18}$$

with:

F5 = transformation rate from NOnit to N2Onit $[\mu g\ N \cdot g^{-1}\ soil \cdot h^{-1}]$
r2gas = first-order transformation constant $[h^{-1}]$ {0.8}
NOnit = nitric oxide concentration $[\mu g\ N \cdot g^{-1}\ soil]$
E4nit = relative nitric oxide reductase (nitrification) activity [–]

Instead of modelling the full enzyme system (as in Section 2.3.2), only the relative enzyme activities of the two reductase enzymes (E3nit and E4nit) are modelled. In the denitrification model developed in Section 2.4, a similar notation of relative enzyme activities is used. The notation is adopted from the model developments by Dendooven and colleagues (Dendooven and Anderson, 1994, 1995b; Dendooven *et al.*, 1994). The relative enzyme activity is set to unity if all conditions are optimum (i.e. substrate, environmental conditions such as pH, temperature, moisture). Adjustments of the optimum enzyme activity (= 1) are calculated for substrate concentrations according to a Michaelis–Menten notation and for environmental factors (temperature and moisture), as described in Section 2.3.4.

The relative enzyme activities adjusted for non-optimal substrate concentrations are calculated for the reduction from NO2nit to NOnit by:

$$E3nit = \frac{E3nit\_max \cdot NO2nit}{E3nit\_Km + NO2nit} \tag{2.3/19}$$

with:

E3nit_max = maximum relative enzyme activity [–] {1}
E3nit_Km = Michaelis–Menten parameter $[\mu g\ N \cdot g^{-1}\ soil]$ {0.1}
NO2nit = nitrite concentration $[\mu g\ N \cdot g^{-1}\ soil]$ {0.1}

and for the reduction from NOnit to N2Onit by:

$$E4nit = \frac{E4nit\_max \cdot NOnit}{E4nit\_Km + NOnit} \tag{2.3/20}$$

with:

E4nit_max = maximum relative enzyme activity [–] {1}
E4nit_Km = Michaelis–Menten parameter $[\mu g\ N \cdot g^{-1}\ soil]$ {0.001}
NOnit = nitric oxide concentration $[\mu g\ N \cdot g^{-1}\ soil]$ {0.001}

So far, we have considered only the production of nitric oxide and nitrous oxide in the soil. However, some of the gas will be emitted to the atmosphere. These emissions can be described by a diffusion process, governed by concentration gradients and diffusivities (the proportionality constants, see Section 1.3). The concept of the basic transport equation has already been introduced in the first chapter (Section 1.3). Here we shall apply the basic concept to the diffusion of gases in soil. The basic notation of the transport equation is Fick's law where (see eqn 1.3/1):

$$F = -D \cdot \frac{dC}{dx} \tag{2.3/21}$$

with:

$D =$ gas diffusivity $[m^2 \cdot h^{-1}]$
$dC =$ concentration difference $[\mu g \cdot m^3]$ (between soil and atmosphere)
$dx =$ diffusion distance [m] (between soil and atmosphere)
$F =$ gas flux assuming planar diffusion $[\mu g \cdot m^{-2} \cdot h^{-1}]$

This notation applies for gas fluxes, *F*, in one dimension, when the area available for flow remains constant.

Gas diffusivities are usually calculated by considering gas diffusivities through air (D_0) measured at normal temperature and pressure (NTP: 273.16 K, 101.3 kPa), and adjusting these for the particular soil conditions. Soil contains solid particles, water and air. Diffusivities in water are approximately four orders of magnitude smaller than diffusivities through air and, for practical reasons, they are generally considered to be zero (Campbell, 1985, pp. 14–15). Diffusivities through soil air are lower than through free air because the diffusion pathways are occupied by solid particles and water.

For calculating the gas diffusivities, the procedure described in Campbell (1985, pp. 14–15) is used. First, the gas diffusivity through air at NTP (D_{0_NTP}) is adjusted for the soil temperature and then multiplied by an air-filled porosity function to adjust for soil tortuosity.

$$D = D_0 \cdot f_\varepsilon \tag{2.3/22}$$

with f_ε, a function of air-filled porosity:

$$f_\varepsilon = b \cdot \phi_g^m$$

The diffusivity at NTP is adjusted for actual temperatures by:

$$D_0 = D_{0\_NTP} \cdot \left(\theta / \theta_0 \right)^n \tag{2.3/23}$$

where:

$\phi_g =$ air filled porosity of soil $[m^3 \text{ air} \cdot m^{-3} \text{ soil}]$
$D_{0\_NTP} =$ diffusivity in air at normal temperature and pressure (NTP) $[m^2 \cdot h^{-1}]$ (note: the units usually used in soil physics are per second; however, to avoid confusion, the units are expressed as per hour)

$\theta =$ actual temperature [K] (note: in order to convert °C to Kelvin we have to add 273.16 to the °C value)
$\theta_0 =$ Kelvin temperature at 0°C [K] {273.16}
b = constant for the air-filled porosity function [–] {0.9}
m = constant for the air-filled porosity function [–] {2.3}
n = constant for temperature adjustments [–] {1.75}

Adjustments for non-normal air pressures are not considered here but can easily be calculated with the procedure described in Campbell (1985, pp. 14–15). The diffusivities for nitric oxide (Diff_NO) and nitrous oxide ('Diff_N2O') are calculated according to the above method as:

$$D_0 = D_{0\,-NTP} \cdot \left(\theta/\theta_0\right)^n \cdot b \cdot \phi_g^m$$

or in the notation as used later in the ModelMaker model:

$$\text{Diff\_NO} = \text{Diff0\_NO} \cdot \left(\frac{\text{Temp} + \text{Kelvin}}{\text{Kelvin}}\right)^{\text{Diff\_n}} \cdot \text{Diff\_b} \cdot (\text{porosity} - \text{VWC})^{\text{Diff\_m}} \qquad (2.3/24)$$

$$\text{Diff\_N2O} = \text{Diff0\_N2O} \cdot \left(\frac{\text{Temp} + \text{Kelvin}}{\text{Kelvin}}\right)^{\text{Diff\_n}} \cdot \text{Diff\_b} \cdot (\text{porosity} - \text{VWC})^{\text{Diff\_m}} \qquad (2.3/25)$$

where:

Diff0_N2O = diffusivity of N_2O through air at NTP (0°C, 101.3 kPa) $[m^2 \cdot h^{-1}]$ {0.05148}
Diff0_NO = diffusivity of NO through air at NTP (0°C, 101.3 kPa) $[m^2 \cdot h^{-1}]$ {0.05148} (note: at the moment the same diffusivity is used for N_2O and NO)
Temp = soil air temperature [0°C]
Kelvin = Kelvin temperature corresponding to 0°C [K] {273.16}
Diff_n = parameter for temperature adjustment [–] {1.75}
Diff_b = parameter for moisture adjustment [–] {0.9}
Diff_m = parameter for moisture adjustment [–] {2.3}
porosity = total porosity of soil [m^3 pores $\cdot$ m^{-3} soil]; this can be calculated with the relationship: $1 - \frac{\text{dens\_bulk}}{\text{dens\_solid}}$, where 'dens_bulk' is the measured bulk density of the soil (usually between 900 and 1500 kg m^{-3} soil) and 'dens_solid' is the density of the solid material (usually around 2650 kg m^{-3}).
VWC = volumetric water content [m^3 water $\cdot$ m^{-3} soil]

For calculating the gas flux, the following assumptions are made:

- The site of gas production is at a well-defined depth in the soil.
- Gas flux occurs only between the site of production in soil and the soil surface (only upwards).

These assumptions represent a simplified system and do not reflect real field situations. However, under these considerations, it is possible to calculate gas fluxes out of the soil without modelling a multilayered soil profile. Such a multilayered system for gas flux would be similar to the notation described in Chapter 3 for heat movement in soil.

The emission rate of NOnit_rate and N2Onit_rate (both expressed in dimensions of $\mu g\ N\ g^{-1}$ soil h^{-1}) can be calculated as (eqn 2.3/21):

$$\text{NOnit\_rate} = -\text{Diff\_NO} \cdot \frac{\text{NOnit\_soil} - \text{NOair}}{\text{Diff\_dist}} \quad (2.3/26)$$

and:

$$\text{N2Onit\_rate} = -\text{Diff\_N2O} \cdot \frac{\text{N2Onit\_soil} - \text{N2Oair}}{\text{Diff\_dist}} \quad (2.3/27)$$

with:

NOnit_soil =	concentration of NO in soil [$\mu g\ N \cdot m^{-3}$]
NOair =	concentration of NO in air [$\mu g\ N \cdot m^{-3}$]
N2Onit_soil =	concentration of N_2O in soil [$\mu g\ N \cdot m^{-3}$]
N2Oair =	concentration of N_2O in air [$\mu g\ N \cdot m^{-3}$]
Diff_dist =	diffusion distance between the site of gas production and the soil surface [m] {0.1} (note: it is assumed that the diffusion distance is 10 cm, which corresponds to an emission from a 20-cm-thick soil layer)

It should be noted that for upward movement the flux has a negative value and for downward movement a positive value. However, we consider the flux from the 'soil atmosphere perspective', where an upward flux (out of the soil) would cause an increase and therefore a positive flux into the atmosphere. The parameter Diff_dist defines the diffusion distance and is therefore crucial for a realistic calculation of the gas emission. At the moment, we do not have any indication what value we should use for this parameter. This should be kept in mind while developing the gas diffusion model.

The air concentrations of NO (NOair) and N_2O (N2Oair) are approximately 0.1 and 310 ppbv (parts per billion on a volume basis), respectively (Conrad, 1995). Via the ideal gas law, they can be converted into dimensions of $\mu g\ N\ m^{-3}$:

$$\text{NOair} = \frac{\text{NOair\_ppb} \cdot 10^{-9} \cdot \text{Vair} \cdot \text{Pressure}}{\text{R} \cdot (\text{T\_air} + \text{Kelvin})} \cdot \text{molwt\_N} \cdot 1000000 \quad (2.3/28)$$

$$\text{N2Oair} = \frac{\text{N2Oair\_ppb} \cdot 10^{-9} \cdot \text{Vair} \cdot \text{Pressure}}{\text{R} \cdot (\text{T\_air} + \text{Kelvin})} \cdot \text{molwt\_N} \cdot 2 \cdot 1000000 \qquad (2.3/29)$$

with:

NOair_ppb =	NO mixing ratio in air [ppbv] {0.1}
N2Oair_ppb =	N_2O mixing ratio in air [ppbv] {310}
Vair =	volume of air [m^3] {1}
Pressure =	air pressure [Pa] {101300}
R =	gas constant [$J \cdot mol^{-1} \cdot K^{-1}$] {8.314}
T_air =	air temperature [°C]
Kelvin =	Kelvin temperature at 0°C [K] {273.16}

The soil concentration of nitric oxide (NOnit) and nitrous oxide (N2Onit) (in dimensions of $\mu g\ N\ g^{-1}$ soil) is converted into units of $\mu g\ N\ m^{-3}$ soil (NOnit_soil and N2Onit_soil) via the transformation:

$$\text{NOnit\_soil} = \text{NOnit} \cdot \text{dens\_bulk} \cdot 1000000 \qquad (2.3/30)$$

and:

$$\text{N2Onit\_soil} = \text{N2Onit} \cdot \text{dens\_bulk} \cdot 1000000 \qquad (2.3/31)$$

where:

dens_bulk = bulk density [$g \cdot cm^{-3}$] {1}

Note: multiplication of NOnit or N2Onit (units: $mg\ N\ g^{-1}$) by the soil bulk density dens_bulk (units $g\ cm^{-3}$) yields NOnit_soil or N2Onit_soil expressed in $\mu g\ N\ cm^{-3}$. To obtain $\mu g\ N\ m^{-3}$, we have to multiply by 1,000,000.

This calculation assumes a uniform distribution of the gas produced over all soil constituents (i.e. over solid, water and air parts), so that this calculated concentration is also present in the gas phase, where the diffusion process takes place.

It might be more realistic to assume that all the gas produced is immediately transferred into the soil water phase and then equilibrated via a solubility expression with the gas phase (e.g. Heinke and Kaupenjohann, 1997). The exchange between the water and gas phase of a certain gas could, for instance, be calculated with the Bunsen coefficient (Müller, 1996). However, this conversion is not considered for the current model development.

Finally, we are able to define the soil NOnit and N2Onit concentrations with the differential equations:

$$\frac{d\text{NOnit}}{dt} = \text{F4} - \text{F5} - \text{NOnit\_rate\_c} \qquad (2.3/32)$$

with:

NOnit_init = initial 'NOnit' concentration [$\mu g\ N \cdot g^{-1}$ soil] {0.01}

and:

$$\frac{d\text{N2Onit}}{dt} = \text{F5} - \text{N2Onit\_rate\_c} \tag{2.3/33}$$

with:

N2Onit_init = initial N2Onit concentration [μg · N · g^{-1} soil] {0.01}

where the NOnit_rate_c and N2Onit_rate_c correspond to NOnit_rate and N2Onit_rate (eqn 2.3/26 and 2.3/27) but are expressed in dimensions of μg N g^{-1} soil h^{-1} to be compatible with the F4 and F5 transformation rates.

To obtain cumulative emission, the rates NOnit_rate and N2Onit_rate are integrated in separate compartments (indicated by the '_e' notation):

$$\frac{d\text{NOnit\_e}}{dt} = \text{NOnit\_rate} \tag{2.3/34}$$

with:

NOnit_e_init = initial NOnit_e [μg N · m^{-3}] {0}

and:

$$\frac{d\text{N2Onit\_e}}{dt} = \text{N2Onit\_rate} \tag{2.3/35}$$

with:

N2Onit_e_init = initial N2Onit_e [μg N · m^{-3}] {0}

The NO2nit component (eqn 2.3/5) has to be updated by subtracting the flow F4 (NO2nit to NOnit) from NO2nit. The mathematical notation for NO2nit is now:

$$\frac{d\text{NO2nit}}{dt} = (\text{F2} \cdot \text{k\_E1}) - \text{F3} - \text{F4} \tag{2.3/36}$$

Application in ModelMaker

For the calculation of soil diffusivities, we need soil and air temperatures and soil volumetric water contents. These environmental parameters change with time and are best supplied to the model via an additional lookup component. Therefore a <Lookup Table> 'Inputs' has been inserted. When using data sets which are available in an ASCII format (e.g. data from an automatic weather station), it is also possible to use a <Lookup File> to hold the input values. Such a file can be altered by any commercially available spreadsheet software. The data set which is held in the <Lookup table> 'Inputs' consists of random numbers for volumetric water contents (range: 0–0.7) and soil and air temperature (range 5–30°C). Some models have the additional notation *Mod…real.mod*. These models are identical to the models with the same number but are supplied with hourly climate data from May 1997 collected by the automatic weather station at the Environmental Research Station, Linden, Germany.

Since the input data have to be available in many parts of the model, it is advisable to read the inputs into globally available variables (green/black hatching in *Mod2–2b.mod*).

The addition to the basic nitrification model is presented in Fig. 2.8 (*Mod2–2b.mod*).

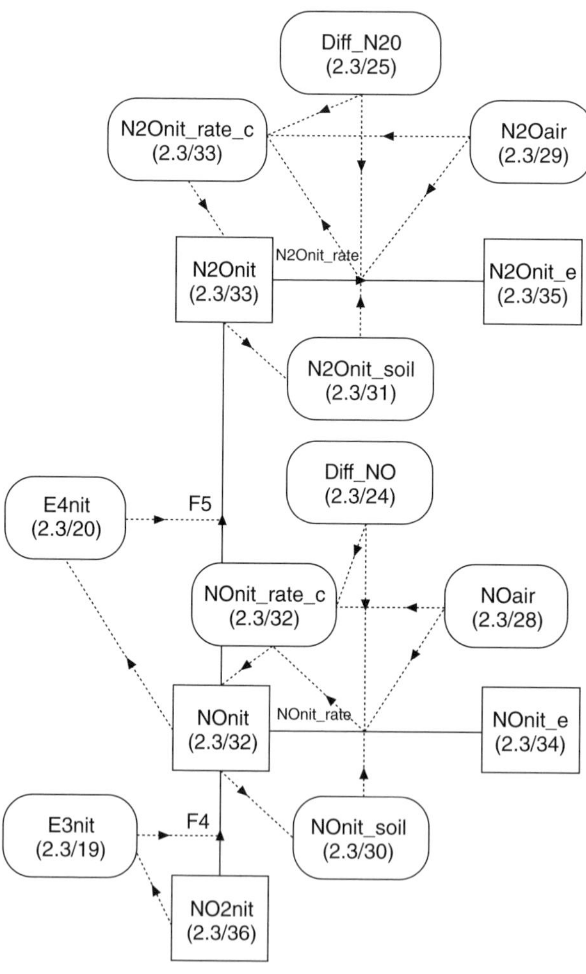

Fig. 2.8. Extension of the basic nitrification model for modelling gaseous N emission from soil (*Mod2-2b.mod*).

First, the necessary components and necessary links between them have to be inserted (Fig. 2.8). After defining the parameters and the definitions of the various equations (see notation in this section), the extended model should be similar to the one described in *Mod2–2b.mod*.

In addition, a user choice has been added which allows the specification of two options: a constant as opposed to a variable environment. When choosing the constant environment option, fixed values for VWC, T_air and T_soil are used, whereas the 'variable environment-option' uses the data supplied via the <Lookup Table> 'Inputs'.

To avoid negative air-filled porosities, the calculation of the N_2O and NO diffusivities (Diff_NO and Diff_N2O) are made conditional. In the case where VWC $\geq$ porosity, a fixed minimum air-filled porosity Poro_min is used. This avoids the model attempting to calculate the power of a negative value which is not defined with Diff_m values < 1 (eqns 2.3/24 and 2.3/25). This example illustrates the importance of checking the program code very carefully. Even if no procedural or mathematical errors are present, situations could occur where the model might use data constellations which would lead to run-time errors. This kind of error sometimes occurs when extreme data values are used. To avoid potential problems, it is advisable to create conditional definitions for those components where problems of this kind are anticipated. However, these additional declarations should be as realistic as possible. For instance, in the extreme case, when the volumetric water content (VWC) of the soil is higher than the total porosity, it can be assumed that the air-filled porosity is close to zero (a value for Poro_min of 0.01 is used).

The extended version of the model *Mod2–2b.mod* can now be used to predict not only the inorganic nitrogen concentration but also the possible gaseous N losses via nitrification. However, the model parameters have not been calibrated using real data and therefore cannot be used to simulate field situations. Furthermore, important processes, such as denitrification and other N transformations have not been considered. However, a sensitivity analysis (as described in the previous section) can provide additional insight as to how the change of certain parameter settings will affect the overall model behaviour. One example presented below is the emission of NOnit and N2Onit. Both are dependent on the transformation rate F4 (from NO2nit to NOnit). A key parameter is the first-order transformation constant r1gas, which has been changed during the sensitivity analysis from 0.001 to 0.005 h^{-1} (Fig. 2.9).

A time conversion option via the parameter Time_convert has also been added in the notation. However, changing this parameter to a value other than 1 should be done very carefully because the input data have to be in a corresponding format as well. Generally, if another time step is desired, all rates (with notations per time) have to be changed accordingly. It is a good idea to set up the program with the desired time step in mind in order to avoid errors related to time conversions.

The current diffusion model for calculating N gas emission is already reasonably advanced. Other researchers have used a simpler first-order approach

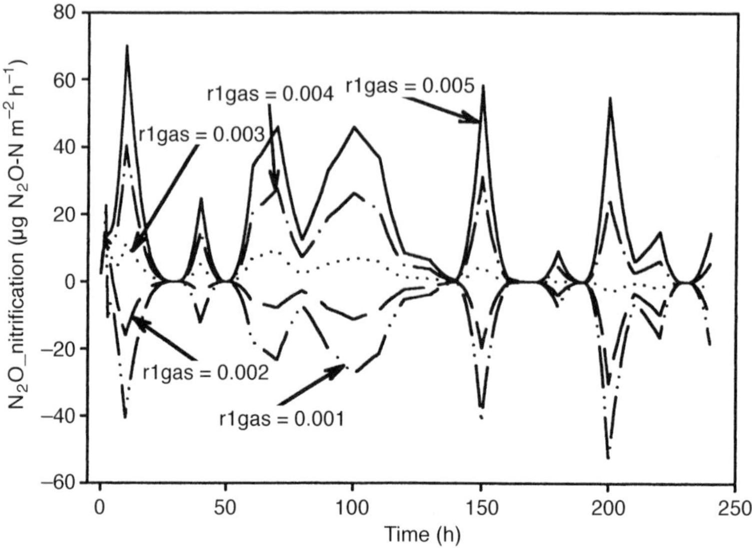

Fig. 2.9. Sensitivity analyis for N_2O emission rate (N2Onit_rate) with varying first-order transformation constants 'r1gas' (0.001–0.005 h^{-1}) (*Mod2-2b.mod*).

(e.g. Cho and Mills, 1979). A first-order gas emission notation could easily be incorporated into the current model by exchanging the equations in NOnit_rate and N2Onit_rate (eqns 2.3/26 and 2.3/27) with:

$$\text{NOnit\_rate} = \text{kNOnit} \cdot \text{NOnit\_soil} \cdot 2 \cdot \text{Diff\_dist} \quad (2.3/37)$$

and:

$$\text{N2Onit\_rate} = \text{kN2Onit} \cdot \text{N2Onit\_soil} \cdot 2 \cdot \text{Diff\_dist} \quad (2.3/38)$$

where:

kNOnit = first-order rate constant for NO (nitrification) emission [h^{-1}] {0.03}
kN2Onit = first-order rate constants for N_2O (nitrification) emission [h^{-1}] {0.002}

In the calculation, it is assumed that the gas is emitted by a layer of soil which is twice as deep as the Diffusion distance (Diff_dist) (remember we defined the diffusion distance to be half the distance of the soil layer where the gas is produced).

The first-order diffusion notation is included into the current notation after the adjustments for soil temperature have been considered (Section 2.3.4

Mod2–2d.mod). The model *Mod2–2d.mod* also contains an <Independent Event> 'Choice', which allows the user to choose between the first-order rate or diffusion emission approach. The calculations in N2Onit_rate and NOnit_rate are made conditional via the <define> 'emission_choice', which holds the value of the user choice (diffusion = 1, first-order = 2).

2.3.4 Influence of soil moisture and soil temperature on nitrification

The final addition to the nitrification model will consider the adjustment of N dynamics for soil temperature and moisture. In the last section we have already introduced the environmental variables; volumetric water content (VWC), air (T_air) and soil temperature (T_soil) and their influence on NO and N_2O diffusivities. However, changing moisture and temperature conditions will also affect the various transformation rates in the model.

Different moisture and temperature optima exist for the various enzymatically catalysed transformation rates. Suboptimal conditions above and below the optimum conditions will result in lower activities. Sometimes, conditions are reached (e.g. frost or drought) where microbial activity will entirely stop.

Over the last century, numerous researchers have investigated the effects of moisture and temperature on nitrification. Instead of defining direct functional dependencies (e.g. Arrhenius equation, Section 1.4.4), the influence of the environmental variables is often introduced into a model by response functions for soil moisture and temperature, which take on values between 0 and 1 (0 = no activity, 1 = optimum activity). These functions can be applied to the various parts of the model where a moisture and/or temperature influence exists.

Soil moisture function

Soil volumetric water contents (VWC) are often measured. However, the effect a certain VWC has on microbial transformations is dependent on the soil type (a high VWC in a clay soil might have the same effect as a low VWC in a sandy soil). The best soil moisture definition is the soil water potential notation, because it expresses the VWC in units of energy (see Chapter 4 for further explanation). However, this notation requires additional information on the 'soil water characteristic', which is often not available. Another expression which we shall use here is the water-filled porosity (WFP) notation, which is often used to define soil moisture relationships for microbial processes (e.g. Granli and Bøckman, 1994). The water-filled porosity (%) is a measure of the extent to which the soil is occupied by water. Very often, one WFP function is created for all the transformations in a system. However, this might not be realistic enough, because different transformations might respond in a different way to soil moisture. For instance, the transformation from NH_4^+–N to NO_3^-–N operates in moist but not saturated conditions, whereas gaseous N emission (NO and N_2O) can occur under almost saturated conditions.

Water-filled porosity (WFP) is defined as:

$$WFP = \frac{VWC}{porosity} \tag{2.3/39}$$

where:

$$porosity = 1 - \frac{dens\_bulk}{dens\_solid} \tag{2.3/40}$$

with:

VWC = volumetric water content [m^3 water · m^{-3} soil]
dens_bulk = soil bulk density [kg · m^{-3} soil] {between 700 and 1500}
dens_solid = density of the solid phase [kg · m^{-3} solid] {2650}

The general shape of a water-filled porosity function can be described in the following way:

- At low WFP values, the microbial activity is zero or very low (fWFP = 0 or near 0).
- An increase in WFP above the first stage will result in an increase in activity (fWFP increases from 0 to 1).
- Usually one value or a range of values exists where the microbial activity is optimal (fWFP = 1).
- Towards saturation (WFP towards 1), the activity might decrease again (fWFP decreasing from 1 to 0 or a value > 0).
- Depending on the microbial process, there might be a range of values towards saturation, where the microbial activity is zero again.

Very often, continuous functions are defined which calculate specific WFP function values (fWFP) (e.g. Parton *et al.*, 1996). The functions developed here define the fWFP values in a piecewise fashion. Following the general description above, a linear regression will be defined for each segment of the fWFP curve. For a piecewise function, WFP values are needed where the activity is zero, increasing, at an optimum and decreasing again. For the increasing and decreasing parts, linear segments are defined with parameters obtained via linear regression of WFP against fWFP data sets.

The general shape of a piecewise linear response function is presented in Fig. 2.10. The activity of microbes at a particular water-filled porosity (WFP) is specified by the function fWFP between 0 and 1 (f is an indicator for a function).

The mathematical notation of the fWFP response function is:

$$\text{fWFP} = \left| \begin{array}{c} 0 \\ \text{Iinc} + \text{Kinc} \cdot \text{WFP} \\ 1 \\ \text{Idec} + \text{Kdec} \cdot \text{WFP} \\ 0 \end{array} \right| \text{ if } \left| \begin{array}{c} \text{WFP} \leq \text{A} \\ \text{A} < \text{WFP} \leq \text{B} \\ \text{B} < \text{WFP} \leq \text{C} \\ \text{C} < \text{WFP} \leq \text{D} \\ \text{D} < \text{WFP} \end{array} \right| \qquad (2.3/41)$$

where:

Iinc = intercept of linear regression for increasing activity [–]
Kinc = slope of linear regression for increasing activity [–]
Idec = intercept of linear regression for decreasing activity [–]
Kdec = slope of linear regression for decreasing activity [–]

The piecewise continuous function in Fig. 2.10 was drawn using the parameter combination:

A = 0.3
B = 0.55
C = 0.7
D = 0.9
Iinc = −1.2
Kinc = 4
Idec = 4.5
Kdec = −5

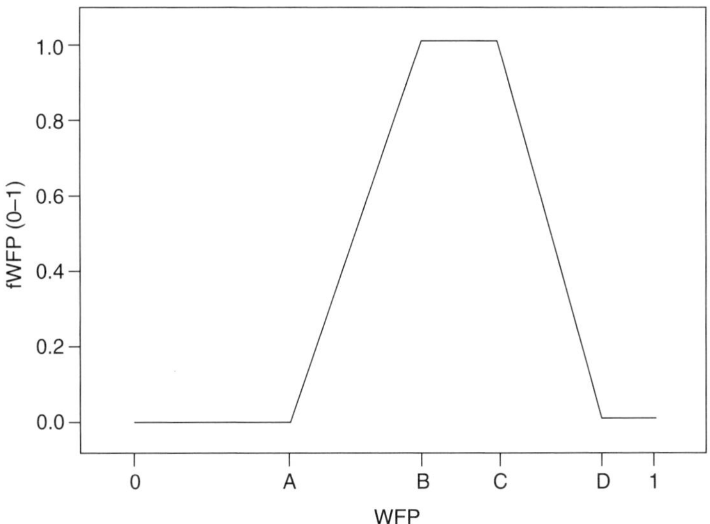

Fig. 2.10. Piecewise continuous water-filled porosity function (fWFP, eqn 2.3/41).

The fWFP function for some processes might be defined differently. For instance, instead of the initial segment, where the function obtains a value of 'zero', a linear increasing response might be more appropriate (e.g. fWFP_MIN in Section 2.5.8).

Temperature function

Adjustment of the microbial processes in response to temperature is enabled via a continuous temperature function. Similar to the water-filled porosity function, it is possible to define temperature values where the activity is zero, optimal and decreasing again, once the temperature rises beyond an optimum value. Usually, at extreme temperatures (high and low), microbial activity will entirely stop.

Instead of a piecewise continuous fit, a simplified version of the continuous temperature function described by Thornley (1998, p. 54) is used:

$$fT = \frac{(T - Tmin)^{qfT} \cdot (Tmax - T)}{(Topt - Tmin)^{qfT} \cdot (Tmax - Topt)} \quad (2.3/42)$$

where:

$$Topt = \frac{Tmin + qfT \cdot Tmax}{1 + qfT} \quad (2.3/43)$$

and:

$$fT = 0 \quad \text{if} \quad T < Tmin \quad \text{or} \quad T > Tmax$$

with:

T = actual temperature [°C]
Tmin = lowest temperature where activity is just zero [°C]
Tmax = highest temperature where activity is just zero [°C]
Topt = temperature where activity is optimal (fT =1) [°C]

In the original 'Thornley' notation a difference was made between Tref (where fT = 1, a reference temperature) and Topt'(original notation Tmax), where maximum activity was reached. However, assuming that fT = 1 at Topt, the original 'Thornley' notation can be simplified so that the factor qfT can be calculated by rearranging eqn 2.3/43.

$$qfT = \frac{Topt - Tmin}{Tmax - Topt} \quad (2.3/44)$$

To calculate the temperature function, only three parameters have to be supplied (Tmin, Tmax and Topt). An example of a temperature function with parameters Tmin = −10°C; Tmax = 50°C; Topt = 30°C is presented in Fig. 2.11.

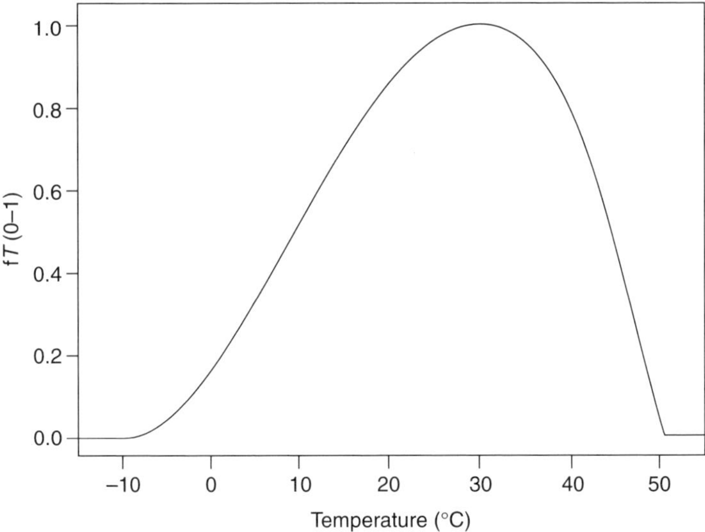

Fig. 2.11. Temperature function (fT) (eqn 2.3/42).

Application in ModelMaker

The volumetric water contents as well as soil and air temperatures were already supplied to the nitrification model via the <Lookup Table> 'Inputs'. Specific water-filled porosity functions for the oxidation of NH_4^+–N to NO_2^-–N (fWFP_NO2nit), oxidation from NO_2^-–N to NO_3^-–N (fWFP_NO3), reduction from NO_2^-–N to NO (fWFP_NOnit) and reduction from NO to N2Onit (fWFP_N2Onit) are defined (according to eqn 2.3/41) with the following parameters:

	_NO2nit	_NO3	_NOnit	_N2Onit
WFP_A	0.087	0.087	0.1	0.1
WFP_B	0.54	0.54	0.5	0.5
WFP_C	0.69	0.69	0.8	0.8
WFP_D	1	1	1	1
Iinc	−0.19	−0.19	−0.25	−0.25
Kinc	2.2	2.2	2.5	2.5
Idec	2.6	2.6	5	5
Kdec	−2.33	−2.33	−5	−5

Data for the WFP functions were taken from Parton *et al.* (1996) and from the theoretical gas emission model by Davidson (1991).

Temperature function (fT)

Only one temperature function (fT_NIT) is defined for all oxidation and reduction processes in the nitrification sequence. The temperature function is equal to the one described in Fig. 2.11 (parameters: fmin_NIT = −10°C; fmax_NIT = 50°C; fopt_NIT = 30°C). The extension 'NIT' is added to indicate that this function is the nitrification function (other functions for denitrification = DEN and for mineralization = MIN will also be introduced later). Data for the fT_NIT function were taken from Parton *et al.* (1996). The WFP and temperature functions are applied to all components in the nitrification model where moisture and/or temperature might have an influence on the model dynamics.

With two environmental functions, two options exist and can be applied:

1. Both functions will be multiplied with each other.
2. Only the function with the lowest value will be applied.

With three or more functions, it would also be possible to multiply with one function and then multiply with the minimum of the other two functions (e.g. Frissel and van Veen, 1981). A good knowledge of the system is helpful to decide how the various functions should be applied to the system.

In the current nitrification model, an option is introduced which prompts the user for a decision as to how the functions fT and fWFP should be treated in the model. The 'GetValue' statement in the <independent event> 'User_choice' is expanded to:

```
GetValue("Environmental adjustment: 1 = NONE, 2 = adjust for
        pH, 3 = adjust for pH and fT,fWFP (minimum), 4 = adjust
        for pH and fT, fWFP (multiply)","Environmental
        adjustment",function_adj,function_adj=1 or
        function_adj=2 or function_adj=3 or function_adj=4);
```

The following components are adjusted in the nitrification model: a1, a2, E3nit and E3nit. They are made conditional and calculated according to the value of the <define> 'function_adj'.

If, instead of the diffusion model, a first-order notation is used to calculate N gas emissions, then the additional variables kNO and kN2O are included, which calculate the adjustments for soil moisture and soil temperature of the emission rate constants. The emission rate constant under optimal conditions is defined by the parameters kNO_max and kN2O_max.

The final nitrification model, including the adjustments for temperature and WFP, is presented in the model *Mod2–2c.mod, Mod2–2c_real.mod* (only diffusion models) or *Mod2–2d.mod* (includes also the first-order emission calculation). Note that some of the parameter settings in the models vary from the original parameters (e.g. in *Mod2–2d.mod*: r1gas = 0.008, r2gas = 2).

Running the model

Some calculations in the final model, especially the ones related to gas emission via diffusion, are very non-linear, which might cause problems in numerical stability. Due to this non-linearity, it might happen that the model output becomes erratic and breaks down. To avoid this, a very small output step is chosen (e.g. <Stop value> and <Output steps> are equal or at least close to each other). Another approach is to change some of the model definitions. With the model *Mod2–2d.mod*, it is possible to choose the alternative first-order emission approach (described above), which is more stable in the calculations than the diffusion approach.

Overall, it should be kept in mind that this model is not suitable for direct simulations of either field or laboratory applications. The purpose of this model was to develop and present conceptual ideas and show how they can be transferred into a numerical model of nitrification. This model should only be seen as one example of how nitrification, with its numerous transformations, may be mathematically described.

One of the aspects which has not been included in the current model is the inhibition due to high substrate concentrations (Laudelout *et al.*, 1974, 1976; Leggett and Iskandar, 1981). An outline showing how this additional aspect can be included is presented in the final section.

2.3.5 Further additions to the nitrification model

High nitrite and ammonium concentrations may have an inhibitory effect on the ammonium and nitrite oxidation rates. Leggett and Iskandar (1981) introduced an inhibition notation which is based on the models by Laudelout and colleagues (e.g. Laudelout *et al.*, 1974, 1976).

Inhibition notations can be included by expanding the current Michaelis–Menten notations (e.g. E1nit, E2nit).

If the original Michaelis–Menten notation is:

$$V = \frac{V_{max} \cdot S}{K_m + S} \tag{2.3/45}$$

then the inhibition on the rate, *V*, can be modelled by including an inhibition notation $\left(1 + \frac{S}{I_S}\right)$ resulting in:

$$V = \frac{V_{max} \cdot S}{(K_m + S) \cdot \left(1 + \frac{S}{I_S}\right)} \tag{2.3/46}$$

with:

I_S = inhibition constant (with the same units as S)

The additional inhibition notation in the denominator of the above equation has the following effect:

- At low substrate concentrations the inhibition notation approaches the value 1 and the Michaelis–Menten notation is therefore similar to the original one (without inhibition notation).
- With increasing substrate concentration, the inhibition notation takes on values higher than unity, leading to a reduction of the transformation rate (V).

A specific formulation of an inhibition notation, which also includes the effect of H^+ concentration on nitrous acid production, is described in detail in the publication by Leggett and Iskandar (1981). This last addition is not incorporated into the current nitrification model but could easily be inserted.

2.4 Denitrification

2.4.1 Conceptual denitrification model

Numerous review papers and descriptions of the denitrification process have been published (e.g. Delwiche, 1981; Firestone, 1982; Tiedje, 1982; Fillery, 1983; Conrad, 1996b). They all agree that denitrification is a microbiological process which is carried out by a large number of microorganisms that have adapted to growing under anaerobic conditions by using N oxides (e.g. NO_3^-–N, NO_2^-–N) as their electron acceptors.

Most of the denitrifiers are able to reduce the N oxides in a reduction sequence from NO_3^-–N right through to N_2. However, there are organisms which lack the last reductive step and finish the reductive sequence with N_2O (e.g. Fillery, 1983). It is now generally accepted that the full denitrification sequence proceeds from NO_3^-–N to N_2 with the intermediates NO_2^-–N, NO and N_2O (Conrad, 1996b).

There is some debate regarding which kinetics best describes the process of denitrification. Models have been developed using all three kinetic types: zero-order, first-order (e.g. Focht, 1974; Kohl *et al.*, 1976) and Michaelis–Menten kinetics (e.g. Betlach and Tiedje, 1981). However, it seems that none of these kinetic types can by itself satisfactorily describe denitrification dynamics. Observations provide evidence that, for example, high concentrations of nitrate can inhibit N_2O reductase and therefore lead to high N_2O emission (Blackmer and Bremner, 1978) or that, under low nitrate concentrations, the production of N_2 might commence earlier than predicted with a kinetic chain reaction (see eqn 1.4/15). Based on these observations, Cho and Mills (1979) proposed a denitrification model which includes a competitive Michaelis–Menten notation. Each reductive step is modelled by considering the concentration of each electron acceptor in relationship to the concentration of all electron acceptors, which can be seen as an overall denitrification activity.

The model developed here is based on the paper by Cho and Mills (1979). The basic model will be extended by including the intermediate NO (which they did not consider) and a dynamic denitrification enzyme system based on the work by Dendooven and colleagues (Dendooven and Anderson, 1994, 1995b; Dendooven *et al.*, 1994).

The basic conceptual model of the denitrification process is presented in Fig. 2.12.

2.4.2 Basic denitrification model

The denitrification model is based on transformations described by Michaelis–Menten kinetics. The Michaelis–Menten notation for the various transformations in the denitrification sequence following the description in Section 1.4.6 is defined by:

$$
\begin{aligned}
&E_{NO3} + NO3 \underset{k_{2\_NO3}}{\overset{k_{1\_NO3}}{\rightleftarrows}} (E_{NO3} \cdot NO3) \xrightarrow{k_{3\_NO3}} NO2den + E_{NO3} \\
&E_{NO2den} + NO2den \underset{k_{2\_NO2den}}{\overset{k_{1\_NO2den}}{\rightleftarrows}} (E_{NO2den} \cdot NO2den) \xrightarrow{k_{3\_NO2den}} NOden + E_{NO2den} \\
&E_{NOden} + NOden \underset{k_{2\_NOden}}{\overset{k_{1\_NOden}}{\rightleftarrows}} (E_{NOden} \cdot NOden) \xrightarrow{k_{3\_NOden}} N2Oden + E_{NOden} \\
&E_{N2Oden} + N2Oden \underset{k_{2\_N2Oden}}{\overset{k_{1\_N2Oden}}{\rightleftarrows}} (E_{N2Oden} \cdot N2Oden) \xrightarrow{k_{3\_N2Oden}} N2 + E_{N2Oden}
\end{aligned}
\tag{2.4/1}
$$

where the terms in parentheses (i.e. $(E_{NO3} \cdot NO3)$, $(E_{NO2den} \cdot NO2den)$, $(E_{NO} \cdot NOden)$, $(E_{N2Oden} \cdot N2Oden)$) represent the specific enzyme–substrate complexes (*X* in eqn 1.4/16).

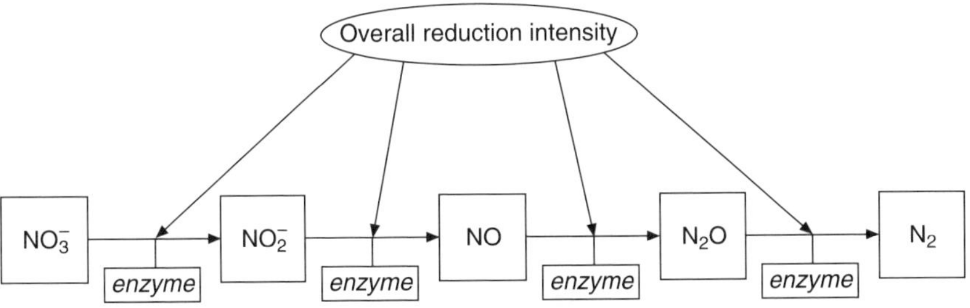

Fig. 2.12. Conceptual denitrification model (based on Cho and Mills, 1979).

Assuming a steady state (enzyme concentration is constant), the enzyme–substrate complexes can be defined as (see eqn 1.4/17):

$$\left(E_{NO3} \cdot NO3\right) = \frac{k_{1\_NO3}}{k_{2\_NO3} + k_{3\_NO3}} \cdot E_{NO3} \cdot NO3$$

$$\left(E_{NO2den} \cdot NO2den\right) = \frac{k_{1\_NO2den}}{k_{2\_NO2den} + k_{3\_NO2den}} \cdot E_{NO2den} \cdot NO2den$$

(2.4/2)

$$\left(E_{NOden} \cdot NOden\right) = \frac{k_{1\_NOden}}{k_{2\_NOden} + k_{3\_NOden}} \cdot E_{NOden} \cdot NOden$$

$$\left(E_{N2Oden} \cdot N2Oden\right) = \frac{k_{1\_N2Oden}}{k_{2\_N2Oden} + k_{3\_N2Oden}} \cdot E_{N2Oden} \cdot N2Oden$$

The changes in the various components in the denitrification sequence can now be defined by the differential equations (see eqn 1.4/20):

$$\frac{dNO3}{dt} = -k_{3\_NO3} \cdot \left(E_{NO3} \cdot NO3\right)$$

$$\frac{dNO2den}{dt} = k_{3\_NO3} \cdot \left(E_{NO3} \cdot NO3\right) - k_{3\_NO2den} \cdot \left(E_{NO2den} \cdot NO2den\right)$$

$$\frac{dNOden}{dt} = k_{3\_NO2den} \cdot \left(E_{NO2den} \cdot NO2den\right) - k_{3\_NOden} \cdot \left(E_{NOden} \cdot NOden\right)$$

$$\frac{dN2Oden}{dt} = k_{3\_NOden} \cdot \left(E_{NOden} \cdot NOden\right) - k_{3\_N2Oden} \cdot \left(E_{N2Oden} \cdot N2Oden\right)$$

$$\frac{dN2}{dt} = -k_{3\_N2Oden} \cdot \left(E_{N2Oden} \cdot N2Oden\right)$$

(2.4/3)

where the '$k_{3...}$' parameters represent the rate constants for the various reduction steps.

The overall denitrification intensity, Rden, can be calculated as:

$$\begin{aligned} Rden = {} & k_{3\_NO3} \cdot \left(E_{NO3} \cdot NO3\right) + k_{3\_NO2den} \cdot \left(E_{NO2den} \cdot NO2den\right) \\ & + k_{3\_NOden} \cdot \left(E_{NOden} \cdot NOden\right) + k_{3\_N2Oden} \cdot \left(E_{N2Oden} \cdot N2Oden\right) \end{aligned} \qquad (2.4/4)$$

This notation, defined by Cho and Mills (1979), corresponds to the sum of the production rate of $NO_2^- - N$, NO, N_2O and N_2 due to the reduction of the precursors. It corresponds, therefore, to the overall activity of the denitrifiers. For the basic model it is assumed that the overall denitrification intensity is constant.

Equation 2.4/4 can be rewritten according to eqn group 2.4/2. This involves multiplying both sides of eqn 2.4/2 by the 'k_3 … ' constants and rewriting eqn 2.4/4 by combination with eqn 2.4/2:

$$\begin{aligned} \mathrm{Rden} = & \frac{k_{1\_NO3} \cdot k_{3\_NO3} \cdot E_{NO3}}{k_{2\_NO3} + k_{3\_NO3}} \cdot \mathrm{NO3} \\ & + \frac{k_{1\_NO2den} \cdot k_{3\_NO2den} \cdot E_{NO2den}}{k_{2\_NO2den} + k_{3\_NO2den}} \cdot \mathrm{NO2den} \\ & + \frac{k_{1\_NOden} \cdot k_{3\_NOden} \cdot E_{NOden}}{k_{2\_NOden} + k_{3\_NOden}} \cdot \mathrm{NOden} \\ & + \frac{k_{1\_N2Oden} \cdot k_{3\_N2Oden} \cdot E_{N2Oden}}{k_{2\_N2Oden} + k_{3\_N2Oden}} \cdot \mathrm{N2Oden} \end{aligned} \qquad (2.4/5)$$

In a steady state, it is possible to simplify eqn 2.4/5 by introducing the following constants:

$$\begin{aligned} \mathrm{k1den} &= \frac{k_{1\_NO3} \cdot k_{3\_NO3} \cdot E_{NO3}}{k_{2\_NO3} + k_{3\_NO3}} \\ \mathrm{k2den} &= \frac{k_{1\_NO2den} \cdot k_{3\_NO2den} \cdot E_{NO2den}}{k_{2\_NO2den} + k_{3\_NO2den}} \\ \mathrm{k3den} &= \frac{k_{1\_NOden} \cdot k_{3\_NOden} \cdot E_{NOden}}{k_{2\_NOden} + k_{3\_NOden}} \\ \mathrm{k4den} &= \frac{k_{1\_N2Oden} \cdot k_{3\_N2Oden} \cdot E_{N2Oden}}{k_{2\_N2Oden} + k_{3\_N2Oden}} \end{aligned} \qquad (2.4/6)$$

The factors k1den, k2den, k3den, k4den are regarded as competitive weighting factors with dimensions of time^{-1}.

The final notation for the overall denitrification intensity is calculated as the sum of all single reduction intensities:

$$\mathrm{Denom} = \mathrm{k1den} \cdot \mathrm{NO3} + \mathrm{k2den} \cdot \mathrm{NO2den} + \mathrm{k3den} \cdot \mathrm{NOden} + \mathrm{k4den} \cdot \mathrm{N2Oden} \qquad (2.4/7)$$

The unit of the denitrification intensity (Denom) is in μg N g^{-1} soil h^{-1}, which can easily be derived from the units of the N oxides (μg N g^{-1} soil) and the dimensions for the competitive weighting factors (h^{-1}).

To avoid an arbitrary increase or decrease in the denitrification rate which might occur in response to the accumulation of intermediates (NO_2^-−N, NO, N_2O), the various reduction rates of eqn group 2.4/3 are constrainted by multiplying all of them by a constant denitrification intensity (Rden, eqn 2.4/4) and division by eqn 2.4/7 (calculated denitrification intensity). For details of this operation, the interested reader is referred to the publication by Cho and Mills (1979). The change in the electron acceptor concentrations with time is finally defined by the differential equations:

$$\frac{d\mathrm{NO3}}{dt} = -\mathrm{D1}$$

$$\frac{d\mathrm{NO2den}}{dt} = \mathrm{D1} - \mathrm{D2}$$

$$\frac{d\mathrm{NOden}}{dt} = \mathrm{D2} - \mathrm{D3} \qquad (2.4/8)$$

$$\frac{d\mathrm{N2Oden}}{dt} = \mathrm{D3} - \mathrm{D4}$$

$$\frac{d\mathrm{N2Oden}}{dt} = \mathrm{D3} - \mathrm{D4}$$

with:

$$\mathrm{D1} = \frac{\mathrm{k1den} \cdot \mathrm{NO3} \cdot \mathrm{Rden}}{\mathrm{Denom}}$$

$$\mathrm{D2} = \frac{\mathrm{k2den} \cdot \mathrm{NO2den} \cdot \mathrm{Rden}}{\mathrm{Denom}}$$

$$\mathrm{D3} = \frac{\mathrm{k3den} \cdot \mathrm{NOden} \cdot \mathrm{Rden}}{\mathrm{Denom}} \qquad (2.4/9)$$

$$\mathrm{D4} = \frac{\mathrm{k4den} \cdot \mathrm{N2Oden} \cdot \mathrm{Rden}}{\mathrm{Denom}}$$

where:

Rden, Denom = constant, calculated denitrification intensity [μg N · g^{-1} soil · h^{-1}]

k1den = competitive weighting factor for NO_3^- reductase [h^{-1}] {1}
k2den = competitive weighting factor for NO_2^- reductase [h^{-1}] {2}
k3den = competitive weighting factor for NO reductase [h^{-1}] {0.5}
k4den = competitive weighting factor for N_2O reductase [h^{-1}] {2}
NO3 = nitrate concentration [$\mu g\ N \cdot g^{-1}$ soil]
NO2den = nitrite (denitrification) concentration [$\mu g\ N \cdot g^{-1}$ soil]
NOden = nitric oxide (denitrification) concentration [$\mu g\ N \cdot g^{-1}$ soil]
N2Oden = nitrous oxide (denitrification) concentration [$\mu g\ N \cdot g^{-1}$ soil]

Initial values for the various electron acceptors:

NO3init = initial nitrate concentration [$\mu g\ N \cdot g^{-1}$ soil] {110}
NO2den_init = initial nitrite (denitrification) concentration [$\mu g\ N \cdot g^{-1}$ soil] {0}
NOden_init = initial NO (denitrification) concentration [$\mu g\ N \cdot g^{-1}$ soil] {0}
N2Oden_init = initial N_2O (denitrification) concentration [$\mu g\ N \cdot g^{-1}$ soil] {0}
N2init = initial N_2O (denitrification) concentration [$\mu g\ N \cdot g^{-1}$ soil] {0}

Application in ModelMaker

Before further model development takes place, it is advisable to compile the current model definitions as the first stage in the development of the denitrification model. This includes the notations described by eqn 2.4/7–2.4/9. Five <compartments> linked via flows D1 to D4 and two <variables> Rden and Denom have to be added to the model (Fig. 2.13, *Mod2–3a.mod*).

The calculations are constrained by a constant denitrification intensity, whereby the <variable> 'Rden' is defined by the <parameter> 'Rmax' (set to 1, as in Cho and Mills, 1979). However, in anticipation of a model extension which introduces a variable 'Rden' notation, the <variable> has already been introduced at this stage. After adding the parameters and the mathematical notations into the various components, the model should be similar to the one provided in the file: *Mod2–3a.mod*.

The model <run options> are set to: <Stop value> = 240 h, <steps> = 240 which corresponds to a simulation period of 10 days with outputs every hour.

Change the <Stop Value> to 2400 (i.e. 100 days' simulation) and observe what happens to the inorganic nitrogen.

After running the model for a long period (this current model for approximately 500 h), all available electron acceptors have been reduced to N_2 and any further reduction will result in negative N concentrations. This feature is related to the various rates of the reduction system in relationship to the overall denitrification intensity (set to a constant value). This basic model, where the denitrification intensity, irrespective of the electron acceptor concentration, does not change,

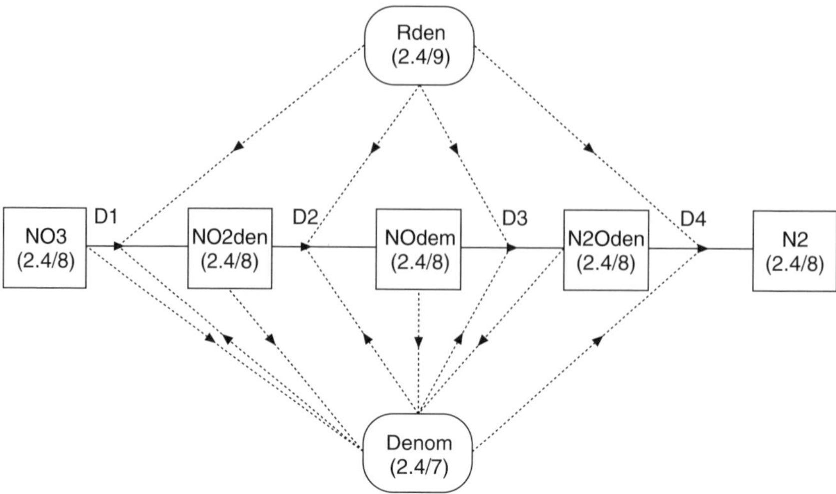

Fig. 2.13. Graphical display of the basic denitrification model (*Mod2-3a.mod*).

may cause such an artefact. To avoid this behaviour and the generation of unrealistic scenarios, the simulation should be stopped before all electron acceptors are depleted. The safety feature included in the current model should therefore keep track of the total N oxide concentrations and stop the model run once the concentration drops below a predefined minimum N oxide concentration. The total N oxide concentration is calculated with the <variable> 'Nsum':

$$\text{Nsum} = \text{NO3} + \text{NO2den} + \text{NOden} + \text{N2Oden} \qquad (2.4/10)$$

The model continues to run while the value of Nsum is still above a predefined stop condition 'Stopcond' (e.g. see program at the end of the paper by Cho and Mills, 1979):

$$\text{Stopcond} = 0.005 \cdot \text{NO3init} \qquad (2.4/11)$$

The comparison between Nsum and Stopcond is carried out in the <Independent event> 'Finish' with the statement:

```
if(NSum<Stopcond)
  ModelExit(Stopcond);
```

The model run will stop immediately (irrespective of the <Stop Value> in <Run Options>) once the above statement is true (i.e. Nsum <Stopcond). This addition to the first stage of model development is introduced into *Mod2–3b.mod*. Such an artefact in the model behaviour is related to problems in the model definition. Clearly, a concentration-related dynamic denitrification intensity and substrate-related enzyme concentrations of the various reductions steps are needed for a

more realistic model behaviour. Therefore the next section is devoted to the denitrification enzyme dynamics.

2.4.3 Denitrification enzyme dynamics

Mathematical description

The original notation of the denitrification model by Cho and Mills (1979) has been extended by Dendooven and Anderson (1995b) to include enzyme characteristics of the various reductase enzymes. They used relative enzyme activities (values between 0 = no activity and 1 = optimum activity). However, they made no attempt to describe the various reduction enzymes dynamically, as would be needed for long-term simulation periods.

A relative concentration or activity parameter has the advantage of being simple to add to the present model and will provide well-defined upper and lower bounds of the activity state of a system. One good example, which will be studied in the context of the C/N transformation model (Section 2.5.7), is the activity parameter (output between 0 and 1), which divides entire microbial populations into active and dormant ones (e.g. Blagodatsky and Richter, 1998).

The enzyme dynamics introduced into the current denitrification model are similar to the notation of the reduction enzymes leading to NO and N_2O production during nitrification (see Section 2.3.3, eqns 2.3/19 and 2.3/20). It is assumed that the activity of the reductase changes in response to the availability of N oxides according to a Michaelis–Menten notation. For the four reductase steps within the denitrification sequence, the relative enzyme activities are defined by:

$$
\begin{aligned}
\text{E1den} &= \frac{\text{E1den\_max} \cdot \text{NO3}}{\text{E1den\_km} + \text{NO3}} \\
\text{E2den} &= \frac{\text{E2den\_max} \cdot \text{NO2den}}{\text{E2den\_km} + \text{NO2den}} \\
\text{E3den} &= \frac{\text{E3den\_max} \cdot \text{NOden}}{\text{E3den\_km} + \text{NOden}} \\
\text{E4den} &= \frac{\text{E4den\_max} \cdot \text{N2Oden}}{\text{E4den\_km} + \text{N2Oden}}
\end{aligned}
\qquad (2.4/12)
$$

with:

E1den_max = maximum relative nitrate reductase activity [–] {1}
E2den_max = maximum relative nitrite reductase activity [–] {1}
E3den_max = maximum relative nitric oxide reductase activity [–] {1}
E4den_max = maximum relative nitrous oxide reductase activity [–] {1}

E1den_km = Michaelis–Menten parameter (nitrate reductase) [$\mu g\ N \cdot g^{-1}$ soil] {5}
E2den_km = Michaelis–Menten parameter (nitrite reductase) [$\mu g\ N \cdot g^{-1}$ soil] {5}
E3den_km = Michaelis–Menten parameter (nitric oxide reductase) [$\mu g\ N \cdot g^{-1}$ soil] {2}
E4den_km = Michaelis–Menten parameter (nitrous oxide reductase) [$\mu g\ N \cdot g^{-1}$ soil] {2}

The relative enzyme concentrations will influence the various reduction rates and are therefore multiplied with the flows D1 to D4 (eqn 2.4/8) (see also Dendooven and Anderson, 1995b). The updated flow rates between the compartments are:

$$\begin{aligned} D1 &= \frac{k1den \cdot NO3 \cdot Rden}{Denom} \cdot E1den \\ D2 &= \frac{k2den \cdot NO2den \cdot Rden}{Denom} \cdot E2den \\ D3 &= \frac{k3den \cdot NOden \cdot Rden}{Denom} \cdot E3den \\ D4 &= \frac{k4den \cdot N2Oden \cdot Rden}{Denom} \cdot E4den \end{aligned} \qquad (2.4/13)$$

Apart from the single reduction rates, the overall denitrification intensity Rden is affected as well. It is assumed that the overall sum of the N oxides Nsum will influence the overall denitrification intensity according to a Michaelis–Menten notation. The maximum growth rate for a denitrifier is first calculated in a similar fashion to the growth rates a1 and a2 for the two nitrifier groups (see eqns 2.3/11 and 2.3/12).The optimum growth rate Vden_max is adjusted for non-optimal substrate concentrations (Nsum) and expressed in a relative way by dividing it by the optimum growth rate Vden_max. In the last section, the denitrification intensity was set to Rmax which may also be seen as the denitrification intensity under optimum conditions. Multiplying it by the relative growth rate adjusts this upper bound of the denitrification intensity to non-optimal conditions. The mathematical notation is:

$$Rden = \frac{\left(\dfrac{Vden\_max \cdot Nsum}{K_m\_den + Nsum}\right)}{Vden\_max} \cdot Rmax \qquad (2.4/14)$$

with:

$$\text{Vden\_max} = \frac{\ln(2)}{\text{tgenden}} - \text{bden} \tag{2.4/15}$$

(For the derivation of this equation see the similar notation for the maximum nitrifier growth rates (a1 and a2) in Section 2.3.2)

where:

tgenden = generation time for denitrifiers under optimum conditions [h] {2}
bden = degradation parameter of denitrifiers [h^{-1}] {0.005}
Rmax = maximum denitrification intensity [$\mu g\ N \cdot g^{-1}\ soil \cdot h^{-1}$] {1}

The overall microbial activity of the denitrification system and the dynamics of the various reduction steps is related to the adjustment of the maximum denitrification intensity Rmax. A parameter such as Rmax can therefore be seen as a 'master-parameter' with which an entire system can easily be adjusted.

Application in ModelMaker

The enzyme dynamics and the variable overall denitrifier intensity are incorporated into the model *Mod2–3c.mod* (Fig. 2.14).

It should be noted that the safety feature incorporated into *Mod2–3b.mod* is not needed any longer because the dynamic enzyme system incorporated into model *Mod2–3c.mod* ensures that the N concentrations cannot drop below zero. This is a feature which results from the notation for the relative enzymes E1den to

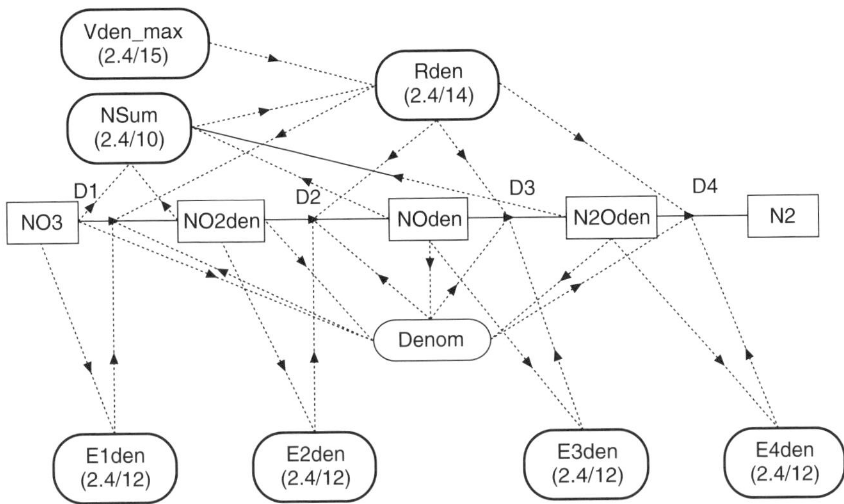

Fig. 2.14. Denitrification model including a dynamic enzyme system (*Mod2-3c.mod*) (bold = additional notation, see also Fig. 2.13).

E4den which will decrease with decreasing N concentrations and therefore reduce the flows D1 to D4 accordingly.

2.4.4 *Denitrification and gaseous N emission*

Emissions of NO and N_2O are calculated in exactly the same way as presented in Section 2.3.3 for emission during nitrification. Therefore the equations are not repeated here. It only has to be kept in mind that, instead of the notation '…nit …', the notation '…den…' is used to indicate that these are emissions during denitrification (e.g. NOnit_rate here: NOden_rate). However, during denitrification, dinitrogen (N2) is also emitted. Dinitrogen represents the final product of the denitrification sequence and cannot be reduced further. It is therefore assumed that all N_2 produced during denitrification is potentially available for emission from soil. Dinitrogen is the main constituent of our atmosphere (approx. 78%) and, to calculate the emission from soil, it is assumed that the concentration difference between the soil and air concentration is given by the production rate in the soil and the diffusivity of N_2, which is taken to be the same as for N_2O. Therefore the diffusion equation for N2 can be described by an equation similar to eqns 2.3/26 and 2.3/27:

$$\text{N2rate} = \text{Diff\_N2} \cdot \frac{\text{N2soil}}{\text{Diff\_dist}} \tag{2.4/16}$$

where:

Diff_N2 = diffusivity of N_2 (similar to eqn 2.3./24) $[m^2 \cdot h^{-1}]$
N2soil = soil N_2 production above ambient $[g\ N \cdot m^{-3}]$
Diff_dist = Diffusion distance [m] {0.1}

The diffusivity Diff_N2 is calculated by (see eqn 2.3/24):

$$\text{Diff\_N2} = \text{Diff0\_N2} \cdot \left(\frac{\text{Temp} + \text{Kelvin}}{\text{Kelvin}}\right)^{\text{Diff\_n}} \cdot \text{Diff\_b} \cdot (\text{porosity} - \text{VWC})^{\text{Diff\_m}} \tag{2.4/17}$$

with:

Diff0_N2 = diffusivity of N_2 through air at NTP (0°C, 101.3 kPa) $[m^2 \cdot h^{-1}]$ {0.05148}

All other parameters are the same as in eqn 2.3/24 (Section 2.3.3).

The cumulative N_2 emission (N2e) is determined by:

$$\text{N2e} = \text{N2\_rate} \tag{2.4/18}$$

where:

N2e = cumulative N_2 emission [$\mu g\ N \cdot m^{-2}$]

Again, conversion of the rate N2_rate from $\mu g\ N\ m^{-2}\ h^{-1}$ into units of $\mu g\ N\ g^{-1}\ h^{-1}$ (N2_rate_c) is needed and N2_rate is subtracted from N2.

In addition to the gaseous N emission calculated by diffusion, a simpler approach using a first-order emission notation, is calculated as described by Cho and Mills (1979). The first-order emission calculation for NOden, N2Oden and N2 are calculated in the same way as already described at the end of Section 2.3.3 (for NOnit and N2Onit; eqns 2.3/37 and 2.3/38). The emission rates with the first-order notation are calculated as:

$$\begin{aligned} &\text{NOden\_rate} = \text{kNOden} \cdot \text{NOden\_soil} \cdot 2 \cdot \text{Diff\_dist} \\ &\text{N2Oden\_rate} = \text{kN2Oden} \cdot \text{N2Oden\_soil} \cdot 2 \cdot \text{Diff\_dist} \\ &\text{N2rate} = \text{kN2} \cdot \text{N2soil} \cdot 2 \cdot \text{Diff\_dist} \end{aligned} \tag{2.4/19}$$

with:

kNO, kN2Oden and kN2 being the first-order emission constants which in the current model are defined by the parameters:

kNOden_max = maximum emission rate constant for NOden emission [h^{-1}] {0.2}
kN2Oden_max = maximum emission rate constant for N2Oden emission [h^{-1}] {0.08}
kN2_max = maximum emission rate constant for N2 emission [h^{-1}] {0.2}

Adjustments of the emission rate constants for soil moisture and temperature are carried out according to the notation presented in Section 2.4.5.

Application in ModelMaker

In order to calculate diffusivities, the <Independent Event> 'Choice' and climate data are supplied to the model via the <Lookup table> 'Inputs' (similar to *Mod2–2b.mod*). Two 'GetValue' statements are inserted into the <independent event> 'Choice', which prompt the user for a decision as to which environmental adjustment and type of emission should be calculated:

```
GetValue("Constant environment = 1; variable environment =
        2","Environment",environ_choice,environ_choice=1 or
        environ_choice=2);
GetValue("Diffusion calculation = 1; first-order calculation =
        2","Emission
        calculation",emission_choice,emission_choice=1 or
        emission_choice=2);
```

The notation for the emission of NOden and N2Oden is similar to the one presented in Fig. 2.8. In addition, the notation for the N_2 emission according to eqns 2.4/16–2.4/18 must be inserted. The graphical outline of the notations which are added to the model *Mod2–3c.mod* is presented in Fig. 2.15.

After inserting all the notations and parameters, the model can be executed. However, in order to obtain outputs which represent reality more closely, some of the parameters must be adjusted. The actual procedure for finding a suitable parameter set may involve the testing of numerous parameter combinations; this can be a very cumbersome task. As already mentioned in the previous section, the maximum denitrification intensity Rmax can act as a 'master parameter' which is able to scale down all transformation processes at once. However, it is often the combination of parameters which determines the overall behaviour of the model. In the case of the denitrification model, the most important parameters are the competitive weighting factors k1den to k4den and their relationship to each other. In addition it was found that the values of the Michaelis–Menten parameters E1den_km to E4den_km are essential in determining the dynamics of the reductase activities and therefore the overall behaviour of the model. It is often possible to obtain parameter settings from investigations reported in the literature. However, very often parameters are not reported or cannot even be measured, so that it is sometimes a tedious task of 'trial and error' to find a suitable parameter configuration.

Tools provided with ModelMaker, such as the <Sensitivity Analysis> or <Model> <Optimize...>, can simplify the task of finding suitable parameter sets. Dendooven and Anderson (1995b) used an optimization routine to find suitable enzyme characteristics and substrate affinities for their model. A similar approach can be carried out with the ModelMaker tool <Model><Optimize...>.

In *Mod2–3d.mod* some model parameters have been altered so that the emission rates of NO, N_2O and N_2 would come into a range which is more in line with actual observations. Generally, the emissions of N_2O would range between 0 and approximately 2000 μg N m^{-2} h^{-1}. The following parameters were changed:

E3den_km =	1e-08 [μg N · g^{-1} soil]
E4den_km =	0.25 [μg N · g^{-1} soil]
K1den =	0.05 [h^{-1}]
K2den =	10 [h^{-1}]
K1den =	5000 [h^{-1}]
K1den =	200000 [h^{-1}]
Rmax =	0.15

While running the model, it is a good idea to try out other parameter combinations and observe the model outcome, which will increase the understanding of how the model works.

To obtain a more 'stable' notation for the calculation of gas emissions, the first-order emission approach has been inserted as well. However, a real comparison between the two emission calculations (diffusion or first-order) can only be

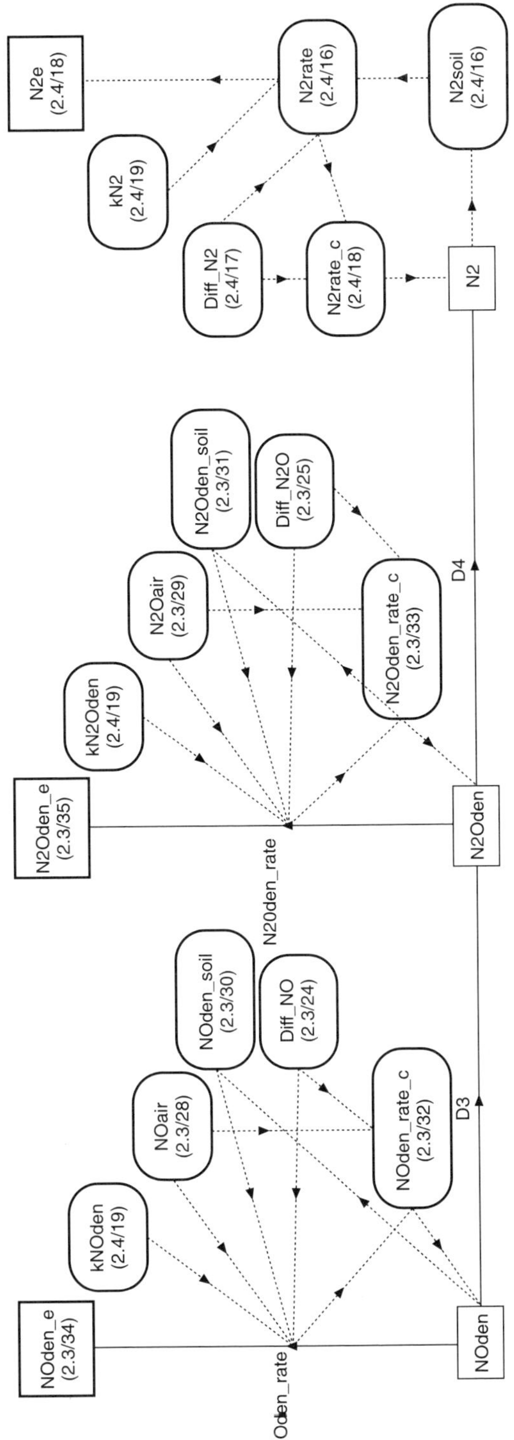

Fig. 2.15. Notation for the gaseous N emissions within the denitrification model (see *Mod2-3d.mod*).

made once the emission constants (kNOden, kN2Oden and kN2) have been adjusted for soil moisture and soil temperature (as is already done for the diffusivities). The adjustment of the model for soil moisture and soil temperature is described in the next section.

2.4.5 Influence of soil moisture and soil temperature

Various processes within the denitrification sequence are adjusted for soil moisture and soil temperature, with relationships similar to the ones described for nitrification (Section 2.3.4).

Soil moisture

Separate water-filled porosity functions (according to eqn 2.3/41) are created for the four reduction steps (i.e. fWFP_NO3 (nitrate), fWFP_NO2den (nitrite), fWFP_NOden (nitric oxide) and fWFP_N2Oden (nitrous oxide)). The parameters for the water-filled porosity functions are:

	_NO3	_NO2den	_NOden	_N2Oden
WFP_A	0.37	0.524	0.1	0.5
WFP_B	1	0.607	0.8	0.85
WFP_C	–	0.655	0.9	1
WFP_D	–	0.705	1	1
Iinc	0; −0.4843	−6.199	−0.143	−1.4286
Kinc	0.1618; 1.474	11.847	1.4286	2.857
Idec	–	13.39	10	–
Kdec	–	−18.939	−10	–

In some cases the WFP functions differ from the basic shape described in Fig. 2.10 (e.g. fWFP_NO3 has two linear increasing and no decreasing segments; once the maximum is reached for both fWFP_NO3 and fWFP_N2Oden, it will stay at the optimum until saturation).

Dendooven and Anderson (1994, 1995a) found that, after the onset of anaerobic conditions, the derepression (i.e. adjustment to the new conditions) of enzymes only occurs after some time. The derepression time (i.e. the time between the onset of anaerobic conditions and when the enzymes are fully adjusted to the new conditions) is dependent on the length of the anaerobic conditions. For instance, the derepression time of nitrous oxide reductase is approximately 15–40 h, during which the enzyme activity would remain at the previous state. If the enzyme activity of the precursors to the nitrous oxide reductase were to be derepressed earlier, an increased N_2O production would be expected during this derepression time. To model the derepression dynamics of the various enzymes, the following assumptions were made:

- Each increase in soil moisture creates new anaerobic zones in the soil according to the fractional soil moisture increase.
- The enzyme activity remains at the same level until the enzymes are fully derepressed.
- Enzyme activities are only adjusted to a higher level after the derepression time is over and only when the soil moisture content at that time is at least as high as at the beginning of the derepression period.
- Decreasing soil moisture will lead to an immediate adjustment of the enzyme activity.

In order to model the effect of the derepression dynamics, the current values of the water-filled porosity WFP, as well as the values before the derepression time WFP_P, have to be compared (i.e. fH2O_NO3 and fH2O_NO3P for nitrate reductase, fH2O_NO2den and fH2O_NO2denP for nitrite reductase, fH2O_NOden and fH2O_NOdenP for nitric oxide reductase and fH2O_N2Oden and fH2O_N2OdenP for nitrous oxide reductase). The 'fH2O…' functions are calculated according to eqn 2.3/41 using the current WFP values and the 'fH2O…P' functions use water-filled porosity values before the derepression time.

The 'fWFP…' functions take on the value of either the 'fH2O…' or 'fH2O…P' function according to the conditional expression:

if

$$\begin{aligned} &WFP \geq WFP\_P \\ &fWFP = fH2O\_P \end{aligned} \qquad (2.4/20)$$

else

$$fWFP = fH2O\_P$$

where:

WFP = current water-filled porosity of soil [–]
WFP_P = water-filled porosity before the derepression time [–]
fH2O = response function calculated with 'WFP' [–]
fH2O_P = response function calculated with 'WFP_P' [–]

(Note: each reduction step is characterized by a separate response function.)

The adjustment to the new level is only carried out if the water-filled porosity (WFP) of the soil is at least as high as before the derepression time. Otherwise the water-filled porosity function is calculated by using the current WFP value.

Derepression times for the various reduction steps are supplied to the model with the parameters:

DeRep_NO3 = time for derepression of nitrate reductase after the onset of anaerobic conditions [h] {0}
DeRep_NO2den = time for derepression of nitrite reductase after the onset of anaerobic conditions [h] {5}

DeRep_NOden = time for derepression of nitric oxide reductase after the onset of anaerobic conditions [h] {0}
DeRep_N2Oden = time for derepression of nitrous oxide reductase after the onset of anaerobic conditions [h] {16}

Data for the WFP functions are taken from Dendooven and Anderson (1994, 1995a), Parton *et al.* (1996) and the theoretical model by Davidson (1991).

Temperature function
Temperature adjustment of the denitrification activity is modelled via the temperature response function described in Section 2.3.4 (eqn 2.3/42). The temperature function (fT_DEN) is calculated using the parameters:

Tmin_DEN = lowest temperature where denitrification activity is just zero [°C] {−15}
Tmax_DEN = highest temperature where denitrification activity is just zero [°C] {75}
Topt_DEN = temperature where denitrification activity is optimal [°C] {30}

Data for the temperature function are taken from Parton *et al.* (1996) and Nömmik (1956).

Application in ModelMaker
The components which must be included in the previous model for calculation of soil moisture and temperature functions are presented in *Mod2–3e.mod* (Fig. 2.16).

In order to model the derepression dynamics, we shall have to include <delays> into the model which supply the program with the WFP_P values before the derepression times ('DeRep... '). Water-filled porosity functions are calculated for the current WFP and the WFP before the derepression time (WFP_P); the <variables> 'fWFP... ' are made conditional according to eqn 2.4/20.

The soil temperature function (fT_DEN) is defined in one variable. In the denitrification model, the following mathematical expressions are adjusted for soil temperature and soil moisture:

E1den, E2den, E3den, E4den, and the first-order emission constants kNOden, kN2Oden and kN2.

As already described for the nitrification model, the user can decide at the beginning of each simulation run (<Independent event> 'Choice'; <Define> 'function_adj') how the two functions should be applied (no adjustments = 1, taking the minimum of both functions = 2 or multiplying the functions by each other = 3). An adjustment for soil pH is not considered in the current model but could easily be incorporated according to the procedure described in Section 2.3.3 for the nitrification model.

The final model is presented in the model *Mod2–3e_real.mod*. It was decided to use more realistic environmental parameters to avoid calculation problems

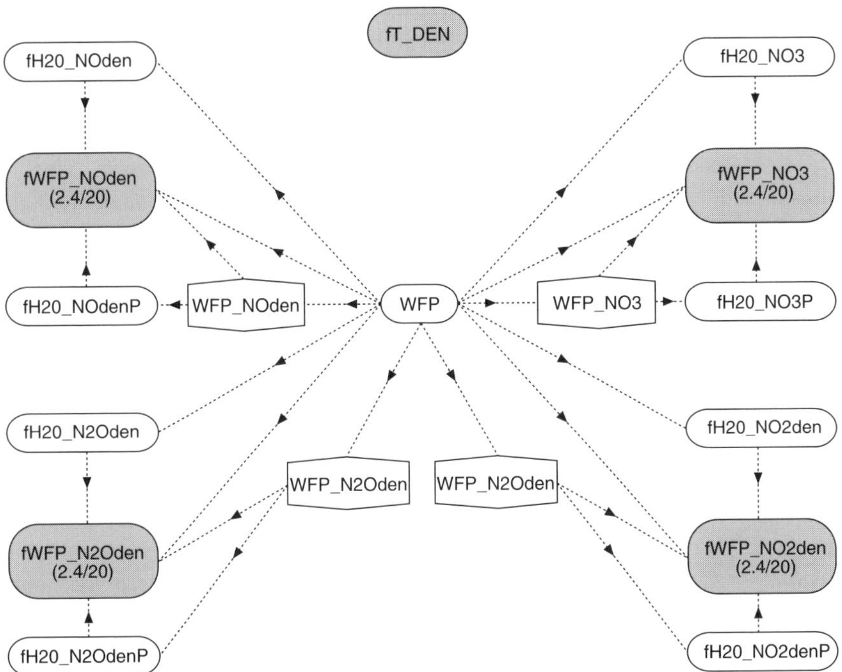

Fig. 2.16. Notation for the water-filled porosity functions (including derepression dynamics) and soil temperature function in the denitrification model (*Mod2-3e.mod*).

caused by erratic input data. It should always be kept in mind that the quality of input data also determines the stability of the simulation runs.

One's own data sets can easily be exchanged with the data currently held in the <Lookup Table> 'Inputs' or by supplying the model with data via an ASCII file held in a <Lookup File>. Usually this task can be performed via a spreadsheet package (e.g. Excel, Lotus, Quattro Pro) which is able to read and write data to files in an ASCII format.

In the final model *Mod2–3e.mod*, some of the parameters have been changed. These are: Rmax = 0.5, N2init = 0.001 and kN2_max = 0.1. The denitrification model can now be used for simulations. The various user choices make it easy to compare various simulation scenarios.

The current model (*Mod2–3e.mod*) is set up for a simulation period of 10 days. During this time, only a small change in the soil N concentration occurs; nevertheless, the N gas emissions may be high. In field situations, we often observe reasonably quick changes in the inorganic N concentrations. This discrepancy between simulations and observations points to the fact that, apart from denitrification, there are other N transformation processes operating in soil (e.g. N uptake by plants, N immobilization). Therefore, the denitrification model alone (like the nitrification model alone) is not sufficient to model N transformations in soil. The two processes of nitrification and denitrification are just two of several other processes which are carried out by the larger part of the living microbial biomass. These other processes include the decomposition of plant residue as well as the mineralization and build-up of various soil organic-matter fractions. Most N transformations in soil are linked to C transformations. Therefore, a model for C/N transformation is essential for understanding the N cycle in soil as well as the dynamics of the processes of nitrification and denitrification.

2.5 C/N Transformations in Soil Organic Matter

2.5.1 Conceptual model

An overview of C/N transformations in soil has already been presented in the introductory section (Section 2.1). An excellent overview of present models dealing with C/N transformations is presented in the book by Powlson *et al.* (1996): *Evaluation of Soil Organic Matter Models Using Existing Long-Term Datasets*. Most of the models have been developed over the last 25 years and some of the more well known include:

ROTHC, a carbon turnover model developed by Jenkinson and colleagues at Rothamsted experimental station (e.g. Coleman and Jenkinson, 1996).
CENTURY, a turnover model developed for predicting long-term changes by Parton and colleagues (e.g. Parton, 1996).
DAISY, a C/N turnover model developed by Jensen and colleagues (e.g. Mueller *et al.*, 1996).
VVV, a C/N turnover model based on the developments by van Veen and colleagues in the early 1980s (e.g. Gunnewiek, 1996).
DNDC, a C/N turnover model with the emphasis of predicting gaseous N emissions developed by Li and colleagues (e.g. Li, 1996).
HURLEY PASTURE MODEL, an ecosystem model with a soil and litter submodel, developed by Thornley and colleagues (e.g. Thornley, 1998).
ECOSYS, an ecosystem model which includes a submodel dealing with C and N transformations, mainly developed by Grant (e.g. Grant *et al.*, 1993).

The Rothamsted model, originally developed by Jenkinson and Rayner (1977), was one of the first widely used C transformation models. However, the basic concepts in modelling transformation processes in soil are similar for all of the above-mentioned models. Most of them divide the various organic constituents into plant residues and soil organic matter (SOM) components. These organic constituents are then further subdivided into fractions characterized by various transformation rates (e.g. fast, slow, very resistant). Jenkinson and Rayner (1977) found that the time required to degrade the various soil organic-matter fractions may vary from a few days to several thousand years. This large range of turnover characteristics of organic matter is partly due to chemical and physical stabilization, related to clay content and other chemical compounds in soil (van Veen *et al.*, 1984). To incorporate this feature into a model, three or more soil organic-matter fractions are considered which undergo degradation at characteristic transformation rates.

As already pointed out, the characteristics and dynamics of the microbial biomass are essential for the understanding of turnover processes in soil. Models vary considerably in the way they treat the action of the microbial biomass. One group of models may only consider the actual transformation without physically passing material through a microbial biomass pool (e.g. Jenkinson and Rayner, 1977; Parton, 1996), while other models put emphasis on the material passing through the microbial biomass before it is allocated to other soil organic-matter fractions (e.g. van Veen *et al.*, 1984). While the first group of models is easier to model, the second group of models represents the transformations in a more process-orientated way.

Another difference is the treatment of the concomitant flows of C and N. While most of the models assume that the C/N ratios of the various parts of the system will not change (e.g. Parton, 1996), there are other models which will calculate the C:N ratios in a dynamic way according to N mineralization–immobilization characteristics in relation to C transformations (e.g. van Veen *et al.*, 1984). Since net mineralization and immobilization are functions of the C:N ratios of newly formed biomass, a variable C:N ratio of the biomass and the other parts of the system will allow for optimal flexibility of the model. Discrepancies between calculated and observed C:N ratios of various soil organic-matter fractions would point to problems with the model concept.

An important feature of process-orientated models (e.g. van Veen *et al.*, 1984) is how they treat the growth/decay and dynamics of the microbial biomass. Depending on substrate characteristics (availability of C and N) and environmental conditions (freezing, drying, etc.), the microbial biomass may be active or inactive (Hunt, 1977). Degradation of microbial biomass may also happen at various speeds due to the quality of the substrates (e.g. fast or difficult to degrade) or to the capacity of the soil to protect a certain microbial population from degradation (van Veen *et al.*, 1984). The dynamics of the microbial biomass is best characterized by a dynamic response to influencing factors (Blagodatsky and Richter, 1998).

Finally, some loss of carbon in the form of CO_2 occurs, which is a result of the maintenance and growth of the microbial biomass. The portion of the C which is lost as carbon dioxide (CO_2–C) depends on the efficiency with which biosynthesis occurs (Frissel and van Veen, 1981).

The model developed in this section is based on the division of organic material proposed by van Veen *et al.* (1984), with the following assumptions:

- Inputs of crop residues have a fixed C:N ratio.
- C:N ratios of organic material vary according to the actual C/N flows within the system.

- All degraded material passes through the microbial biomass pool (except the organic-matter transformations due to physical and chemical stabilization).
- N immobilization by the biomass is dependent on the rate of C flowing into the biomass and the C:N ratio of the microbial biomass.

In most models the transformation rates are calculated according to first-order kinetics. However, it was found that Michaelis–Menten kinetics describes the transformation kinetics in a more realistic way (Frissel and van Veen, 1981; van Veen *et al.*, 1984). During the course of the model development, both kinetics will be used, thereby enabling direct comparison.

The compilation of these ideas enables a conceptual model to be formulated (Fig. 2.17).

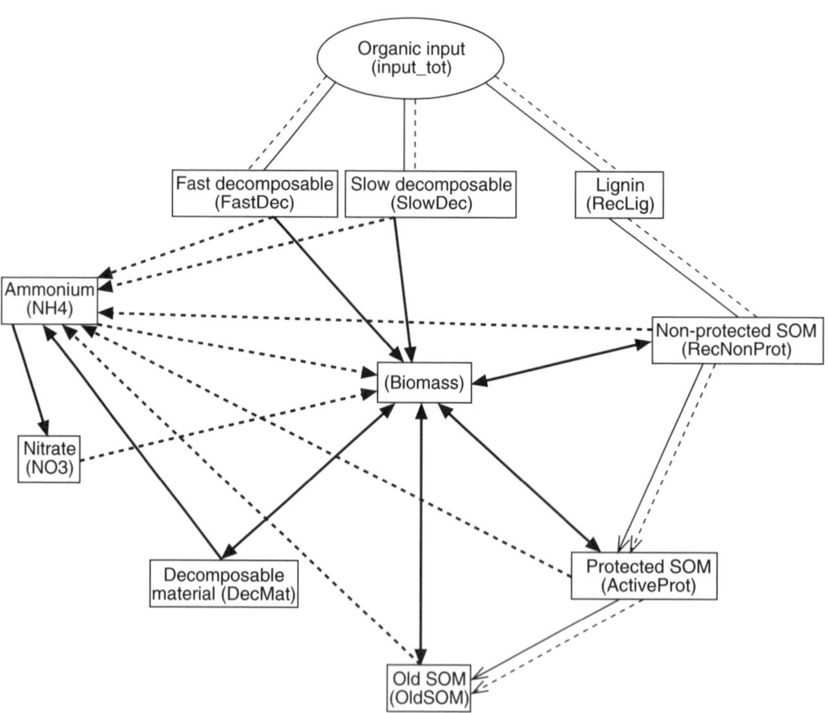

Fig. 2.17. Conceptual C/N transformation model (based on van Veen *et al.*, 1984) (thick solid lines = C transformation, thick dashed lines = N transformation, thin solid lines = C flow (physical, chemical stabilization); thin dashed lines = N flow (physical, chemical stabilization), without arrow = direct incorporation without transformation) (in parentheses = names as they appear in the model).

2.5.2 C transformation concept

The C/N transformation model developed in this section will be by far the most complex model which we have developed so far. This is already obvious if we compare the apparent complexity of the conceptual model for C/N transformations (Fig. 2.17) with the conceptual models for nitrification (Fig. 2.5) and denitrification (Fig. 2.12). However, the apparent complexity as outlined in Fig. 2.17 will diminish once the basic concept of the C/N transformations is understood because the transformation processes of the various organic substrates with the biomass all follow the same principle. The first step in the model development will therefore be the derivation of the basic exchange mechanism between the biomass ('Biomass') and one substrate pool (SOM1 = soil organic-matter pool 1).

First, we shall assume that the substrate contains only carbon and that the degradation of the SOM1 as well as the biomass follows first-order kinetics. The microbial biomass takes up the carbon of the SOM1 pool for biosynthesis. Part of the transformed SOM1 material is incorporated into the biomass. The rest is emitted as carbon dioxide (CO2_SOM1). Assuming that the microbes are working with an efficiency for biosynthesis defined by Y_SOM1 (e.g. Y_SOM1 = 0.5) means that only half of the C flowing through the microbes is incorporated into their bodies, the rest being lost as CO_2.

The gross rate of C flowing through the microbial biomass (Cgross_SOM1) is defined by (e.g. van Veen and Paul, 1981):

$$\begin{aligned} \text{Cgross\_SOM1} &= \text{CO2\_SOM1} \cdot \left\{ 1 + \left(\frac{\text{Y\_SOM1}}{1 - \text{Y\_SOM1}} \right) \right\} \\ &= \text{CO2\_SOM1} \cdot (1 - \text{Y\_SOM1}) \end{aligned} \qquad (2.5/1)$$

This equation can easily be derived by introducing the notation 'Cnet', which determines the rate of C actually incorporated into the microbial biomass (note: to simplify the derivation the extension '_SOM1' is omitted):

$$\text{Cnet} = \text{Cgross} \cdot \text{Y}$$

(net flow is equal to the gross rate multiplied by the efficiency)

Exchanging 'Cgross' with 'Cnet' + 'CO2' and also replacing 'Cgross' with $\frac{\text{Cnet}}{\text{Y}}$, we obtain:

$$\frac{\text{Cnet}}{\text{Y}} = \text{Cnet} + \text{CO2}$$

'CO2' can be defined by rearranging:

$$\text{CO2} = \frac{\text{Cnet}}{\text{Y}} - \text{Cnet} \quad \text{or} \quad \text{CO2} = \text{Cnet} \cdot \left(\frac{1}{\text{Y}} - 1 \right)$$

and finally:

$$CO2 = Cnet \cdot \left(\frac{1-Y}{Y}\right) \quad \text{and} \quad Cnet = CO2 \cdot \left(\frac{Y}{1-Y}\right)$$

'Cgross' can now be defined as:

$$Cgross = CO2 + CO2 \cdot \left(\frac{Y}{1-Y}\right) \quad \text{(or eqn 2.5 / 1)}$$

The processes of soil organic-matter degradation, biomass growth, CO_2 loss and biomass decay occur simultaneously. Assuming that the entire microbial biomass is involved in the decomposition of the soil organic-matter pool (SOM1) and that the transformation follows first-order kinetics, the notation for the gross transformation rate (Cgross_SOM1) is:

$$Cgross\_SOM1 = k\_SOM1 \cdot SOM1 \cdot \frac{Biomass}{Ctot} \qquad (2.5/2)$$

with:

Cgross_SOM1 = gross rate of C flow from soil organic matter to the biomass [μg C · g^{-1} soil · day^{-1}]
Ctot = total C in organic substrate (in this case it is equal to SOM1; note: it does not include Biomass carbon) [μg C · g^{-1} soil]
SOM1 = soil organic matter [μg C · g^{-1} soil]
k_SOM1 = first-order degradation constant [day^{-1}] {0.01; 0.001} (T = 100; 1000 days)
Biomass = fraction of biomass involved in the degradation of component SOM1 (assuming that the entire biomass is involved in the degradation of SOM1) [μg C · g^{-1} soil]

The degradation constant (k) is calculated as the inverse of the turnover time (*T*), (i.e. k = 1/*T*; e.g. Jenkinson and Rayner, 1977). Since most of the transformations are considered to be long-term changes (a few days to several hundred years), time units of 'days' rather than 'hours' are used in the C/N transformation model.

The microbial biomass is considered as one pool, despite the fact that this pool is made up of numerous microbial populations of possibly quite different characteristics. To allow for microbial diversity but still use only one microbial biomass pool, the entire biomass is divided into subgroups according to the fraction of the specific C substrate which the biomass subgroup is feeding on and set in relation to the overall C pool size (e.g. Frissel and van Veen, 1981). For instance, the size of the biomass subgroup which is considered to take part in the degradation of the C substrate pool SOM1 is calculated as:

$$\text{SOM1} \cdot \frac{\text{Biomass}}{\text{Ctot}}$$

with:

Ctot = sum of C in all substrate pools [μg C • g soil^{-1}]

This partitioning of the biomass is important, if more than one substrate pool is considered.

The microbial biomass is also continuously decaying. Assuming that the portions of the biomass which have grown on a particular substrate are also degrading at various speeds, it is possible to use the division of microbial biomass defined above in connection with substrate-specific decay rates. In the case of only one substrate, the overall decay of the biomass (Bio_SOM1) is defined by:

$$\text{Bio\_SOM1} = \text{k\_BioSOM1} \cdot \text{SOM1} \cdot \frac{\text{Biomass}}{\text{Ctot}} \qquad (2.5/3)$$

with:

k_BioSOM1 = decay rate constant of the biomass (feeding on 'SOM1') [day^{-1}] {0.02} (T = 50 days)

In the more general case with several C pools, the overall biomass degradation flux is partitioned into subflows, which enter particular soil organic-matter pools. In the first simple example, the entire C flow from the biomass will enter the SOM1 pool.

Finally, we can define the dynamics of the Biomass and SOM1 pool by the differential equations (combining eqns 2.5/1–2.5/3):

$$\frac{d\text{Biomass}}{dt} = (\text{Cgross\_SOM1} - \text{CO2\_SOM1}) - \text{Bio\_SOM1} \qquad (2.5/4)$$

with:

Biomass_init = initial amount of biomass [μg C • g^{-1} soil] {200}

and:

$$\frac{d\text{SOM1}}{dt} = \text{Bio\_SOM1} - \text{Cgross\_SOM1} \qquad (2.5/5)$$

with:

SOM1_init = initial amount of soil organic-matter [μg C • g^{-1} soil] {300}

Application in ModelMaker

The C transformation concept in soil just presented (eqn 2.5/1–2.5/5) is modelled in *Mod2–4a.mod* (Fig. 2.18). After inserting the <Compartments> (2), <Variables> (2), <Flows> (2) and <Influences> (5), the graphical output of the model should be similar to the one presented in Fig. 2.18.

The diagram looks slightly different from the ones we have considered so far. This is mainly a result of the two opposing flows between the <compartment> 'Biomass' and 'SOM1'. However, this notation is needed because there is a constant exchange between the two pools. The graphical representation of the model within ModelMaker would be easier to understand if the two opposing <flows> were separate, but this is not possible with the current ModelMaker version (version 3.03). Changing the degradation parameters and observing the effects on the two C pools (Biomass and SOM1) might be helpful in order to gain more insight into the dynamics of the C model even with this simple system (this can conveniently be carried out with a sensitivity analysis, as described in Section 2.3.3).

Before the model is extended from a single- to a multi-substrate model (as in Fig. 2.17), the concept of N transformations in soil is needed.

2.5.3 N transformation concept

Most organic substrates in the soil consist of carbon and nitrogen. Therefore, while microorganisms feed on the organic substrates to obtain carbon for their body

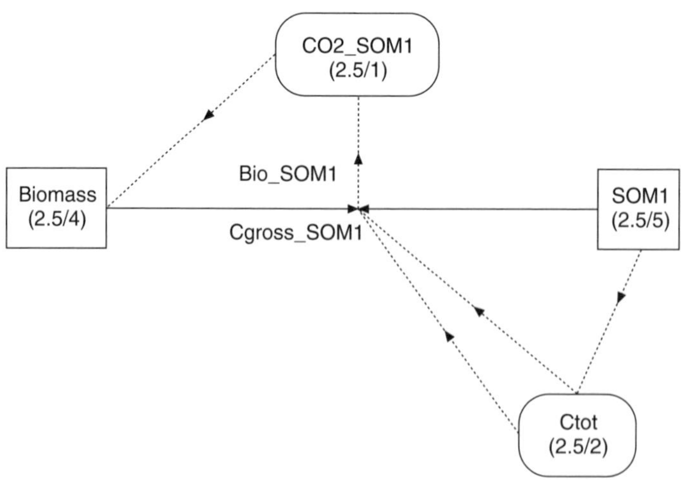

Fig. 2.18. C transformation concept (*Mod2-4a.mod*).

structure, a concomitant N mineralization flux occurs because the degraded organic material will release some nitrogen which enters the inorganic N pool of the soil. While carbon is taken up, the N requirement of the microbial biomass is met by immobilizing inorganic N in relation to the C:N ratio of the biomass. It is generally assumed that mineralization releases nitrogen in the form of ammonium ($NH_4^+{-}N$) which is then available for other N transformations, such as nitrification to nitrate (NO3) (Section 2.3).

To model the C and corresponding N transformations, the concept presented by van Veen *et al.* (1984) is adopted. In their notation, the substrate degradation constants (e.g. k_SOM1) apply to N mineralization. The body structures of microbes are characterized by specific C:N ratios, which will govern the allocation of the assimilated C. Therefore the C transformation rates (e.g. Cgross_SOM1_C) are obtained by multiplication with the C:N ratio of the biomass (CN_Bio) (Frissel and van Veen, 1981). The equation for gross C transformation (eqn 2.5/2) is extended to:

$$\mathrm{Cgross\_SOM1\_C} = \mathrm{k\_SOM1} \cdot \mathrm{SOM1} \cdot \frac{\mathrm{Biomass}}{\mathrm{Ctot}} \cdot \mathrm{CN\_Bio} \qquad (2.5/6)$$

with:

$$\mathrm{CN\_Bio} = \frac{\mathrm{Biomass\_C}}{\mathrm{Biomass\_N}}$$

(C:N ratio of microbial biomass [–]; eqns 2.5/4 and 2.5/14)

Note: k_SOM1 is here considered as a first-order rate constant for N mineralization (e.g. Frissel and van Veen, 1981).

The corresponding N decomposition flux which enters the ammonium pool (NH4) is calculated by dividing the gross C transformation rate (Cgross_SOM1_C) by the C:N ratio of the substrate (e.g. CN_SOM1):

$$\mathrm{Cgross\_SOM1\_N} = \frac{\mathrm{Cgross\_SOM1\_C}}{\mathrm{CN\_SOM1}} \qquad (2.5/7)$$

with:

$$\mathrm{CN\_SOM1} = \frac{\mathrm{SOM1\_C}}{\mathrm{SOM1\_N}}$$

(C:N ratio of soil organic-matter [$\mu g\ C \cdot \mu g^{-1}\ N$]; eqns 2.5/5 and 2.5/15)

Nitrogen which is needed by the microbes is mainly taken up from the ammonium pool (NH4) but also to a lesser extent from the nitrate pool (NO3) (Jansson *et al.*, 1955). The potential N immobilization flux (N_imm_pot) can be calculated by:

$$N\_imm\_pot = \frac{Cgross\_SOM1\_C - CO2\_SOM1}{CN\_Bio} \qquad (2.5/8)$$

With this notation, the N immobilization flow is directly proportional to the net C transformation rate. If not enough N is present, generally an approach is used where the C flow slows down to match the available N (Frissel and van Veen, 1981; van Veen *et al.*, 1984; Parton, 1996). However, recent investigations provide evidence that microbes may continue to assimilate organic C even under conditions when the inorganic N concentration would not be sufficient to match the C transformation rate. Under these conditions, it seems possible that microbes grow tissue with a relatively wide C:N ratio (Blagodatsky and Richter, 1998). To allow for such an N concentration effect, the potential N immobilization rate (N_imm_pot, eqn 2.5/8) is adjusted by a Michaelis–Menten-type approach. In addition, the overall immobilization flux is split into NH4 and NO3 immobilization. The immobilization rates for ammonium (NH4_Bio) and nitrate (NO3_Bio) are defined by:

$$NH4\_Bio = N\_imm\_pot \cdot fNH4im \cdot \frac{N\_NH4 \cdot NH4}{kn\_NH4 + NH4} \qquad (2.5/9)$$

and:

$$NO3\_Bio = N\_imm\_pot \cdot (1 - fNH4im) \cdot \frac{N\_NO3 \cdot NO3}{kn\_NO3 + NO3} \qquad (2.5/10)$$

where:

N_NH4 = Michaelis–Menten constant for NH_4^+–N immobilization [–] {5}
kn_NH4 = saturation constant for NH_4^+–N immobilization [μg N · g^{-1} soil] {4}
N_NO3 = Michaelis–Menten constant for NO_3^-–N immobilization [–] {5}
kn_NO3 = saturation constant for NO_3^-–N immobilization [μg N · g^{-1} soil] {4}
fNH4im = fraction of total immobilization flux taken up from NH_4^+–N [–] {0.8}

The Michaelis–Menten-type constants N_NH4 and N_NO3 can be considered as dimensionless N immobilization adjustment parameters.

Since the emphasis here is on modelling C/N transformations in soil, the nitrification–denitrification system is only considered in a very basic way (for more details, see Sections 2.3 and 2.4). The dynamics of ammonium (NH4) is modelled with the differential equation:

$$\frac{d\text{NH4}}{dt} = \text{Cgross\_SOM1\_N} - \text{NH4\_Nloss} - \text{NH4\_NO3} - \text{NH4\_Bio} \tag{2.5/11}$$

where:

NH4_Nloss = k_Nloss_nit · NH4	(N loss during nitrification)
NH4_NO3 = k_nit · NH4	(nitrification rate)

with:

NH4_init =	initial NH_4^+−N concentration [μg N · g^{-1} soil] {10}
k_Nloss_nit =	first-order rate constant for N loss during nitrification [day^{-1}] {0.001; 0.01}
k_nit =	first-order nitrification rate constant [day^{-1}] {0.005}

The dynamics of nitrate (NO3) is defined by:

$$\frac{d\text{NO3}}{dt} = \text{NH4\_NO3} - \text{NO3\_Nloss} - \text{NO3\_Bio} \tag{2.5/12}$$

where:

NO3_Nloss = k_Nloss_den · NO3	(N loss during denitrification)

with:

NO3_init =	initial NO_3^-−N concentration [μg N · g^{-1} soil] {5}
k_Nloss_den =	first-order rate constant for N loss during denitrification [day^{-1}] {0.001; 0.01}

In Section 2.5.2, we defined the flow of C resulting from the degradation of the microbial biomass as Bio_SOM1 (eqn 2.5/3) (we shall now call it Bio_SOM1_C). In addition, an N flow occurs, which is defined by the C flow and the C:N ratio of the microbial biomass:

$$\text{Bio\_SOM1\_N} = \frac{\text{Bio\_SOM1\_C}}{\text{CN\_Bio}} \tag{2.5/13}$$

Finally, we are able to define the nitrogen pools for biomass (Biomass_N) (eqns 2.5/9, 2–5/10 and 2.5/13):

$$\frac{d\text{Biomass\_N}}{dt} = \text{NH4\_Bio} + \text{NO3\_Bio} - \text{Bio\_SOM1\_N} \tag{2.5/14}$$

where:

$$\text{Biomass\_N\_init} = \frac{\text{Biomass\_C\_init}}{\text{CN\_Bio\_init}}$$

(definition for the initial microbial biomass N)

with:

CN_Bio_init = initial C:N ratio of the biomass [μg C $\cdot$ μg^{-1} N] {8}

and soil organic matter (SOM1_N) (eqns 2.5/7 and 2.5/13):

$$\frac{d\text{SOM1\_N}}{dt} = \text{Bio\_SOM1\_N} - \text{Cgross\_SOM1\_N} \qquad (2.5/15)$$

where:

$$\text{SOM1\_N\_init} = \frac{\text{SOM1\_C\_init}}{\text{CN\_SOM1\_init}}$$

(definition for the initial N content in SOM1)

with:

CN_SOM1_init = initial C:N ratio of the soil organic matter (SOM1) [μg C $\cdot$ μg^{-1} N] {11}

Application in ModelMaker

The first model (*Mod2–4a.mod*) is extended according to eqns 2.5/6 to 2.5/15. To better distinguish the components associated with C and N transformations, the notations '_C' and '_N' have been added. Some components have to be inserted, as presented in Fig. 2.19. In order to keep the graphical output simple, some of the components were made globally available. These are the <Flow> 'Cgross_SOM1' and the <Variables> 'CN_Bio' and 'CN_SOM1' (components which are globally available are indicated by grey hatching).

The model *Mod2–4b.mod* contains a full description of the concepts which will be applied in the next section when developing the basic C/N transformation model. Therefore the model *Mod2–4b.mod* should only be seen as a case-study to illustrate mathematical tools needed in C/N transformations. The current model does not include any input of organic C material and the C:N ratios of both the biomass and the soil organic matter will continuously decline, because carbon is lost from but not added to the system. Furthermore, the model does not include the influence of substrate-specific transformation rates. This can be achieved by introducing a Michaelis–Menten notation for the various transformation rates.

If the model were run long enough, the system would come to a point where all C in the substrate is depleted. Depending on substrate availability, microbes are able to change into a state of dormancy or non-activity. A dynamic representation

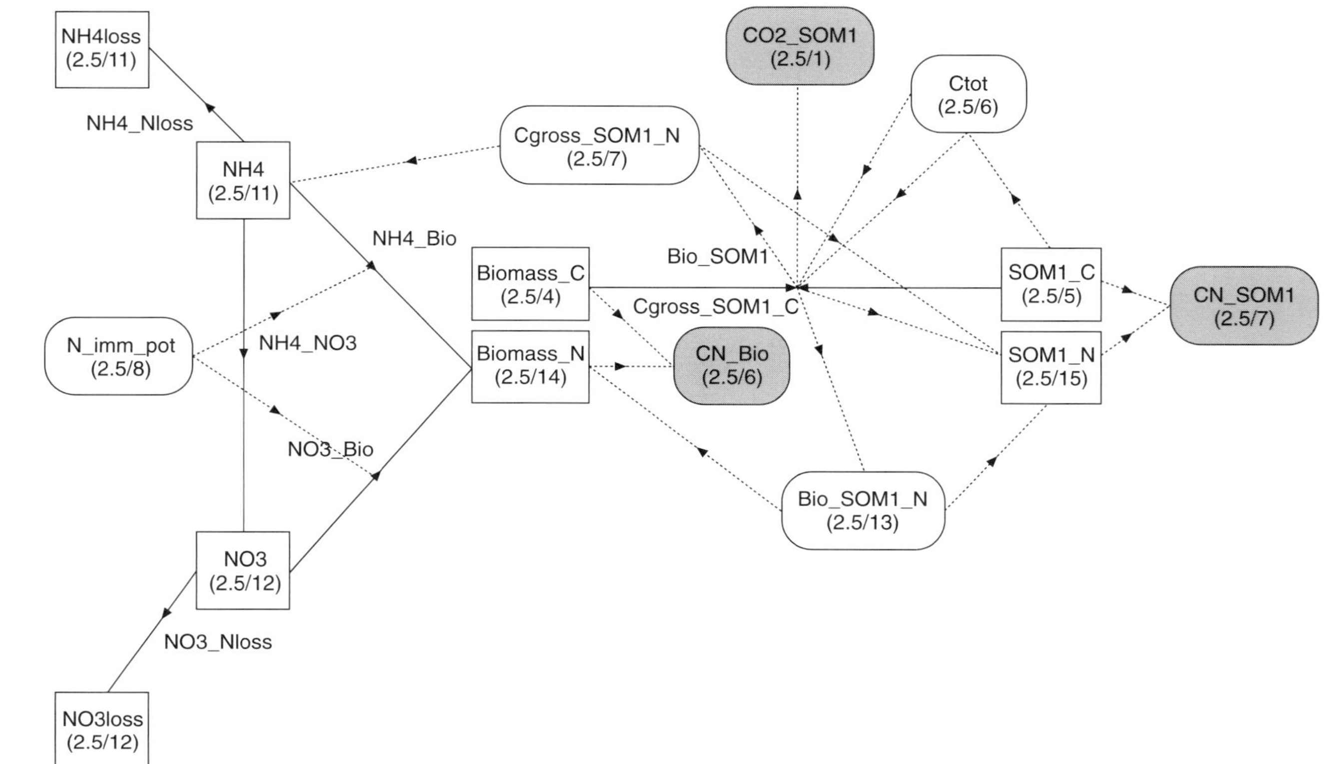

Fig. 2.19. C and N transformation concept (*Mod2-4b.mod*).

of the state of activity of microbes is introduced later (e.g. Hunt, 1977; Blagodatsky and Richter, 1998).

2.5.4 *Basic C/N transformation model*

The basic concepts for C and N transformations in soil presented in the last section will be extended in this section to a multi-substrate system to create a basic C/N transformation model. A key concept in this model is that all microbially mediated transformations always pass through the microbial biomass pool before they are allocated to other soil organic-matter fractions. The substrate and soil organic-matter pools to be considered are presented in the conceptual model (Fig. 2.17).

The organic substrates and inputs are grouped into organic inputs (e.g. plant residues) and soil organic-matter fractions. These two main groups are then further divided into subgroups of similar transformation characteristics (Fig. 2.17). The organic inputs are grouped into:

- fast decomposable organic inputs (e.g. proteins);
- slow decomposable organic inputs (e.g. cellulose);
- lignin fraction.

The total organic input is usually supplied to the model either via a user-defined input section (e.g. <Lookup File> or <Lookup Table>) or via an additional plant growth model (e.g. Thornley, 1988; see also Chapter 6 in this book). In the model developed here, we assume that we know the total daily organic input (Input_tot), which is supplied via a <Lookup File> 'Input' to the model.

The units of Input_tot are in μg dry matter g^{-1} soil day^{-1}. In order to express Input_tot in units of μg C g^{-1} soil day^{-1}, we need to know the fraction of C in the dry matter (f_CinOrgInp) and, in order to further allocate the total organic input (Input_tot) to subgroups of identical transformation characteristics, we need to know the fractions of fast decomposable organic inputs (FastDec), slow decomposable organic inputs (SlowDec) and lignin (RecLig). C inputs into the three organic input fractions are calculated by:

into FastDec:

$$\text{Input\_tot} \cdot \text{f\_CinOrgInp} \cdot \text{f\_FastDec}$$

into SlowDec:

$$\text{Input\_tot} \cdot \text{f\_CinOrgInp} \cdot \text{f\_SlowDec} \qquad (2.5/16)$$

into RecLig:

$$\text{Input\_tot} \cdot \text{f\_CinOrgInp} \cdot \text{f\_RecLig}$$

with:

f_CinOrgInp = fraction of C in total organic inputs [$g\ C \cdot g^{-1}$ DM] {0.5}
f_FastDec = fraction of fast decomposable organic inputs in total organic C inputs [–] {0.25}
f_SlowDec = fraction of slow decomposable organic inputs in total organic C inputs [–] {0.6}
f_RecLig = fraction of lignin in total organic C inputs [–] {0.15}

(Note: the terms for the flows are directly written into the differential equations FastDec, SlowDec and RecLig (see eqn group 2.5/28).

The N contents of the three fractions are usually well known and the C:N ratios of the three input fractions are well-defined, e.g.

CN_FastDec = C:N ratio of fast decomposable organic inputs [$\mu g\ C \cdot \mu g^{-1}$ N] {7}
CN_SlowDec = C:N ratio of slow decomposable organic inputs [$\mu g\ C \cdot \mu g^{-1}$ N] {20}
CN_RecLig = C:N ratio of lignin organic input fraction [$\mu g\ C \cdot \mu g^{-1}$ N] {150}

While the fast and slow decomposable fractions will undergo direct microbial transformations, the RecLig fraction will immediately enter the pool of recalcitrant non-protected soil organic-matter (Rec_NonProt) and decompose together with this soil organic-matter fraction (according to van Veen *et al.*, 1984).

The soil organic matter is divided into four subpools (Fig. 2.17):

- decomposable material (DecMat);
- recalcitrant non-protected soil organic matter (RecNonProt);
- active protected soil organic matter (ActiveProt);
- old soil organic matter (OldSOM).

Flows resulting from decomposing microbial biomass can be grouped into easy and more difficult mineralizable products. The soil organic fraction which contains easily mineralizable microbial products and plant products (e.g. root exudates) is contained in the pool of decomposable material (DecMat). This soil organic-matter pool is the most active one, with the fastest turnover time of all SOM fractions. Recalcitrant microbial and plant material (products which are more difficult to mineralize) are contained in the recalcitrant non-protected soil organic-matter pool (RecNonProt). Large differences in turnover rates of some soil organic compounds have been explained by a physical protection from degradation by clay minerals or other soil constituents (van Veen *et al.*, 1984; Parton, 1996; see also Section 2.1). Therefore, a small fraction of the microbial products will enter the active protected soil organic matter (ActiveProt). However, this pool can also be supplied by chemical stabilization processes from the RecNonProt pool. This ActiveProt pool has a turnover time of several hundred years. The soil organic-matter fraction with the slowest turnover time is the old organic matter (OldSOM), which is sometimes considered as an inert pool (e.g. Jenkinson and

Parry, 1989). This pool is only supplied via the action of chemical stabilization and protection mechanisms from material in the ActiveProt pool. The slow turnover time of this pool (more than 1000 years), as well as the slow rate of supply, will make this pool very inactive. The participation of this pool in the overall SOM dynamics is only evident after simulations of several thousand years.

According to the above description of the basic C/N transformation model, together with the concepts described in Sections 2.5.2 and 2.5.3, it is possible to describe the various transformations mathematically.

The gross C flows into the biomass from transformations of all six C substrate pools (recall that the RecLig pool is part of the RecNonProt pool) (according to eqn 2.5/6) are:

$$\mathrm{FastDec\_Bio} = \mathrm{k\_FastDec} \cdot \mathrm{FastDec} \cdot \frac{\mathrm{BioTot}}{\mathrm{Ctot}} \cdot \mathrm{CN\_Bio}$$

$$\mathrm{SlowDec\_Bio} = \mathrm{k\_SlowDec} \cdot \mathrm{SlowDec} \cdot \frac{\mathrm{BioTot}}{\mathrm{Ctot}} \cdot \mathrm{CN\_Bio}$$

$$\mathrm{DecMat\_Bio} = \mathrm{k\_DecMat} \cdot \mathrm{DecMat} \cdot \frac{\mathrm{BioTot}}{\mathrm{Ctot}} \cdot \mathrm{CN\_Bio} \qquad (2.5/17)$$

$$\mathrm{RecNonProt\_Bio} = \mathrm{k\_RecNonProt} \cdot \mathrm{RecNonProt\_C} \cdot \frac{\mathrm{BioTot}}{\mathrm{Ctot}} \cdot \mathrm{CN\_Bio}$$

$$\mathrm{ActiveProt\_Bio} = \mathrm{k\_ActiveProt} \cdot \mathrm{ActiveProt} \cdot \frac{\mathrm{BioTot}}{\mathrm{Ctot}} \cdot \mathrm{CN\_Bio}$$

$$\mathrm{OldSOM\_Bio} = \mathrm{k\_OldSOM} \cdot \mathrm{OldSOM} \cdot \frac{\mathrm{BioTot}}{\mathrm{Ctot}} \cdot \mathrm{CN\_Bio}$$

where:

$$\mathrm{Ctot} = \mathrm{FastDec} + \mathrm{SlowDec} + \mathrm{DecMat} + \mathrm{RecNonProt} + \mathrm{ActiveProt} + \mathrm{OldSOM}$$

with:

BioTot = Biomass carbon pool (= Biomass_C) [μg C · g^{-1} soil]
CN_Bio = C:N ratio of biomass [μg C · μg^{-1} N]
k_FastDec = first-order N mineralization rate constant (FastDec) [day^{-1}] {0.1} (T = 10 days)
k_SlowDec = first-order N mineralization rate constant (SlowDec) [day^{-1}] {0.08} (T = 12.5 days)
k_DecMat = first-order N mineralization rate constant (DecMat) [day^{-1}] {0.08} (T = 12.5 days)
k_RecNonProt = first-order N mineralization rate constant (RecNonProt) [day^{-1}] {0.0001} (T = 27 years)

k_ActiveProt = first-order N mineralization rate constant (ActiveProt) [day^{-1}] {0.000008} (T = 342 years)
k_OldSOM = first-order N mineralization rate constant (OldSOM) [day^{-1}] {0.000002} (T = 1370 years)

The flows of CO_2 associated with the various C transformations are calculated in analogy to eqn 2.5/1:

$$\begin{aligned}
&\text{CO2_FastDec} = \text{FastDec_Bio} \cdot (1 - \text{Y_FastDec}) \\
&\text{CO2_SlowDec} = \text{SlowDec_Bio} \cdot (1 - \text{Y_SlowDec}) \\
&\text{CO2_DecMat} = \text{DecMat_Bio} \cdot (1 - \text{Y_DecMat}) \\
&\text{CO2_RecNonProt} = \text{RecNonProt_Bio} \cdot (1 - \text{Y_RecNonProt}) \\
&\text{CO2_ActiveProt} = \text{ActiveProt_Bio} \cdot (1 - \text{Y_ActiveProt}) \\
&\text{CO2_OldSOM} = \text{OldSOM_Bio} \cdot (1 - \text{Y_OldSOM})
\end{aligned} \qquad (2.5/18)$$

with:

Y_FastDec = biosynthetic efficiency for microbes associated with turnover of the substrate pool FastDec [–] {0.6}
Y_SlowDec = biosynthetic efficiency for microbes associated with turnover of the substrate pool SlowDec [–] {0.65}
Y_DecMat = biosynthetic efficiency for microbes associated with turnover of the substrate pool DecMat [–] {0.6}
Y_RecNonProt = biosynthetic efficiency for microbes associated with turnover of the substrate pool RecNonProt [–] {0.7}
Y_ActiveProt = biosynthetic efficiency for microbes associated with turnover of the substrate pool ActiveProt [–] {0.8}
Y_OldSOM = biosynthetic efficiency for microbes associated with turnover of the substrate pool OldSOM [–] {0.8}

The use of variable biosynthetic efficiencies is related to considerations by Parton *et al.* (1987), who used higher efficiencies for microbes feeding on substrate with longer turnover times.

The N mineralization flows, which are defined by the C flows (eqn 2.5/17) divided by the C:N ratios of the various substrates, are (for C:N ratios, see eqn group 2.5/29):

associated with FastDec decomposition:

$$\frac{\text{FastDec\_Bio}}{\text{CN\_FastDec}}$$

associated with SlowDec decomposition:

$$\frac{\text{SlowDec\_Bio}}{\text{CN\_SlowDec}}$$

associated with DecMat decomposition:

$$\frac{\text{DecMat\_Bio}}{\text{CN\_DecMat}}$$

associated with RecNonProt decomposition:

$$\frac{\text{RecNonProt\_Bio}}{\text{CN\_RecNonProt}} \tag{2.5/19}$$

associated with ActiveProt decomposition:

$$\frac{\text{ActiveProt\_Bio}}{\text{CN\_ActiveProt}}$$

associated with OldSOM decomposition:

$$\text{RecLig\_RecNonProt} = \text{RecLig}$$

The flow of the lignin fraction (RecLig) into the RecNonProt_C pool is defined by:

$$\text{RecLig\_RecNonProt} = \text{RecLig} \tag{2.5/20}$$

with the accompanying N flow:

$$\frac{\text{RecLig\_RecNonProt}}{\text{CN\_RecLig}} \tag{2.5/21}$$

with:

CN_RecLig = C:N ratio of the lignin fraction [μg C $\cdot$ μg^{-1} N] {150}

The carbon flow related to chemical stabilization from the RecNonProt_C to the ActiveProtC pool is defined by:

$$\text{RecNonProt\_ActiveProt} = \text{k\_stabRecAct} \cdot \text{RecNonProt\_C} \tag{2.5/22}$$

with the accompanying N flow:

$$\frac{\text{RecNonProt\_ActiveProt}}{\text{CN\_RecNonProt}} \tag{2.5/23}$$

with:

k_stabRecAct = first-order chemical stabilization constant for flow from RecNonProt to the ActiveProt pool [day^{-1}] {0.00001}

The C flow from the ActiveProt to the OldSOM pool is given by:

$$\text{ActiveProt\_OldSOM} = \text{k\_stabActOld} \cdot \text{ActiveProt\_C} \tag{2.5/24}$$

with the N flow:

$$\frac{\text{ActiveProt\_OldSOM}}{\text{CN\_ActiveProt}} \tag{2.5/25}$$

with:

k_stabActOld = first-order chemical stabilization constant for flow from the ActiveProt to the OldSOM pool [day^{-1}] {0.000003}

After considering the various flows from the substrate pools to the microbial biomass and between pools due to stabilization processes, we shall consider the C and N flows associated with the degradation of microbial biomass. It is assumed that the subgroups of the microbial biomass also have variable decay characteristics, dependent upon the substrate pools they are feeding on (i.e. a group of microbes involved in degradation of decomposable material with a fast turnover time will also decay faster than a group of microbes living on protected soil organic matter). This feature can be incorporated into the model by the definition of microbe-specific decay-rate constants.

The overall biomass decay rate will be distributed into the pools DecMat, RecNonProt and ActiveProt. The C flows into the pools are defined (by analogy to eqn 2.5/3) by:

$$\text{Bio\_DecMat} = (X) \cdot \frac{\text{BioTot}}{\text{Ctot}} \cdot \text{f\_DecMat} \tag{2.5/26}$$

$$\text{Bio\_RecNonProt} = (X) \cdot \frac{\text{BioTot}}{\text{Ctot}} \cdot \text{f\_RecNonProt}$$

$$\text{Bio\_ActiveProt} = (X) \cdot \frac{\text{BioTot}}{\text{Ctot}} \cdot \text{f\_ActiveProt}$$

where:

$$X = \left(\begin{array}{l} \text{k\_BioFastDec} \cdot \text{FastDec} + \text{k\_BioSlowDec} \cdot \text{SlowDec} \\ + \text{k\_BioDecMat} \cdot \text{DecMat\_C} + \text{k\_BioRecNonProt} \cdot \text{RecNonProt\_C} \\ + \text{k\_BioActiveProt} \cdot \text{ActiveProt\_C} + \text{k\_BioOldSOM} \cdot \text{OldSOM} \end{array} \right)$$

with:

f_DecMat = fraction of total microbial degradation flux entering the DecMat pool [–] {0.4}

f_RecNonProt = fraction of total microbial degradation flux entering the RecNonProt pool [–] {0.5}

f_ActiveProt = fraction of total microbial degradation flux entering the ActiveProt pool [–] {0.1}

k_BioFastDec = first-order decay constant for microbial biomass associated with transformations of the substrate pool FastDec [day^{-1}] {0.07} k_BioSlowDec = first-order decay constant for microbial biomass associated with transformations of the substrate pool SlowDec [day^{-1}] {0.07}

k_BioDecMat = first-order decay constant for microbial biomass associated with transformations of the substrate pool DecMat [day^{-1}] {0.07}

k_BioRecNonProt = first-order decay constant for microbial biomass associated with transformations of the substrate pool RecNonProt [day^{-1}] {0.07}

k_BioActiveProt = first-order decay constant for microbial biomass associated with transformations of the substrate pool ActiveProt [day^{-1}] {0.009}

k_BioOldSOM = first-order decay constant for microbial biomass associated with transformations of the substrate pool OldSOM [day^{-1}] {0.001}

The N flows associated with the C flows (eqn 2.5/26) entering the N pools, Bio_DecMat_N, Bio_NonProt_N, Bio_ActiveProt_N, are obtained by division of the C flows by the C:N ratio of the biomass, i.e.:

$$\text{into DecMat\_N:} \quad \frac{\text{Bio\_DecMat}}{\text{CN\_Bio}}$$

$$\text{into RecNonProt\_N:} \quad \frac{\text{Bio\_RecNonProt}}{\text{CN\_Bio}} \tag{2.5/27}$$

$$\text{into ActiveProt\_N:} \quad \frac{\text{Bio\_ActiveProt}}{\text{CN\_Bio}}$$

Finally, the differential equations for the C and N pools can be defined as a result of the flows in and out of the various compartments. The notations for the differential equations defining the various C pools are:

$$\frac{d\text{FastDec}}{dt} = (\text{Input\_tot} \cdot \text{f\_CinOrgInp} \cdot \text{f\_FastDec}) - \text{FastDec\_Bio}$$

(eqns 2.5/16, 2.5/17)

with:

FastDec_init = initial C concentration in substrate pool FastDec [$\mu g\ C \cdot g^{-1}$ soil] {1}

$$\frac{d\text{SlowDec}}{dt} = (\text{Input\_tot} \cdot \text{f\_CinOrgInp} \cdot \text{f\_SlowDec}) - \text{SlowDec\_Bio}$$

(eqns 2.5/16, 2.5/17)

with:

SlowDec_init = initial C concentration in substrate pool SlowDec [μg C · g^{-1} soil] {1}

$$\frac{d\text{DecMat\_C}}{dt} = \text{Bio\_DecMat} - \text{DecMat\_Bio}$$

(eqns 2.5/17, 2.5/26)

with:

DecMat_Cinit = initial C concentration in substrate pool DecMat [μg C · g^{-1} soil] {50}

$$\frac{d\text{RecNonProt\_C}}{dt} = \begin{pmatrix} \text{RecLig\_RecNonProt} + \text{Bio\_RecNonProt} \\ -\text{RecNonProt\_Bio} - \text{RecNonProt\_ActiveProt} \end{pmatrix}$$

(eqns 2.5/17, 2.5/20, 2.5/22, 2.5/26)

with:

RecNonProt_Cinit = initial C concentration in substrate pool RecNonProt [μg C · g^{-1} soil] {300}

$$\frac{d\text{ActiveProt\_C}}{dt} = \begin{pmatrix} \text{RecNonProt\_ActiveProt} + \text{Bio\_ActiveProt} \\ -\text{ActiveProt\_Bio} - \text{ActiveProt\_OldSOM} \end{pmatrix}$$

(eqns 2.5/17, 2.5/22, 2.5/24, 2.5/26)

with:

ActiveProt_Cinit = initial C concentration in substrate pool ActiveProt' [μg C · g^{-1} soil] {600}

$$\frac{d\text{OldSOM\_C}}{dt} = \text{ActiveProt\_OldSOM} - \text{OldSOM\_Bio}$$

(eqns 2.5/17, 2.5/24)

with:

OldSOM_Cinit = initial C concentration in substrate pool OldSOM [μg C · g^{-1} soil] {700}

(2.5/28)

The notations for the differential equations defining the soil organic matter N pools are (recall that an N-pool definition for the FastDec and SlowDec compartments is not needed, due to the fixed C:N ratios of these pools and the fact that no N will enter these compartments):

$$\frac{d\text{DecMat\_N}}{dt} = \frac{\text{Bio\_DecMat}}{\text{CN\_Bio}} - \frac{\text{DecMat\_Bio}}{\text{CN\_DecMat}}$$

(eqns 2.5/17, 2.5/26)

where:

$$\text{CN\_DecMat} = \frac{\text{DecMat\_C}}{\text{DecMat\_N}}$$

(eqn 2.5/28)

with:

$$\text{DecMat\_Ninit} = \frac{\text{DecMat\_Cinit}}{\text{CN\_DecMat\_init}}$$

CN_DecMat_init = initial C:N ratio of substrate pool DecMat $[\mu g\ C \cdot \mu g^{-1}\ N]$ {8}

$$\frac{d\text{RecNonProt\_N}}{dt} = \left(\begin{array}{l} \dfrac{\text{RecLig\_RecNonProt}}{\text{CN\_RecLig}} + \dfrac{\text{Bio\_RecNonProt}}{\text{CN\_Bio}} \\ - \dfrac{\text{RecNonProt\_Bio}}{\text{CN\_RecNonProt}} - \dfrac{\text{RecNonProt\_ActiveProt}}{\text{CN\_RecNonProt}} \end{array} \right)$$

(eqns 2.5/17, 2.5/21, 2.5/23, 2.5/27)

where:

$$\text{CN\_RecNonProt} = \frac{\text{RecNonProt\_C}}{\text{RecNonProt\_N}}$$

(eqn 2.5/28)

with:

$$\text{RecNonProt\_Ninit} = \frac{\text{RecNonProt\_Cinit}}{\text{CN\_RecNonProt\_init}}$$

CN_RecNonProt_init = initial C:N ratio of substrate pool RecNonProt $[\mu g\ C \cdot \mu g^{-1}\ N]$ {11}

$$\frac{d\text{ActiveProt\_N}}{dt} = \left(\begin{array}{l} \dfrac{\text{RecNonProt\_ActiveProt}}{\text{CN\_RecNonProt}} + \dfrac{\text{Bio\_ActiveProt}}{\text{CN\_Bio}} \\ - \dfrac{\text{ActiveProt\_Bio}}{\text{CN\_ActiveProt}} - \dfrac{\text{ActiveProt\_OldSOM}}{\text{CN\_ActiveProt}} \end{array} \right)$$

(eqns 2.5/17, 2.5/23, 2.5/25, 2.5/27)

where:

$$\text{CN\_ActiveProt} = \frac{\text{ActiveProt\_C}}{\text{ActiveProt\_N}}$$

(eqn 2.5/28)

with:

$$\text{ActiveProt\_Ninit} = \frac{\text{ActiveProt\_Cinit}}{\text{CN\_ActiveProt\_init}}$$

CN_ActiveProt_init = initial C:N ratio of substrate pool ActiveProt [μg C $\cdot$ μg^{-1} N] {11}

$$\frac{d\text{OldSOM\_N}}{dt} = \frac{\text{ActiveProt\_OldSOM}}{\text{CN\_ActiveProt}} - \frac{\text{OldSOM\_Bio}}{\text{CN\_OldSOM}}$$

(eqns 2.5/17, 2.5/25)

where:

$$\text{CN\_OldSOM} = \frac{\text{ActiveProt\_C}}{\text{ActiveProt\_N}}$$

(eqn 2.5/28)

with:

$$\text{OldSOM\_Ninit} = \frac{\text{OldSOM\_Cinit}}{\text{CN\_OldSOMot\_init}}$$

CN_OldSOM_init = initial C:N ratio of substrate pool OldSOM [μg C $\cdot$ μg^{-1} N] {11}

(2.5/29)

The N mineralization flows (eqn 2.5/19) will enter the ammonium pool (NH4) (eqn 2.5/11) which is now defined by:

$$\frac{d\text{NH4}}{dt} = \left(\begin{array}{l} \frac{\text{FastDec\_Bio}}{\text{CN\_FastDec}} + \frac{\text{SlowDec}}{\text{CN\_SlowDec}} + \frac{\text{DecMat\_Bio}}{\text{CN\_DecMat}} + \\ + \frac{\text{RecNonProt\_Bio}}{\text{CN\_RecNonProt}} + \frac{\text{ActiveProt\_Bio}}{\text{CN\_ActiveProt}} + \frac{\text{OldSOM\_Bio}}{\text{CN\_OldSOM}} \end{array} \right) - \left(\begin{array}{l} \text{NH4\_NO3} \\ +\text{NH4\_Bio} \\ +\text{NH4\_Nloss} \end{array} \right)$$

(2.5/30)

The notations for the N losses during nitrification and denitrification (eqn 2.5/11) as well as the notation for the nitrate pool (eqn 2.5/12), remain the same.

The notation of the N immobilization rates (eqns 2.5/9 and 2.5/10) remains unchanged. However, the potential immobilization flux (N_imm_pot), calculated according to eqn 2.5/8, is now defined by (see eqns 2.5/17 and 2.5/18):

$$\text{N\_imm\_pot} = \frac{\left\{\begin{bmatrix} (\text{FastDec\_Bio} - \text{CO2\_FastDec}) + (\text{SlowDec\_Bio} - \text{CO2\_SlowDec}) + \\ (\text{DecMat\_Bio} - \text{CO2\_DecMat}) + (\text{RecNonProt\_Bio} - \text{CO2\_RecNonProt}) + \\ (\text{ActiveProt\_Bio} - \text{CO2\_ActiveProt}) + (\text{OldSOM\_Bio} - \text{CO2\_OldSOM}) \end{bmatrix}\right\}}{\text{CN\_Bio}} \tag{2.5/31}$$

Finally, the equation describing the biomass dynamics can be defined. The differential equation for the biomass C pool (Biomass_C) is (eqns 2.5/17; 2.5/18; 2.5/26):

$$\frac{d\text{Biomass\_C}}{dt} = \left\{\begin{bmatrix} \begin{bmatrix} (\text{FastDec\_Bio} - \text{CO2\_FastDec}) + (\text{SlowDec\_Bio} - \text{CO2\_SlowDec}) \\ +(\text{DecMat\_Bio} - \text{CO2\_DecMat}) + (\text{RecNonProt\_Bio} - \text{CO2\_RecNonProt}) \\ +(\text{ActiveProt\_Bio} - \text{CO2\_ActiveProt}) + (\text{OldSOM\_Bio} - \text{CO2\_OldSOM}) \end{bmatrix} \\ -[\text{Bio\_DecMat} + \text{Bio\_RecNonProt} + \text{Bio\_ActiveProt}] \end{bmatrix}\right\} \tag{2.5/32}$$

with:

Biomass_C_init = initial biomass carbon [μg C $\cdot$ g^{-1} soil] {300; 200}

The differential equation for the biomass N pool (Biomass_N) is (eqns 2.5/9; 2.5/10; 2.5/27):

$$\frac{d\text{Biomass\_N}}{dt} = \text{NH4\_Bio} + \text{NO3\_Bio} - \left(\frac{\text{Bio\_DecMat}}{\text{CN\_Bio}} + \frac{\text{Bio\_RecNonProt}}{\text{CN\_Bio}} + \frac{\text{Bio\_ActiveProt}}{\text{CN\_Bio}}\right) \tag{2.5/33}$$

with:

$$Biomass\_N\_init = \frac{Biomass\_C\_init}{CN\_Bio\_init}$$

initial biomass nitrogen [$\mu g\ N \cdot g^{-1}$ soil]

CN_Bio_init = initial C:N ratio of the microbial biomass [$\mu g\ C \cdot \mu g^{-1}\ N$] {8}

Applications in ModelMaker

The basic C/N transformation model (*Mod2–4c.mod*, Fig. 2.20) is already quite complex.

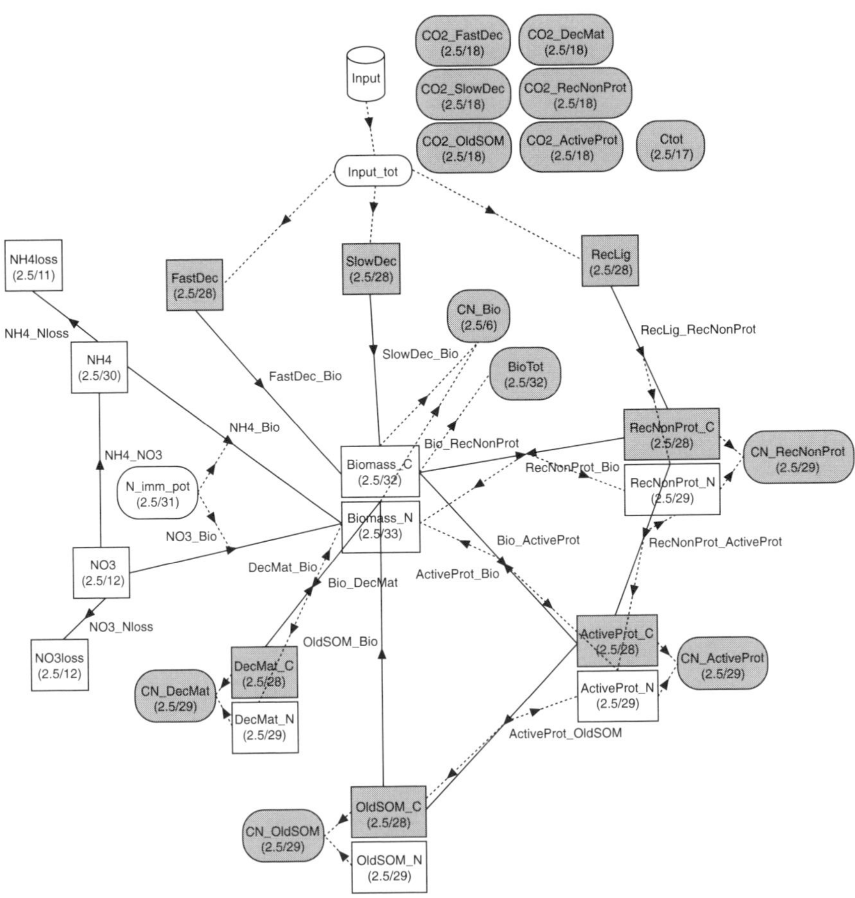

Fig. 2.20. Basic C/N transformation model in ModelMaker (*Mod2-4c.mod*).

Generally, defining components (<variables>, <compartments>, <defines>, etc.) in ModelMaker as <global> should be minimized, because <influences> illustrate the mathematical links within the model. However, if no components were global, all <influences> would have to be in place and this would make the graphical description of a complex model incomprehensible. Therefore, the following components have been made globally available:

<Flows>:

FastDec_Bio, SlowDec_Bio, DecMat_Bio, RecNonProt_Bio, ActiveProt_Bio and OldSOM_Bio

<Variables>:

CN_DecMat, CN_RecNonProt, CN_ActiveProt, CN_OldSOM

CO2_FastDec, CO2_SlowDec, CO2_DecMat, CO2_RecNonProt, CO2_ActiveProt, CO2_OldSOM

Ctot

<Compartments>:

FastDec, SlowDec, DecMat_C, RecNonProt_C, ActiveProt_C, OldSOM_C

An example where too many <Influences> are present can be seen when the *Mod2–4c.mod* (Fig. 2.20) is compared with model *Mod2–4c1.mod* (open this file in ModelMaker). The latter model contains a lot more <influences> because the compartments FastDec, SlowDec, DecMat_C, RecNonProt_C, ActiveProt_C and OldSOM_C have not been made <global>.

<Global> <variables> (e.g. CO2_FastDec, CO2_SlowDec, CO2_DecMat, CO2_RecNonProt, CO2_ActiveProt, CO2_OldSOM and Ctot) can be moved to one side of the graphical display, so that they do not obstruct other model components.

The C/N transformation model *Mod2–4c.mod* can now be used to get a basic understanding of the various transformation processes occurring in soil. To show that the model with the dynamic notations for C:N ratios and accompanying N flows does not behave in a obscure way, it is a good idea to select long simulation periods. Therefore, the run options in <Model> <Run Options> in the model *Mod2–4c.mod* are set to 36,500 days, which equals 100 years and 1200 output steps (= output each month).

One other test to determine if the model behaves realistically is to observe outputs of easily verifiable model components, such as the inorganic N concentrations NH4 and NO3. Model *Mod2–4c.mod* predicts that both concentrations are within 0 and 3 μg N g^{-1} soil, which is within the range of measured inorganic N concentration from unfertilized soil.

However, we should bear in mind that the current model presents only a basic description of C/N transformations in soil. The work by Frissel and van Veen

(1981) provides evidence for instance that the use of Michaelis–Menten rather than first-order kinetics would lead to more realistic simulation results. Therefore, the first extension of the current model will be the addition of Monod-type kinetics (similar to the Michaelis–Menten notation) to describe the various transformations.

2.5.5 Substrate-specific microbial growth and decay rates

Substrate-specific growth and decay rates are generally used in process-orientated C/N models (e.g. Frissel and van Veen, 1981; van Veen *et al.*, 1984; Blagodatsky and Richter, 1998). In this section, a Monod notation is introduced to define substrate-specific transformation rates for the exchange processes with the microbial biomass.

Recall (Section 1.4.2) that a first-order transformation is defined by:

$$v = \frac{dS}{dt} = k \cdot S \qquad (2.5/34)$$

with:

k = first-order rate constant [$time^{-1}$]
S = substrate concentration [$\mu g\ S \cdot g^{-1}$ soil]
v = transformation rate [$\mu g\ S \cdot g^{-1}\ soil \cdot time^{-1}$]

In order to adjust the transformation rate constant (k) for substrate effects, a 'Monod' notation, which is similar to the Michaelis–Menten notation, is introduced (see Panikov, 1995, Section 1.3):

$$k_s = V_{max} \cdot \frac{S}{k_m + S} \qquad (2.5/35)$$

with:

V_{max} = maximum transformation rate [$time^{-1}$]
k_m = Michaelis–Menten constant [$\mu g\ S \cdot g^{-1}$ soil]
k_s = substrate-specific transformation rate variable [$time^{-1}$]

The substrate-specific transformation variable (we can no longer talk about a constant) k_s has to be exchanged with the transformation constant k (eqn 2.5/34) to include the substrate effect in the calculation of the transformation rates (Blagodatsky and Richter, 1998). Therefore, the necessary additions to the current model are the 'ks' variables for each substrate, and associated parameters V_{max} and K_m. The Monod-type notations for substrate-specific transformation rates (according to eqns 2.5/34 and 2.5/35) are defined for the C/N model by:

$$ks\_FastDec = \frac{Vmax\_FastDec \cdot FastDec}{km\_FastDec + FastDec}$$

$$ks\_SlowDec = \frac{Vmax\_SlowDec \cdot SlowDec}{km\_SlowDec + SlowDec}.$$

$$ks\_DecMat = \frac{Vmax\_DecMat \cdot DecMat}{km\_DecMat + DecMat}$$

$$ks\_RecNonProt = \frac{Vmax\_RecNonProt \cdot RecNonProt}{km\_RecNonProt + RecNonProt}$$

$$ks\_ActiveProt = \frac{Vmax\_ActiveProt \cdot ActiveProt}{km\_ActiveProt + ActiveProt}$$

$$ks\_OldSOM = \frac{Vmax\_OldSOM \cdot OldSOM}{km\_OldSOM + OldSOM}$$

(2.5/36)

with:

Vmax_FastDec = maximum transformation rate for mineralization of substrate FastDec [day^{-1}] {0.1}
Vmax_SlowDec = maximum transformation rate for mineralization of substrate SlowDec [day^{-1}] {0.08}
Vmax_DecMat = maximum transformation rate for mineralization of substrate DecMat [day^{-1}] {0.08}
Vmax_RecNonProt = maximum transformation rate for mineralization of substrate RecNonProt [day^{-1}] {0.0001}
Vmax_ActiveProt = maximum transformation rate for mineralization of substrate ActiveProt [day^{-1}] {0.000008}
Vmax_OldSOM = maximum transformation rate for mineralization of substrate OldSOM [day^{-1}] {0.000002}
km_FastDec = half-saturation constant for transformation of substrate FastDec [$\mu g\ S \cdot g^{-1}$ soil] {50}
km_SlowDec = half-saturation constant for transformation of substrate SlowDec [$\mu g\ S \cdot g^{-1}$ soil] {80}
km_DecMat = half-saturation constant for transformation of substrate DecMat [$\mu g\ S \cdot g^{-1}$ soil] {50}
km_RecNonProt = half-saturation constant for transformation of substrate RecNonProt [$\mu g\ S \cdot g^{-1}$ soil] {1000}
km_ActiveProt = half-saturation constant for transformation of substrate ActiveProt [$\mu g\ S \cdot g^{-1}$ soil] {1000}

km_OldSOM = half-saturation constant for transformation of substrate OldSOM [$\mu g\ S \cdot g^{-1}$ soil] {2000}

(Note: at the moment the 'Vmax...' are equal to the first-order constants 'k_ ...' (see constant in eqn 2.5/17); however, to allow for maximum flexibility, the two constants have been defined separately.)

Substrate-specific flows to the microbial biomass are defined by exchanging the first-order constants ('k_...') (eqn 2.5/17) with the 'ks_...' variables (eqn 2.5/36). The flows to Biomass_C are (by analogy to eqn 2.5/17):

$$FastDec\_Bio = ks\_FastDec \cdot FastDec \cdot \frac{BioTot}{Ctot} \cdot CN\_Bio$$

$$SlowDec\_Bio = ks\_SlowDec \cdot SlowDec \cdot \frac{BioTot}{Ctot} \cdot CN\_Bio$$

$$DecMat\_Bio = ks\_DecMat \cdot DecMat \cdot \frac{BioTot}{Ctot} \cdot CN\_Bio$$

$$RecNonProt\_Bio = ks\_RecNonProt \cdot RecNonProt \cdot \frac{BioTot}{Ctot} \cdot CN\_Bio$$

$$ActiveProt\_Bio = ks\_ActiveProt \cdot ActiveProt \cdot \frac{BioTot}{Ctot} \cdot CN\_Bio$$

$$OldSOM\_Bio = ks\_OldSOM \cdot OldSOM \cdot \frac{BioTot}{Ctot} \cdot CN\_Bio$$

(2.5/37)

Furthermore, substrate-specific microbial decay rates (a) are also introduced. The notation described by Blagodatsky and Richter (1998) is applied, which, in the basic notation, is formulated by:

$$a = \frac{a_{max}}{1 + k_a \cdot S} \quad (2.5/38)$$

with:

a_{max} = maximum specific microbial death rate [day^{-1}]
k_a = inhibition constant for microbial death [g soil $\cdot \mu g^{-1}$ C]
S = substrate [$\mu g\ S \cdot g^{-1}$ soil]

Equation 2.5/38 has a similar effect to the inhibition notation described in Section 2.3.5 for nitrification. In general, the higher the inhibition constant (k_a), the higher is the numerical value of the denominator in eqn 2.5/38, therefore reducing the overall value of a. However, if the substrate concentration (S) is very low, the denominator will approach unity and the value of a will be close to a_{max}. Since inhibition constants cannot readily be determined it will be more or less a matter of 'trial and error' to find values which best describe the shape of the substrate-specific death rate (a).

In the theoretical approach by Blagodatsky and Richter (1998), the notation was only developed for one substrate. However, this notation can easily be extended to a multi-substrate system by defining substrate-dependent decay rates. The equations for a (eqn 2.5/38) extended to the entire C/N system are:

$$a\_FastDec = \frac{amax\_FastDec}{1 + ka\_FastDec \cdot FastDec}$$

$$a\_SlowDec = \frac{amax\_SlowDec}{1 + ka\_SlowDec \cdot SlowDec}$$

$$a\_DecMat = \frac{amax\_DecMat}{1 + ka\_DecMat \cdot DecMat\_C}$$

$$a\_RecNonProt = \frac{amax\_RecNonProt}{1 + ka\_RecNonProt \cdot RecNonProt\_C}$$

$$a\_ActiveProt = \frac{amax\_ActiveProt}{1 + ka\_ActiveProt \cdot ActiveProt\_C}$$

$$a\_OldSOM = \frac{amax\_OldSOM}{1 + ka\_OldSOM \cdot OldSOM\_C} \quad (2.5/39)$$

with:

amax_FastDec = maximum specific death rate for microbes associated with the substrate FastDec [day^{-1}] {0.07}

amax _SlowDec = maximum specific death rate for microbes associated with the substrate SlowDec [day^{-1}] {0.07}

amax _DecMat = maximum specific death rate for microbes associated with the substrate Decmat [day^{-1}] {0.07}

amax _RecNonProt = maximum specific death rate for microbes associated with the substrate RecNonProt [day^{-1}] {0.07} amax _ActiveProt = maximum specific death rate for microbes associated with the substrate ActiveProt [day^{-1}] {0.07}

amax _OldSOM = maximum specific death rate for microbes associated with the substrate OldSOM [day^{-1}] {0.07}

The notations for substrate-specific decay rates are introduced into eqn 2.5/26 by exchanging the 'k_Bio...' parameters in the *X* component with the '*a*_ ...' variables (eqn 2.5/39), i.e.:

$$X = \begin{pmatrix} a\_FastDec \cdot FastDec + a\_SlowDec \cdot SlowDec \\ +a\_DecMat \cdot DecMat\_C + a\_RecNonProt \cdot RecNonProt\_C \\ +a\_ActiveProt \cdot ActiveProt\_C + a\_OldSOM \cdot OldSOM \end{pmatrix}$$

(2.5/40)

Application in ModelMaker

The extension of the basic C/N transformation model with the Monod-type notation includes the calculation of transformation rates according to eqns 2.5/37 and 2.5/39. The notation is added to the model as <global> <variables> and is presented in the file *Mod2–4d.mod.*

In addition, an <independent event> 'User_choice' with a <define> 'kin_choice' has been added to the model to allow for user inputs as to whether substrate-specific growth and/or decay rates should be calculated or not. The necessary addition in <Actions> of the <independent event> 'User_choice' is the statement:

```
Kin_choice = GetChoice("Calculate substrate specific growth/decay
       rates? (Y/N)?","Kinetic Choice",Kin_choice);>(Yes = 1,
       No = 0).
```

In order to carry out the user decision, the flows to the Biomass_C (defined by eqns 2.5/17, 2.5/37) and away from the Biomass_C (eqn 2.5/26 in combination with 2.5/39 and 2.5/40) are defined conditionally. The two options are:

```
1: kin_choice = 0  use first-order kinetics
2: kin_choice = 1  use Michaelis-Menten kinetics
```

At the beginning of each simulation, the user has to choose which kind of kinetics should be used for the model calculations.

The current model is set up for 36,500 days (= 100 years) to check if long-term simulations give appropriate results. For instance, the dynamically calculated C:N ratios over a longer time may provide an indication as to how realistic the model simulations are. Run model *Mod2–4d.mod* and observe the output of the C:N ratios in the graph CN_ratios. It appears that the C:N ratios fall in a range which is generally expected in soil (between 5 and 15; the biomass generally has

C:N ratios below 10). However, observing the dynamics of the C pools (graph: C pools), the Biomass_C concentration remains at unrealistically low levels. The Biomass_C and DecMat_C are in a very quick exchange so that a build-up of the Biomass_C pool does not occur. The extension in the next section will, therefore, consider a concept which alleviates this problem, leading to a more realistic simulation of the microbial biomass (Biomass_C).

2.5.6 Protection capacity of microbial biomass

Turnover studies of labelled (^{15}N, ^{14}C) organic material over several years showed that soils with a high clay content generally retained higher proportions of labelled material in the microbial biomass pool than other soils (Ladd *et al.*, 1981). While the turnover rates were similar in all soils, the observed difference in biomass retention was related to an effect of the soil texture. Based on these observations, van Veen *et al.* (1984) introduced a soil texture-specific protection capacity of the microbial biomass. The concept is similar to the chemical and physical protection capacity of soil organic-matter fractions introduced in the previous section. However, instead of defining a fraction of protected biomass, a certain amount of microbial biomass is defined which is resistant to fast decay. Any microbial biomass in excess of this threshold value decays at a fast rate.

The concept of the protection of microbial biomass is incorporated into the current model by altering the notations for microbial decay (eqns 2.5/26, 2.5/39 and 2.5/40). Until the microbial biomass (Biomass_C) has reached the value for protected Biomass (ProtCap), the decay rate of the entire microbial biomass is adjusted by a decay rate reduction factor (ProtCap_k_red). If Biomass_C grows beyond the protected amount, the microbial biomass decays at two rates: the protected part decays at the slow rate and the unprotected part of the biomass decays at the fast rate. The equations describing the decay of microbial biomass (eqns 2.5/26, 2.5/39 and 2.5/40) are defined as conditional equations, according to the amount of biomass. The equations for biomass decay (2.5/26) are now defined by:

If Biomass_C < ProtCap

$$\text{Bio\_DecMat} = (\text{X}) \cdot \frac{\text{BioTot}}{\text{Ctot}} \cdot \text{ProtCap\_k\_red} \cdot \text{f\_DecMat}$$

$$\text{Bio\_RecNonProt} = (\text{X}) \cdot \frac{\text{BioTot}}{\text{Ctot}} \cdot \text{ProtCap\_k\_red} \cdot \text{f\_RecNonProt}$$

$$\text{Bio\_ActiveProt} = (\text{X}) \cdot \frac{\text{BioTot}}{\text{Ctot}} \cdot \text{ProtCap\_k\_red} \cdot \text{f\_ActiveProt}$$

(2.5/41)

If Biomass_C > ProtCap

$$\text{Bio\_DecMat} = (X) \cdot \frac{\text{ProtCap} \cdot \text{ProtCap\_k\_red} + (\text{BioTot} - \text{ProtCap})}{\text{Ctot}} \cdot \text{f\_DecMat}$$

$$\text{Bio\_RecNonProt} = (X) \cdot \frac{\text{ProtCap} \cdot \text{ProtCap\_k\_red} + (\text{BioTot} - \text{ProtCap})}{\text{Ctot}} \cdot \text{f\_RecNonProt}$$

$$\text{Bio\_ActiveProt} = (X) \cdot \frac{\text{ProtCap} \cdot \text{ProtCap\_k\_red} + (\text{BioTot} - \text{ProtCap})}{\text{Ctot}} \cdot \text{f\_ActiveProt}$$

with:

ProtCap = protection capacity of microbial biomass [μg C • g^{-1} soil] {300; 200}
ProtCap_k_red = reduction factor for specific transformation rates [–] {0.01}

Dependent upon whether the transformation rate constants are adjusted for substrate relationships or not, *X* is defined by either eqn 2.5/26 or eqn 2.5/40.

Application in ModelMaker

In order to allow for maximum flexibility in defining simulation scenarios an additional <independent event> 'User_choice' is added to the model which contains the statement:

```
prot_choice=GetChoice("Use Protection Capacity
        (Y/N)?","Protection Capacity choice",Prot_choice);
     if(prot_choice=1){
     ProtCap_choice=ProtCap;
    }
    else{
     ProtCap_choice=0;
    }
```

According to the value of Prot_choice the 'if…else statement' decides whether ProtCap_choice takes on the value of a predefined protection capacity ProtCap or not (ProtCap_choice = 0).

The flows defined in eqn 2.5/41 have to be included into the 'Bio_…' components and made conditional according to the amount of Biomass_C. Together with the conditional statements defined in the last section, four different ways of modelling the transformation rates are now available (eqn 2.5/41 in combination with the *X* variable, calculated either by eqn 2.5/26 or 2.5/40):

```
1: No substrate adjustment, without protection capacity
        (Kin_choice = 0, Prot_Choice = 0)
2: No substrate adjustment, with protection capacity
        (Kin_choice = 0, Prot_Choice = 1)
3: substrate adjustment, without protection capacity
        (Kin_choice = 1, Prot_Choice = 0)
4: substrate adjustment, with protection capacity
        (Kin_choice = 1, Prot_Choice = 1)
```

The extended model is presented in *Mod2–4e.mod.*

The current model has not been tested against 'real data'; however, it is worthwhile looking at some of the simulation results to get an idea as to which 'user-defined' scenario might best describe reality. Scientists who have experience in the measurement of some of the model outputs might have an immediate 'feeling' if the model outputs are in the range of expected values.

In order to illustrate some of the model simulations, the output for the microbial biomass Biomass_C and easily decomposable soil organic matter DecMat_C is presented for all possible user-defined simulation scenarios (Fig. 2.21). The simulations are carried out for 100 years in order to show the model behaviour over a long time period. It can clearly be seen that, where no biomass protection capacity is used, the initial biomass (Biomass_Cinit = 300 mg C g^{-1} soil) declines rapidly. On the other hand, biomass growth is only visible for the scenarios where substrate-specific transformation rates are calculated. The easily decomposable carbon concentrations (DecMat_C) are best described with scenarios including the protection capacity of the microbial biomass. Model outputs for the Biomass_C including the protection capacity and substrate-specific transformation rates, are between 300 and 600 mg C g^{-1} soil, which falls in the range of values reported for a range of ecosystems around the world (Wardle, 1998). This is also in line with the results by van Veen *et al.* (1984), who remarked that the capacity to preserve biomass was the key concept of the successful description of C and N turnover in soil.

While modelling packages such as ModelMaker are helpful in developing models in a very fast and easy way, they are sometimes, especially with extensive code, very slow in carrying out the model calculations. It is therefore advisable, especially if the model should run for long periods, to use a fast computer. For instance, it took roughly 4 min to run the current model *Mod2–4e.mod* for a period of 100 years on a computer equipped with a 100 MHz Pentium processor. A change in the <Model Run Options…> to fewer output steps or a fixed-step size might speed up the calculations considerably.

2.5.7 Physiological state of microorganisms

It is well recognized that microbes can take on different physiological states according to substrate availability and environmental conditions. Hunt (1977), for

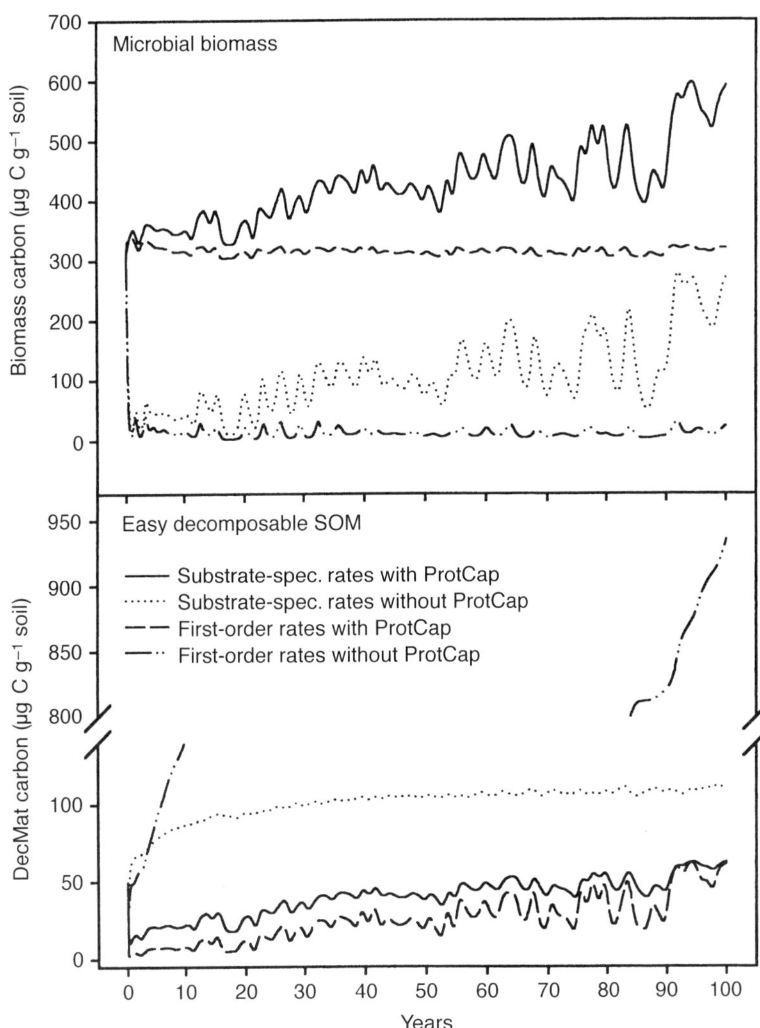

Fig. 2.21. Microbial biomass (Biomass_C) and easy decomposable SOM (DecMat_C) for all User_choice scenarios using the C/N transformation model *Mod2-4e.mod* ('ProtCap' denotes protection capacity).

instance, stressed the point that microbes are present in two physiological states: active or dormant, which in turn influences the dynamics of C and N transformations in soil.

The extension of the model describing the physiological state of microbes is based on theoretical considerations to explain complex phenomena of microbial growth in soil (Panikov, 1996). The theory is based on the division of cell constituents into primary components (P components), which are essential for cell growth (e.g. mRNA, ribosomal proteins, other RNA fractions, enzymes of the primary metabolic pathway), and U components, which are needed for cell survival (e.g. enzymes of secondary metabolism, protective pigments, reserved substances, transport systems of high affinity) (Panikov, 1995, pp. 199–206). Evidence that the P and U components in cells are related to each other in very specific ways led to the development of an 'index of physiological state' (*r*). Based on the Panikov theory, Blagodatsky and Richter (1998) recently proposed a theoretical concept for modelling microbial growth in combination with C and N transformations. The model includes two substrate pools: soluble C and insoluble soil organic matter. Their theoretical concept is used to extend the C/N transformation model.

In order to gain a better understanding of the concept, the model developed by Blagodatsky and Richter (1998) will be examined (Fig. 2.22); once we understand the concept, we shall apply the theory to our C/N model.

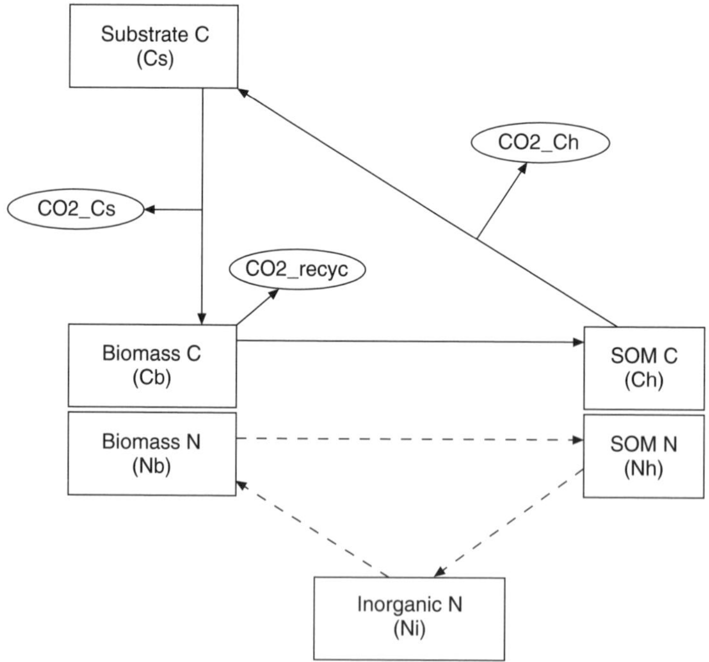

Fig. 2.22. Conceptual C/N model by Blagodatsky and Richter (1998) (solid lines = C transformations; dashed lines = N transformations; SOM = soil organic matter) (see text for further details).

Blagodatsky and Richter (1998) consider three carbon pools in their model: soluble carbon (Cs), Biomass (Cb) and insoluble SOM (Ch). Corresponding N pools are only considered for the biomass pool (Nb) and the insoluble SOM (Nh). In addition, a pool of inorganic nitrogen (Ni) is considered. The biomass only takes up carbon from the Cs pool and all material resulting from degradation of the Cb pool enters the Ch pool. This simplifies the transformation concept, because the entire Cb pool takes part in the transformation processes (recall that in our C/N model we defined substrate-specific biomass fractions (Section 2.5.4). The Ch pool decomposes to soluble carbon Cs, due to the action of the entire microbial biomass but without passing through the Cb pool. Carbon dioxide losses occur during the conversions from the Cs to the Cb pool (CO2_Cs), from the Ch to the Cs pool (CO2_Ch) and during recycling of C material (CO2_recyc). Flows of nitrogen are linked to the C transformations and calculated according to the actual C:N ratio of the biomass. N will flow from the Nb to the Nh pool and by decomposition from the Nh to the Ni pool. N immobilization is calculated as a separate process which is dependent on the microbial biomass (Ch) and the C:N ratio of the biomass (recall that in our C/N model N immobilization is currently calculated according to the actual flows of C into the microbial biomass rather than to the amount of biomass itself).

Substrate-specific transformation rates are calculated for microbial growth ('mu_Cs') (similar to eqn 2.5/35), decay of microbial biomass ('a_Cs') (similar to eqn 2.5/38), decomposition of SOM (q_Ch) (similar to eqn 2.5/35) and N immobilization ('mu_ni') (similar to eqn 2.5/35). Before introducing the 'index of physiological state', the current model is assembled.

Since the emphasis of this model is not the development of new expressions for the C flows but the introduction of the new 'index of physiological state', the notation for the mathematical description of the flows follows the one presented in the previous sections. Note that this notation varies slightly from the mathematical notation used by Blagodatsky and Richter (1998).

The various C flows in the model can be described mathematically by:

$$\begin{aligned} \text{Cs\_Cb} &= \text{mu\_Cs} \cdot \text{Cb} \\ \text{Cb\_Ch} &= \text{a\_Cs} \cdot \text{Cb} \\ \text{Ch\_Cs} &= \text{q\_Ch} \cdot \text{Cb} \end{aligned} \qquad (2.5/42)$$

where:

Cs_Cb = flow from soluble carbon (Cs) to biomass (Cb) [$\mu g\ C \cdot g^{-1}$ soil $\cdot$ day^{-1}]

Cb_Ch = flow from biomass (Cb) to insoluble SOM (Ch) [$\mu g\ C \cdot g^{-1}$ soil $\cdot$ day^{-1}]

Ch_Cs = flow from insoluble SOM (Ch) to soluble carbon (Cs) [$\mu g\ C \cdot g^{-1}$ soil $\cdot$ day^{-1}]

with:

$$mu\_Cs = \frac{mumax \cdot Cs}{ks + Cs}$$

$$a\_Cs = \frac{amax}{1 + ka \cdot Cs} \qquad (2.5/43)$$

$$q\_Ch = \frac{qmax \cdot Ch}{kh + Ch}$$

where:

mu_Cs = specific microbial growth rate [day^{-1}]
a_Cs = substrate-dependent specific microbial decay rate [day^{-1}]
q_Cs = specific rate of soil organic-matter decomposition [day^{-1}]
mumax = maximum specific microbial growth rate [day^{-1}] {0.7}
amax = maximum specific microbial decay rate [day^{-1}] {0.5}
qmax = maximum specific rate of soil organic-matter decomposition [day^{-1}] {0.1}
ks = half-saturation constant for microbial growth [$\mu g\ C \cdot g^{-1}$ soil] {100}
kh = half-saturation constant for organic-matter decomposition [$\mu g\ C \cdot g^{-1}$ soil] {50}
ka = inhibition constant for microbial death rate [μg^{-1} C] {30}

The loss of carbon dioxide associated with the flows Cs_Cb, Ch_Cs and due to death and recycling of the biomass are defined according to eqn 2.5/1:

$$\begin{aligned} CO2\_Cs &= Cs\_Cb \cdot (1 - Ys) \\ CO2\_Ch &= Ch\_Cs \cdot (1 - Yr) \\ CO2\_recyc &= amax \cdot (1 - Yr) \cdot Cb \end{aligned} \qquad (2.5/44)$$

with:

Ys = efficiency of substrate (Cs) uptake [–] {0.6}
Yr = efficiency of microbial biosynthesis [–] {0.6}

The differential equations defining the various C compartments are defined by:

$$\frac{dCs}{dt} = (Ch\_Cs - CO2\_Ch) - Cs\_Cb$$

$$\frac{dCb}{dt} = (Cs\_Cb - CO2\_Cs) - Cb\_Ch - CO2\_recyc \qquad (2.5/45)$$

$$\frac{dCh}{dt} = Cb\_Ch - Ch\_Cs$$

with:

Cs_init = initial substrate (Cs) [μg C $\cdot$ g^{-1} soil] {2000}
Cb_init = initial microbial biomass (Cb) [μg C $\cdot$ g^{-1} soil] {500}
Ch_init = initial insoluble SOM (Ch) [μg C $\cdot$ g^{-1} soil] {200}

The N flow from the microbial biomass to insoluble SOM is defined by:

$$\mathrm{Nb\_Nh} = \mathrm{a\_Cs} \cdot \mathrm{Cb} \cdot \frac{\mathrm{Nb}}{\mathrm{Cb}} \qquad (2.5/46)$$

and the N release during decomposition of insoluble SOM to inorganic nitrogen is given by:

$$\mathrm{Nh\_Ni} = \mathrm{q\_Ch} \cdot \mathrm{Cb} \cdot \frac{\mathrm{Nh}}{\mathrm{Ch}} \qquad (2.5/47)$$

Finally, the N immobilization flux from the inorganic nitrogen pool (Ni) to the biomass (Nb) is described by:

$$\mathrm{Ni\_Nb} = \mathrm{mu\_Ni} \cdot \mathrm{Cb} \qquad (2.5/48)$$

with the specific rate of N immobilization:

$$\mathrm{mu\_Ni} = \frac{\mathrm{nmax} \cdot \mathrm{Ni}}{\mathrm{kn} + \mathrm{Ni}} \qquad (2.5/49)$$

with:

nmax = maximum specific rate of N immobilization [day^{-1}] {0.8}
kn = half-saturation constant for N immobilization [μg N $\cdot$ g^{-1} soil] {40}

The notation for the various differential equations defining the N pools in the model are:

$$\frac{d\mathrm{Nb}}{dt} = \mathrm{Ni\_Nb} - \mathrm{Nb\_Nh}$$

$$\frac{d\mathrm{Nh}}{dt} = \mathrm{Nb\_Nh} - \mathrm{Nh\_Ni} \qquad (2.5/50)$$

$$\frac{d\mathrm{Ni}}{dt} = \mathrm{Nh\_Ni} - \mathrm{Ni\_Nb}$$

with:

$$\mathrm{Nb\_init} = \frac{\mathrm{Cb\_init}}{\mathrm{CN\_Cb\_init}}$$

initial nitrogen in microbial biomass [μg N $\cdot$ g^{-1} soil]

CN_Cb_init = initial C:N ratio of microbial biomass [–] {8}

$$\text{Nh\_init} = \frac{\text{Ch\_init}}{\text{CN\_Ch\_init}}$$

initial nitrogen in insoluble SOM [μg N · g^{-1} soil]
CN_Ch_init = initial C:N ratio of insoluble SOM [–] {11}
Ni_init = initial inorganic nitrogen [μg N · g^{-1} soil] {150}

The basic model described above is presented in *Mod2–4f1.mod* (Fig. 2.23). However, this is only the first stage in the model development and it is not a complete version of the model by Blagodatsky and Richter (1998).

The important step introduced by Blagodatsky and Richter (1998) is the 'index of physiological state' or 'Activity State' parameter (*r*). The parameter 'r' for a single substrate Cs is defined by (see Panikov, 1996):

$$\frac{dr}{dt} = \text{mu\_Cs} \cdot (\text{fr\_Cs} - \text{r}) \qquad (2.5/51)$$

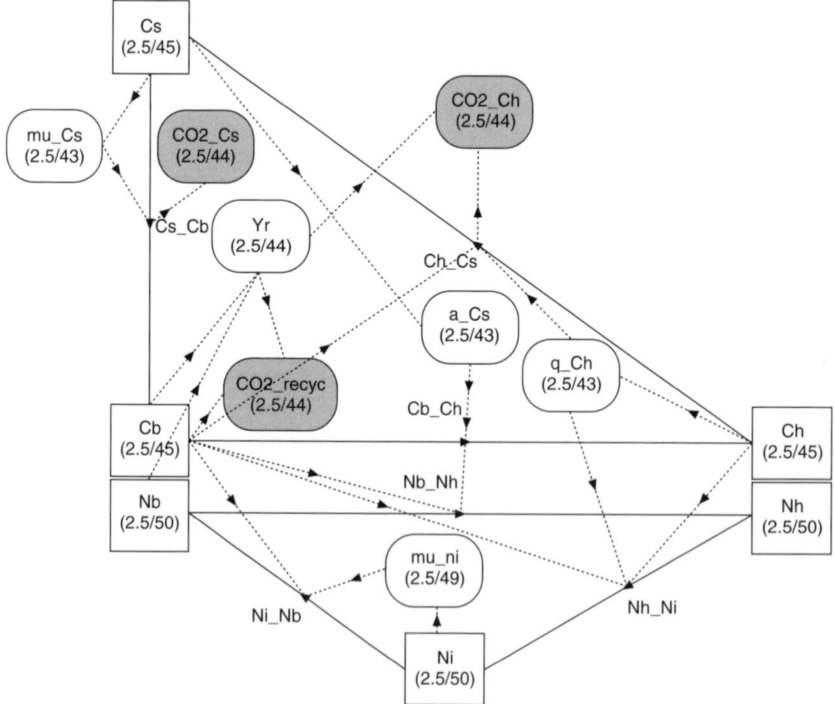

Fig. 2.23. First stage in the development of the model by Blagodatsky and Richter (1998) (*Mod2-4f1.mod*).

where:

r_init = initial 'index of physiological state' [–] {0.3}
mu_Cs = specific microbial growth rate (eqn 2.5/43)
fr_Cs = substrate-specific response function which influences the specific microbial growth rate according to a general inhibition notation (see, for example, Section 2.3.5 for further information on inhibition notations). The inhibition notation is defined by:

$$\text{fr\_Cs} = \frac{\text{Cs}}{\text{krC} + \text{Cs}} \qquad (2.5/52)$$

with:

krC = inhibition constant for C-dependent microbial activity [μg C · g^{-1} soil] {40}

Equation 2.5/51 consists of two terms, the product 'mu_Cs · fr_Cs', which defines the growth rate of new cell components (generally between 0 and 1), and the product 'mu_Cs · r', which defines the rate of *r* component dilution (a negative term) due to microbial growth (Blagodatsky and Richter, 1988). The higher the value of *r* and the higher the response function fr_Cs (high with high substrate concentration), the higher is the resulting dilution of the newly calculated *r* component.

It is not only the carbon substrate but also the amount of inorganic nitrogen which will determine the state of activity. Therefore, a response function, such as the one defined for carbon (eqn 2.5/52) is also defined for available inorganic nitrogen (fr_Ni):

$$\text{fr\_Ni} = \frac{\text{Ni}}{\text{krN} + \text{Ni}} \qquad (2.5/53)$$

with:

krN = inhibition constant for N-dependent microbial activity [μg N · g^{-1} soil] {20}

Either the two response functions are multiplied by each other or the one which is currently at a minimum is used. The latter approach would follow a 'Liebig' approach and will be used in the current model. Therefore, the index *r* with two substrates will be calculated as:

$$\frac{dr}{dt} = \text{mu\_Cs} \cdot (\text{fr\_min} - \text{r}) \qquad (2.5/54)$$

with:

fr_min = min(fr_Cs, fr_Ni)

Finally, a C–N-dependent dynamic calculation of the biosynthetic efficiencies has been included in the model by Blagodatsky and Richter (1998). These workers assume that the biosynthetic efficiency (e.g. eqn 2.5/1) is not constant but linked to the N:C ratio of the microbial biomass (note: they define N:C instead of C:N ratios). With increasing N:C ratios (decreasing C:N ratios), the efficiency increases until it has reached the maximum biosynthetic efficiency. The calculation of the biosynthetic efficiency by Blagodatsky and Richter (1998) is defined by:

$$Yr = Ym - NC\_act \qquad (2.5/55)$$

with:

$$NC\_act = NC\,max - \frac{Nb}{Cb}$$

NC max = 0.15 · (r + 1.6), the maximum N:C ratio in the microbial biomass
Ym = efficiency of microbial biosynthesis [–] {0.6}

In the model by Blagodatsky and Richter (1998), the dynamic biosynthetic efficiency (eqn 2.5/55) is applied to the mineralization of organic matter and the reutilization of microbial biomass (exchanging the parameter in eqn 2.5/44, *Yr* with eqn 2.5/55).

The 'r' index is multiplied by all six flows in the model: Cs_Cb, Cb_Ch, Ch_Cs (eqn 2.5/42), Nb_Nh (eqn 2.5/46), Nh_Ni (eqn 2.5/47), Ni_Nb (eqn 2.5/48).

In addition, the N immobilization flux is multiplied by ' $NCmax - \frac{Nb}{Cb}$ ', (eqn 2.5/55), so that it now reads (previously defined by eqn 2.5/48):

$$Ni\_Nb = mu\_Ni \cdot Cb\_r \cdot NC\_act \qquad (2.5/56)$$

with:

$$Cb\_r = Cb \cdot r$$

$$NC\_act = NC\,max - \frac{Nb}{Cb}$$

Application in ModelMaker

The extension to the model defining the 'r' value is presented in Fig. 2.24. The complete model is presented in *Mod2–4f2.mod*. To simplify the notation in *Mod2–4f2.mod*, a global variable Cb_r = Cb · r has been introduced which alleviates the problems of the numerous flows otherwise needed. In addition a global <variable> $NC\_act = NC\,max - \frac{Nb}{Cb}$ has been defined, which is needed for the calculation of Yr. The variable fr_min evaluates whether the response function for substrate Cs or Ni is at a minimum. This value is then used in the calculation of the *r* value.

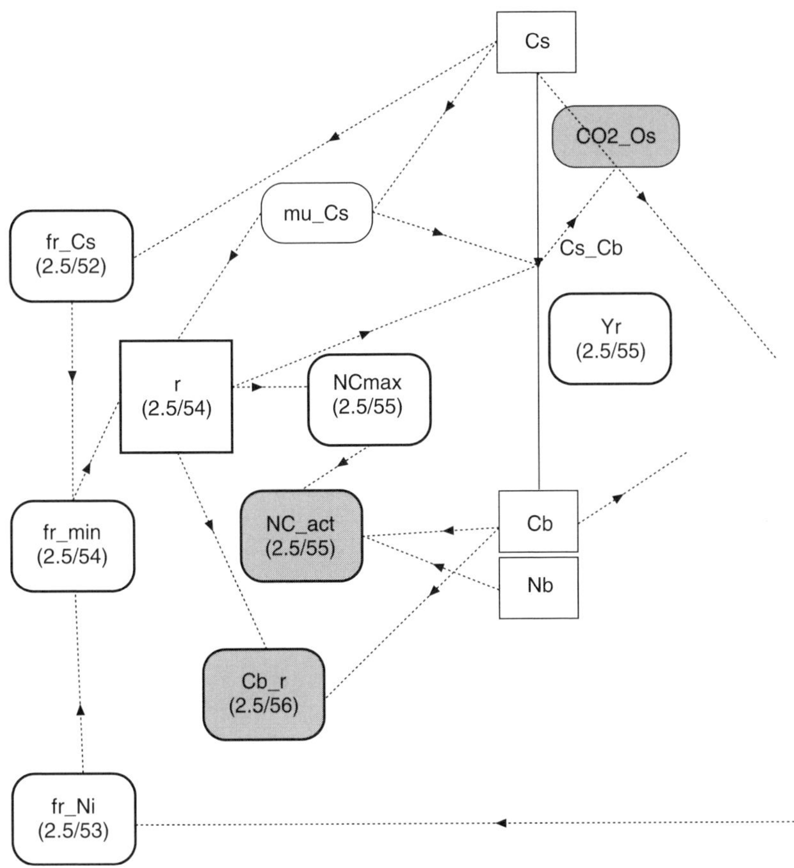

Fig. 2.24. Extension of the model *Mod2-4f1.mod* with the notation of the 'index of physiology' (*r*) (*Mod2-4f2.mod*).

Currently, the model *Mod2–4f2.mod* is set up for a simulation period of 20 days with 240 output steps (i.e. every 2 h). This kind of set-up is an example of how to run a simulation with time steps lower than the time definition in the parameter settings (i.e. parameters are set up on a per day basis but the output is presented on a 2-hourly basis). Recall that in the nitrification model we introduced a constant Time_convert, which scaled down the parameters to suitable time units. The method presented here is probably the safest way to introduce different time steps.

Graphical outputs for the N and C pools, the response functions ('fr_ …') and the *r* value have been defined in *Mod2–4f2.mod*. Once we are familiar with the concept developed in this section for the 'index of physiological state', we can move on and incorporate this new concept into our multi-substrate C/N transformation model.

Including the index of physiology in the C/N model

Recall that the last step in the C/N model development was the introduction of the protection capacity of the microbial biomass (Section 2.5.6). In order to apply the concept described in this section, we have to introduce substrate-specific response functions ('fr…') and *r* values ('r_…'). The response functions for the various substrates are defined according to eqn 2.5/52:

$$fr\_FastDec = \frac{FastDec}{kr\_FastDec + FastDec}$$

$$fr\_SlowDec = \frac{SlowDec}{kr\_SlowDec + SlowDec}$$

$$fr\_DecMat = \frac{DecMat\_C}{kr\_DecMat + DecMat\_C}$$

$$fr\_RecNonProt = \frac{RecNonProt\_C}{kr\_RecNonProt + RecNonProt\_C}$$

$$fr\_ActiveProt = \frac{ActiveProt\_C}{kr\_ActiveProt + ActiveProt\_C}$$

$$fr\_OldSOM = \frac{OldSOM\_C}{kr\_OldSOM + OldSOM\_C}$$

(2.5/57)

with:

kr_FastDec = inhibition constant for C-dependent microbial activity associated with substrate FastDec [$\mu g\ C \cdot g^{-1}$ soil] {2}
kr_SlowDec = inhibition constant for C-dependent microbial activity associated with substrate SlowDec [$\mu g\ C \cdot g^{-1}$ soil] {5}
kr_DecMat = inhibition constant for C-dependent microbial activity associated with substrate DecMat [$\mu g\ C \cdot g^{-1}$ soil] {2}
kr_RecNonProt = constant for C-dependent microbial activity associated with substrate RecNonProt [$\mu g\ C \cdot g^{-1}$ soil] {200}
kr_ActiveProt = inhibition constant for C-dependent microbial activity associated with substrate ActiveProt [$\mu g\ C \cdot g^{-1}$ soil] {400}
kr_OldSOM = inhibition constant for C-dependent microbial activity associated with substrate OldSOM [$\mu g\ C \cdot g^{-1}$ soil] {500}

One response function is defined for total inorganic nitrogen (NH4 + NO3):

$$\text{fr\_Ni} = \frac{\text{NH4} + \text{NO3}}{\text{kr\_Ni} + (\text{NH4} + \text{NO3})} \tag{2.5/58}$$

with:

kr_Ni = inhibition constant for N-dependent microbial activity associated with total inorganic nitrogen (NH4 + NO3) [μg N $\cdot$ g^{-1} soil] {1}

The substrate-dependent *r* values are defined according to eqn 2.5/54 (using eqns 2.5/36 and eqn 2.5/57):

$$\frac{d\text{r\_FastDec}}{dt} = \text{ks\_FastDec} \cdot \{min(\text{fr\_FastDec}, \text{fr\_Ni}) - \text{r\_FastDec}\}$$

$$\frac{d\text{r\_SlowDec}}{dt} = \text{ks\_SlowDec} \cdot \{min(\text{fr\_SlowDec}, \text{fr\_Ni}) - \text{r\_SlowDec}\}$$

$$\frac{d\text{r\_DecMat}}{dt} = \text{ks\_DecMat} \cdot \{min(\text{fr\_DecMat}, \text{fr\_Ni}) - \text{r\_DecMat}\}$$

$$\frac{d\text{r\_RecNonProt}}{dt} = \text{ks\_RecNonProt} \cdot \left\{\begin{matrix} min(\text{fr\_RecNonProt}, \text{fr\_Ni}) \\ -\text{r\_RecNonProt} \end{matrix}\right\}$$

$$\frac{d\text{r\_ActiveProt}}{dt} = \text{ks\_ActiveProt} \cdot \{min(\text{fr\_ActiveProt}, \text{fr\_Ni}) - \text{r\_ActiveProt}\}$$

$$\frac{d\text{r\_OldSOM}}{dt} = \text{ks\_OldSOM} \cdot \{min(\text{fr\_OldSOM}, \text{fr\_Ni}) - \text{r\_OldSOM}\} \tag{2.5/59}$$

with:

r_FastDec_init = initial *r* value for microbial activity associated with substrate FastDec [–] {0.3}
r_SlowDec_init = initial *r* value for microbial activity associated with substrate SlowDec [–] {0.3}
r_DecMat_init = initial *r* value for microbial activity associated with substrate DecMat [–] {0.3}
r_RecNonProt_init = initial *r* value for microbial activity associated with substrate RecNonProt [–] {0.3}
r_ActiveProt_init = initial *r* value for microbial activity associated with substrate ActiveProt [–] {0.3}
r_OldSOM_init = initial *r* value for microbial activity associated with substrate OldSOM [–] {0.3}

The biosynthetic efficiencies are calculated according to eqn 2.5/55 for all substrates undergoing microbial transformations. The definitions are:

$$
\begin{aligned}
&\text{YFastDec} = \text{Y\_FastDec} - \text{NC\_FastDec}\\
&\text{YSlowDec} = \text{Y\_SlowDec} - \text{NC\_SlowDec}\\
&\text{YDecMat} = \text{Y\_DecMat} - \text{NC\_DecMat}\\
&\text{YRecNonProt} = \text{Y\_RecNonProt} - \text{NC\_RecNonProt}\\
&\text{YActiveProt} = \text{Y\_ActiveProt} - \text{NC\_ActiveProt}\\
&\text{YOldSOM} = \text{Y\_OldSOM} - \text{NC\_OldSOM}
\end{aligned}
\qquad (2.5/60)
$$

with: (eqn 2.5/55):

$$
\begin{aligned}
&\text{NC\_FastDec} = 0.15\ (\text{r\_FastDec} + 1.6)\\
&\text{NC\_SlowDec} = 0.15\ (\text{r\_SlowDec} + 1.6)\\
&\text{NC\_DecMat} = 0.15\ (\text{r\_DecMat} + 1.6)\\
&\text{NC\_RecNonProt} = 0.15\ (\text{r\_RecNonProt} + 1.6)\\
&\text{NC\_ActiveProt} = 0.15\ (\text{r\_ActiveProt} + 1.6)\\
&\text{NC\_OldSOM} = 0.15\ (\text{r\_OldSOM} + 1.6)
\end{aligned}
$$

The N immobilization flux has so far been calculated with eqns 2.5/8 and 2.5/31. However, it can also be defined by the derivation in this section (eqn 2.5/56). The potential N immobilization flux for the multi-substrate C/N model can be calculated according to eqns 2.5/28, 2.5/59 and 2.5/60 using:

$$
\text{N\_imm\_pot} = \left\{
\begin{aligned}
&(\text{FastDec} \cdot \text{r\_FastDec} \cdot \text{NC\_FastDec}) +\\
&(\text{SlowDec} \cdot \text{r\_SlowDec} \cdot \text{NC\_SlowDec}) +\\
&(\text{DecMat\_C} \cdot \text{r\_DecMat} \cdot \text{NC\_DecMat}) +\\
&\left(\begin{aligned}&\text{RecNonProt\_C} \cdot \text{r\_RecNonProt} \cdot\\ &\text{NC\_RecNonProt}\end{aligned}\right) +\\
&(\text{ActiveProt\_C} \cdot \text{r\_ActiveProt} \cdot \text{NC\_ActiveProt}) +\\
&(\text{OldSOM\_C} \cdot \text{r\_OldSOM} \cdot \text{NC\_OldSOM})
\end{aligned}
\right\} \cdot \frac{\text{BioTot}}{\text{Ctot}}
\qquad (2.5/61)
$$

To obtain immobilization rates for $NH_4^+{-}N$ and $NH_3^-{-}N$, eqn 2.5/61 is multiplied by the specific N immobilization rate (eqn 2.5/49) where total inorganic nitrogen (Ni) is replaced by either ammonium (NH4) or nitrate (NO3). The fractional N immobilization rates are defined by the factor fNH4im (see eqns 2.5/9 and 2.5/10). In addition, the immobilization rates must be multiplied by the respective *r* values (see eqn 2.5/61). The flows describing the decay of microbial biomass (eqn 2.5/26) are adjusted by exchanging the *X* value with (eqns 2.5/28, 2.5/39, 2.5/59):

$$X = \begin{pmatrix} \text{a\_FastDec} \cdot \text{FastDec} \cdot \text{r\_FastDec} + \\ \text{a\_SlowDec} \cdot \text{SlowDec} \cdot \text{r\_SlowDec} + \\ \text{a\_DecMat} \cdot \text{DecMat\_C} \cdot \text{r\_DecMat} + \\ \text{a\_RecNonProt} \cdot \text{RecNonProt\_C} \cdot \text{r\_RecNonProt} + \\ \text{a\_ActiveProt} \cdot \text{ActiveProt\_C} \cdot \text{r\_ActiveProt} + \\ \text{a\_OldSOM} \cdot \text{OldSOM\_C} \cdot \text{r\_OldSOM} \end{pmatrix} \qquad (2.5/62)$$

Application in ModelMaker

Equations 2.5/57–2.5/62 are added to the previous C/N model (*Mod2–4e.mod*). The notations for the response functions ('fr…', eqns 2.5/57 and 2.5/58) and biosynthetic efficiencies ('Y…', eqn 2.5/60) are defined as <global> <variables>. The compartments defining the various 'indices of physiological state' ('r…', eqn 2.5/59) are added to the existing notation as <compartments>. The new notation for N immobilization (eqn 2.5/61) is added to the already existing <variable> 'N_imm_pot' and the specific rate of N immobilization into the flows NH4_Bio and NO3_Bio. In order to allow the user to decide between the two calculation procedures for immobilization, the user-definable statement:

```
GetValue("Immobilisation calc: Version 1 (van Veen) = 1,
        Version 2 (Blagodatsky) = 2","Immobilisation
        choice",Im_calc, Im_calc=1 or Im_calc=2);
```

is added to the <independent event> 'User_choice'.

To allow the execution of user-defined *r* values <global> <variables> 'rFastDec', 'rSlowDec', 'rDecMat', 'rRecNonProt', 'rActiveProt' and 'rOldSOM' are introduced. These <variables> are conditional and take on values in response to user inputs:

if *r*_calc = 1 (Y), then the dynamically calculated '*r*_…' values are used
if *r*_calc = 0 (N), then the *r* values are set to 1

The user-definable statement which controls *r*_calc is added to the <independent event> 'User_choice':

```
r_calc=GetChoice("Calculate 'index of physiological state'
        (Y/N)?","Activity state",r_calc);
```

The <variables> for calculating the biosynthetic efficiencies (*Y* values, eqn 2.5/60) are also made conditional to allow the calculation according to the value of the <define> Y_calc. The user-definable statement for Y_calc values in the <independent event> 'User_choice' is:

```
Y_calc=GetChoice("Calculate dynamic biosynthetic efficiencies
        (Y-values) (Y/N)?","Biosynthetic efficiency",Y_calc);
```

The final model including the notations described in this section is presented in *Mod2–4g.mod* (Fig. 2.25).

Numerous options are available for running the C/N model. However, some of the combinations do not make sense and are therefore eliminated. For instance, a dynamic *r* calculation only makes sense in combination with substrate-specific transformation and decay rates. Therefore, depending on the user choice for the *r* calculation, the following statement in the Actions of the <independent event> 'User_choice' are inserted:

```
if(r_calc=1){
 Kin_Choice=1
}
```

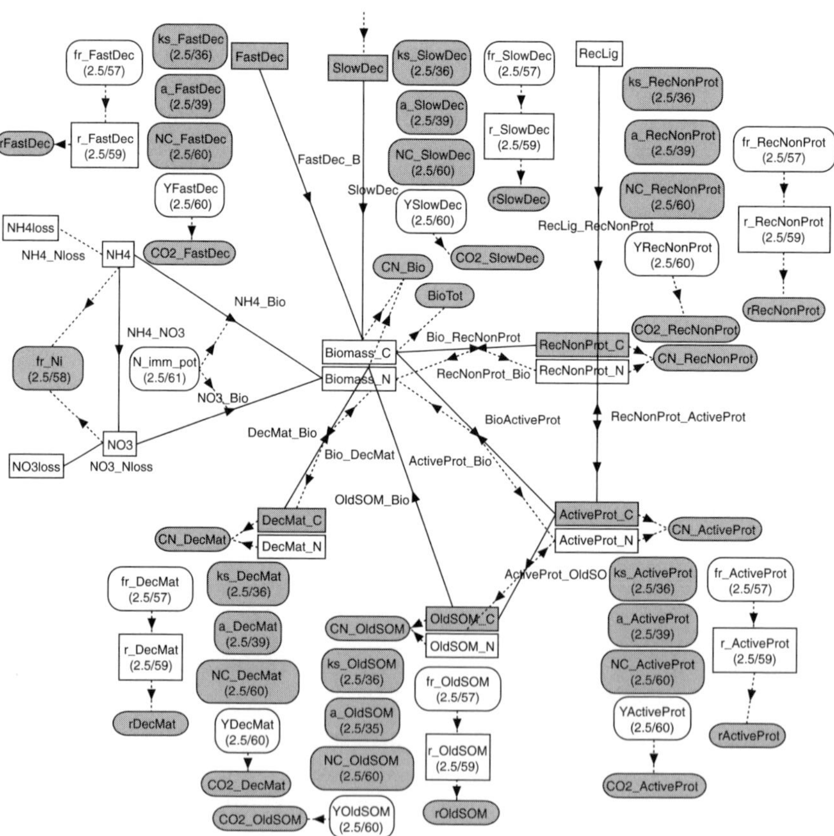

Fig. 2.25. C/N transformation model including the notation for the index of physiological state ('r_..values') and dynamic calculation of biosynthetic efficiencies ('Y...values') (*Mod2-4g.mod*) (bold = new notation; see also Fig. 2.20).

```
else{
 Kin_choice=GetChoice("Calculate substrate specific
        growth/decay rates? (Y/N)?","Kinetic
        Choice",Kin_choice);
}
```

Great care should be exercised when including the various notations in the already existing C/N model (*Mod2–4e.mod*). For instance, once the <variables> for the biosynthetic efficiencies are inserted (*Y* values), the notation for the CO_2 calculations has to be adjusted by exchanging the 'Y_...' parameters with the variables 'Y...' (omitting the underscore!).

Before we look at some of the model outputs, one final addition to the current C/N model is needed: the effects of soil moisture and soil temperature on the various transformation rates.

2.5.8 Influence of soil moisture and soil temperature

Adjustment of the numerous transformation processes within the model for soil moisture and soil temperature is effected by using similar functions to those used in the nitrification and denitrification models. One soil moisture function (fWFP_MIN) similar to eqn 2.3/41 and one soil temperature function (fT_MIN) similar to eqn 2.3/42 are defined. Data for both functions are taken from the publication by Frissel and van Veen (1981).

The parameters for the soil moisture function (fWFP_MIN; eqn 2.3/41) are:

kinc1_MIN =	1.4242	WFP_A_MIN =	0.4
kinc2_MIN =	2.375	WFP_B_MIN =	0.58
Iinc1_MIN =	0	WFP_C_MIN =	0.63
Iinc2_MIN =	−0.3775		
kdec_MIN =	−1.4636		
Idec_MIN =	1.83274		

(Note: two different linear increasing segments are defined for the WFP_A_MIN function.)

The parameters for the soil temperature function (fT_MIN; eqn 2.3/41) are:

Tmin_MIN = −5°C (note: Frissel and van Veen (1981) use 0°C)
Tmax_MIN = 60°C
Topt_MIN = 35°C

Application in ModelMaker

Input values for soil moisture and soil temperature are already present in the <input file> 'Input'. They are read into the <variables> 'VWC' (volumetric water content) and T_soil (soil temperature). Volumetric water content is converted to water-filled porosity (WFP) via eqn 2.3/39.

The soil moisture and soil temperature functions (fWFP_MIN; fT_MIN) are calculated with the respective variables and both functions are combined into an environmental factor ‘fenv’. An additional user choice env_choice selects whether the two environmental parameters are to be multiplied with each other or if the minimum of both functions should be used (see Section 2.3.4 for further information). A ‘GetValue’ statement is added to the <independent event> ‘User_choice’ to allow for various model calculations regarding the influence of the environmental parameters soil moisture and soil temperature:

```
GetValue("Environmental adjustment None = 0, multiplied (fT,
        fWFP)= 1, minimum (fT, fWFP)= 2","Environmental
        Choice",env_choice, env_choice=0 or env_choice=1 or
        env_choice=2);
```

The environmental factor ‘fenv’ is applied to the model by multiplication with the following components:

<Flows>:

FastDec_Bio, SlowDec_Bio, DecMat_Bio, RecNonProt_Bio, ActiveProt_Bio, OldSOM_Bio, Bio_RecNonProt, Bio_DecMat, Bio_ActiveProt, RecNonProt_ActiveProt, ActiveProt_OldSOM

<Variables>

N_imm_pot (only for the ‘Blagodatsky-notation’ because the flows in the other notation are already adjusted)

The notation which has to be added to the current C/N model is presented in Fig. 2.26. The final model is presented in model *Mod2–4h.mod.*

The development of the C/N transformation model is now finished. The C/N transformation model is based on process-orientated considerations; however, this does not mean that the model in its current form is applicable to field situations. Most of the parameters have not been optimized for particular field conditions. Furthermore, the input values are not taken from any particular field investigation. This model therefore has a rather exploratory nature. Nevertheless, the user-definable simulation options, in combination with the possibility to change the numerous parameters, provide a very flexible research model.

The best approach is to just use the model and try out the possible user definable combinations the model offers. The behaviour of the model may best be evaluated when some key parameters are changed. For instance, the parameters related to N loss and N immobilization within the system (i.e. k_Nloss_den, k_Nloss_nit and the immobilization parameters N_NH4, N_NO3, Nmax_NH4, Nmax_NO3) are crucial for the dynamics of the various ‘CN … ratios’. The procedure available in ModelMaker <Model> <sensitivity> (introduced towards the end of Section 2.3.3) is a convenient tool for performing such an analysis.

In Fig. 2.27, one of many examples is presented showing how the model output differs with the two versions for the calculation of N immobilization

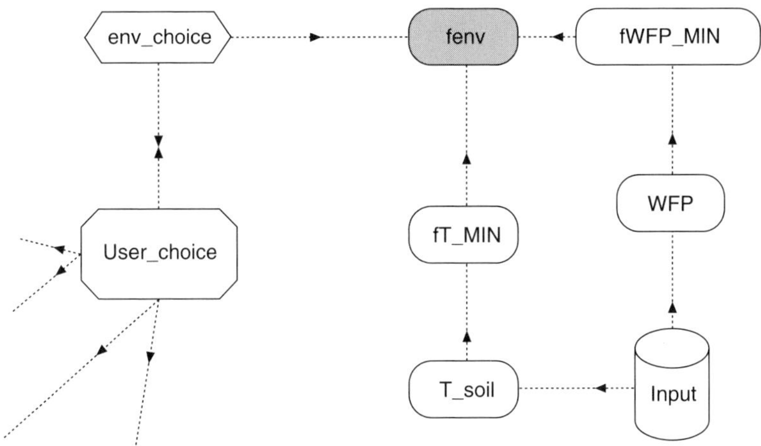

Fig. 2.26. Notation in the C/N model for the environmental factors soil moisture and soil temperature (*Mod2-4h.mod*).

(1 = according to van Veen, 2 = according to Blagodatsky) (Fig. 2.27). Parameters are not altered and user options are set to:

Options	User choice
Calculate 'index of physiological state' ?	YES
Use protection capacity ?	YES
Calculate dynamic biosynthetic efficiencies?	YES
Environmental adjustment:	Option 2 (take minimum)

A simulation over 100 years shows that the outputs for the microbial biomass are similar for the two immobilization options. Therefore, changing between the two N immobilization versions does not affect the output as much as, for instance, the option with or without the microbial protection capacity (see Section 2.5.6, Fig. 2.21). The rather noisy output is related to the inputs from a random-number generator, which was used to generate soil temperature, volumetric water content and organic inputs within a given range.

In another simulation run, the parameters for N loss from the nitrification–denitrification system (k_Nloss_nit and k_Nloss_den) as well as the maximum specific rate of N immobilization (Nmax_NH4 and Nmax_NO3), were changed. This had a considerable effect on the various C:N ratios in the system (not presented here but it can easily be carried out). In order to become familiar with the model and its dynamic behaviour, it is best to try as many options as possible.

An extension one might want to carry out is the combination of the nitrification and denitrification models in Sections 2.3 and 2.4 with the C/N model

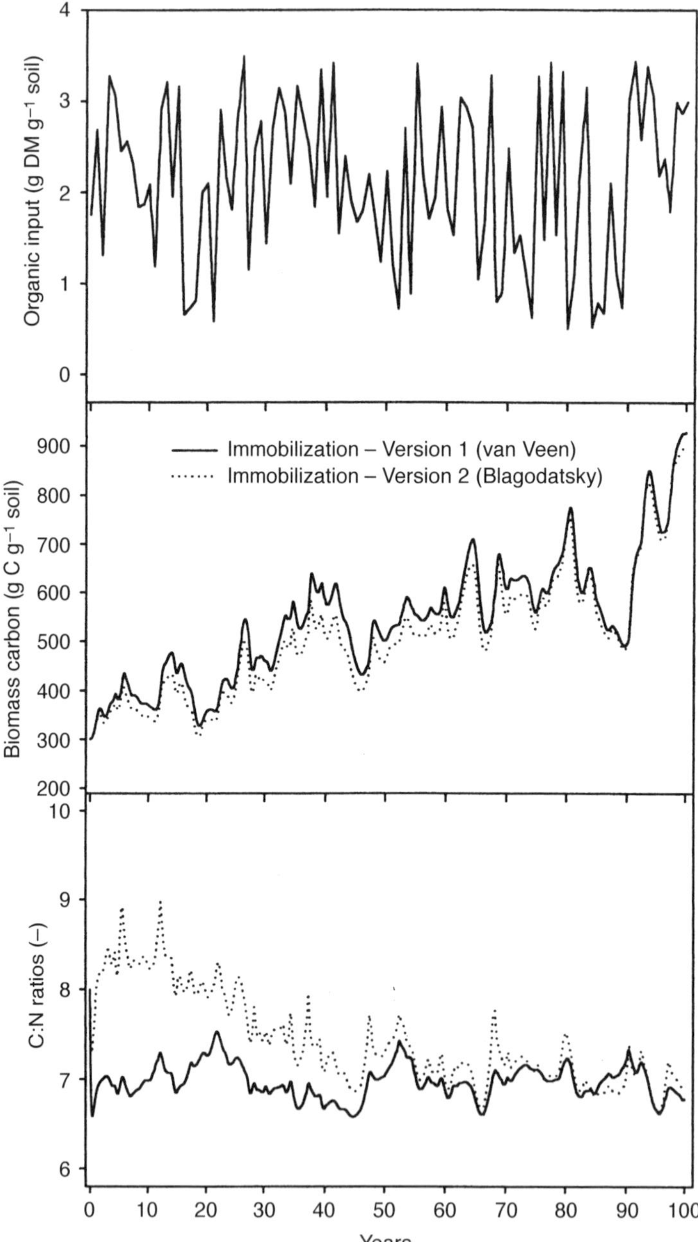

Fig. 2.27. 100-year simulation output of total organic inputs, microbial biomass carbon and C:N ratio of microbial biomass with the C/N transformation model *Mod2-4h.mod* (for parameter settings, see text).

developed here. This can easily be done by copying the appropriate items to one model file.

2.5.9 Future development

Further additions to the C/N transformation model which are not considered here may include a detailed description of what happens during extreme events, such as drying–wetting, freezing–thawing (e.g. Müller *et al.*, 1997). Research has provided evidence that microbial activities under these conditions are particularly high (e.g. drying can cause an accumulation of cellular organic materials, which might lead to enhanced microbial activity) (van Veen *et al.*, 1984).

Soil Temperature 3

3.1 Introduction

Temperature is one of the most important parameters governing numerous processes in the soil and the atmosphere. These include soil physical processes, such as evaporation (Chapter 4), biological processes, such as plant growth (Chapter 6), and microbially mediated transformation processes, such as C/N transformations (Chapter 2). Excellent reviews about soil temperature and heat flow can be found in many textbooks (e.g. Hillel, 1982a; Campbell, 1985, Chapter 4; Monteith and Unsworth, 1990, Chapter 13) and we shall therefore concentrate in this chapter on the essential elements which are needed to understand heat flow in soil and its mathematical modelling.

In the models developed so far, we have regarded soil temperature as an input variable supplied to the models. However, soil temperature changes in a certain pattern, mainly as a result of energy input of the sun to the soil surface, which leads to temperature gradients in the soil and subsequent flows of heat. The general gradient notation under steady-state conditions (i.e. no temperature changes with time) has already been introduced in Section 1.3.

3.2 Soil Temperature Dynamics in Space and Time

The basic equation for heat flow was formulated in the last century and is known as Fourier's law, named after the scientist who first described conduction of heat in solids. In one-dimensional form, this law is written as (see eqn 1.3/1):

$$F_T = -K_T \cdot \frac{dT}{dz} \tag{3.2/1}$$

where:

F_T = flow of heat [$W \cdot m^{-2}$]
K_T = thermal conductivity [$W \cdot m^{-1} \cdot K^{-1}$]
T = temperature [K]
z = distance over which the flow occurs [m]

(Note: values of thermal conductivity (K_T) in K^{-1} equal values expressed in $°C^{-1}$ and values of dT (change in temperature) expressed in K equal values expressed in °C.)

Change of temperature with time (continuity equation)

Soil temperature is never static but changes with time in response to changing environmental conditions. Therefore eqn 3.2/1 has to be expanded to allow for the temperature change both with depth and with time. Consider a simple three-layered soil profile (Fig. 3.1). Supposing that all three layers are characterized by well-defined temperatures T_{i-1}, T_i and T_{i+1} and assuming that the thermal conductivity (K_T) stays constant we can calculate the flows F_T in and out of layer i ($F_{T\_in\_layer\ i}$ and $F_{T\_in\_layer\_i+1}$, following the procedure described in Section 1.3):

$$F_{T\_in\_layer\_i} = -K_T \cdot \frac{T_i - T_{i-1}}{\Delta z}$$

and:

$$F_{T\_in\_layer\_i+1} = -K_T \cdot \frac{T_{i+1} - T_i}{\Delta z}$$

These equations describe the flows of energy (units: W m^{-2} or J s^{-1} m^{-2}) entering and leaving layer i which might cause a corresponding change in the overall energy content of that layer (units: J m^{-3}). The change in energy or heat content of a certain soil volume (e.g. layer i) can be calculated by multiplying the volumetric heat capacity (Ch) of the conducting medium (in our case soil) by the temperature change of this layer (i.e. Ch · dT; units: J m^{-3}). The volumetric heat capacity (Ch, units: J m^{-3} K^{-1}) defines the change in heat content of a unit bulk soil volume per unit change in temperature.

Provided that no other sources or sinks of heat are present in soil layer i and using the principles of energy conservation, then the change in heat flow with distance (i.e. $\frac{dF_T}{dz}$ or $\frac{F_{T\_in\_layer\_i} - F_{T\_in\_layer\_i+1}}{\Delta z}$) equals the change in heat content of the soil layer with time (i.e. $Ch \cdot \frac{dT}{dt}$ or $Ch \cdot \frac{\Delta T}{\Delta t}$).

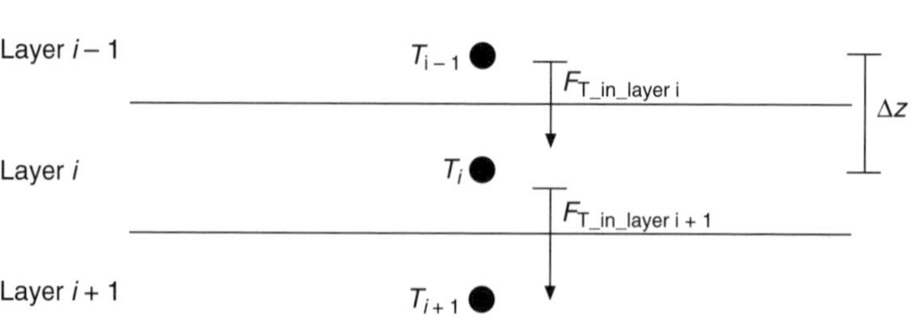

Fig. 3.1. Temperature flow in and out of a soil layer *i* in a theoretical soil profile with three soil layers.

This relationship is defined by the continuity equation for heat flow, which, in mathematical terms, is described by the partial differential equation[1]:

$$Ch \cdot \frac{\partial T}{\partial t} = \frac{\partial F_T}{\partial z} \qquad (3.2/2)$$

or (combining eqns 3.2/1 and 3.2/2):

$$Ch \cdot \frac{\partial T}{\partial t} = \frac{\partial}{\partial z} \cdot \left(K_T \cdot \frac{\partial T}{\partial z} \right) \qquad (3.2/3)$$

where:

Ch = volumetric heat capacity $[J \cdot m^{-3} \cdot K^{-1}]$ or $[J \cdot m^{-3} \cdot °C^{-1}]$
T = temperature [K] or [°C]
K_T = thermal conductivity $[W \cdot m^{-1} \cdot K^{-1}]$

The ratio of the thermal conductivity to the volumetric heat capacity is called the thermal diffusivity (D_T):

$$D_T = \frac{K_T}{Ch} \qquad (3.2/4)$$

Substitution of K_T in eqn 3.2/3 with eqn 3.2/4 yields:

$$\frac{\partial T}{\partial t} = \frac{\partial}{\partial z} \cdot \left(D_T \cdot \frac{\partial T}{\partial z} \right) \qquad (3.2/5)$$

[1] Note: a partial differential equation (indicated by the sign ∂) is a differential equation where the dependent variable (in our case temperature (T) is dependent on more than one independent variable (here depth (*z*) and time (*t*)).

with:

D_T = thermal diffusivity [$m^2 \cdot s^{-1}$]

If the diffusivity D_T stays constant, eqn 3.2/5 simplifies to:

$$\frac{\partial T}{\partial t} = D_T \cdot \frac{\partial^2 T}{\partial z^2} \qquad (3.2/6)$$

The last equation defines the temperature change with time as being proportional to the second-order differential of the temperature change with depth (proportionality constant D_T). Bear in mind (Section 1.2.3) that a differential defines the slope of a function. In the case of the second-order differential, it is the slope of the first-order differential or temperature gradient (eqn 3.2/1). It should therefore be possible to analyse temperature profiles graphically and draw conclusions as to whether certain parts of the soil are currently warming up, cooling down or staying constant. A simplified temperature profile and the temperature gradient, as well as the second-order differential, are presented in Fig. 3.2.

The graphical analysis can be carried out following basic curve-sketching techniques (e.g. Goldstein *et al.*, 1990, Section 2.3).[2] Basically, all temperature profiles curving to the left are currently warming up, all temperature profiles curving to the right are cooling down and all linear segments on a temperature profile are staying constant in temperature (e.g. Hillel, 1982a, p. 171).

3.3 Volumetric Heat Capacity and Thermal Conductivity

The most important thermal properties of soil are the volumetric heat capacity and thermal conductivity of soil. Soil is a composite of many constituents, each characterized by particular thermal properties. Specific volumetric heat capacities (Ch) can be derived by multiplying the specific heat of the soil component (units: $J\ g^{-1}\ K^{-1}$) by the respective density (units: $g\ cm^{-3}$) (note: to obtain the dimensions for volumetric heat capacity in $J\ m^{-3}\ K^{-1}$ we have to multiply the density by 1,000,000 to convert cm^{-3} to m^{-3}) (Table 3.1). The various soil materials are characterized by well-defined thermal conductivities (Table 3.1).

To calculate the overall soil volumetric heat capacity (Ch, units $J\ m^{-3}\ K^{-1}$), the specific heat value of each constituent is multiplied by the respective density and its volumetric fraction in soil and all of the fractional Ch values are summed up:

[2] Note: usually graph-sketching techniques and their analyses are given for the coordinate system where both *x*- and *y*-values are positive. In our case, the *y*-values denote depth, which in mathematical terms are negative *y*-values. Therefore the derivations of the curvature and slope have to be adjusted accordingly.

$$\text{Ch} = \begin{pmatrix} \text{Dens\_water} \cdot \text{C\_water} \cdot \text{VWC} + \\ \text{Dens\_air} \cdot \text{C\_air} \cdot \text{Poro\_air} + \\ \text{Dens\_clay} \cdot \text{C\_clay} \cdot \text{f\_clay} + \\ \text{Dens\_quartz} \cdot \text{C\_quartz} \cdot \text{f\_sand} + \\ \text{Dens\_om} \cdot \text{C\_om} \cdot \text{f\_om} \end{pmatrix} \cdot 10^6 \qquad (3.3/1)$$

where:

Dens_water = density of water [g · cm^{-3}] {see Table 3.1}
Dens_air = density of air [g · cm^{-3}] {see Table 3.1}
Dens_clay = density of clay [g · cm^{-3}] {see Table 3.1}
Dens_quartz = density of quartz or sand [g · cm^{-3}] {see Table 3.1}
Dens_om = density of organic matter [g · cm^{-3}] {see Table 3.1}
C_water = specific heat of water [J · g^{-1} · K^{-1}] {see Table 3.1}
C_air = specific heat of air [J · g^{-1} · K^{-1}] {see Table 3.1}
C_clay = specific heat of clay [J · g^{-1} · K^{-1}] {see Table 3.1}
C_quartz = specific heat of quartz or sand [–] {see Table 3.1}
C_om = specific heat of organic matter [–] {see Table 3.1}
f_clay = volume fraction of clay in soil [–] {0.38}

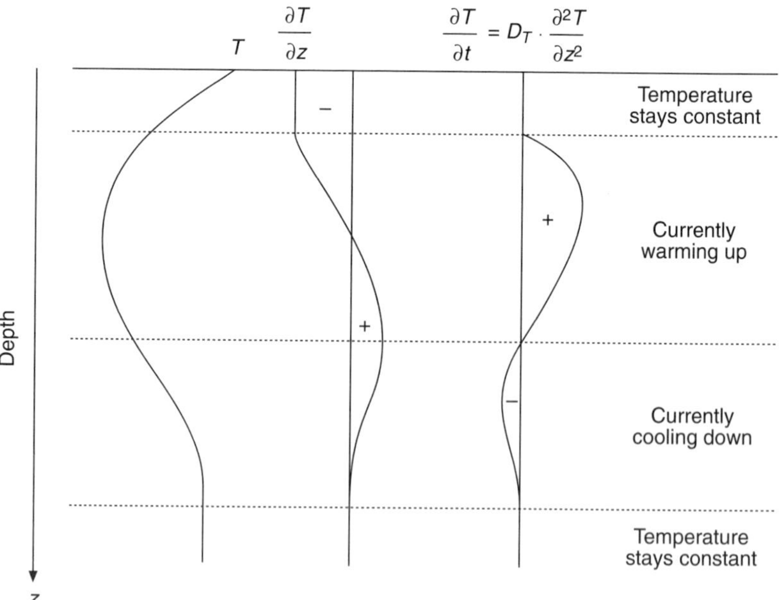

Fig. 3.2. Graphical analysis of a temperature profile (left) and the corresponding first-order (middle) and second-order (right) differentials (modified after Monteith and Unsworth, 1990, p. 226).

f_sand =	volume fraction of sand and silt in soil [–] {0.115}
f_om =	volume fraction of organic matter in soil [–] {0.005}
VWC =	volumetric water content [m^3 water · m^{-3} soil] (from input file)
Poro_air =	air porosity [m^3 air · m^{-3} soil], which is calculated by:

$$\text{Poro\_air} = \text{Poro\_tot} - \text{VWC} \tag{3.3/2}$$

with:

$$\text{Poro\_tot} = 1 - \frac{\text{dens\_bulk}}{\text{dens\_solid}}$$

where:

dens_bulk =	soil bulk density [g · cm^{-3}] {1.30}
dens_solid =	solid density (similar to density of minerals; see Table 3.1) [g · cm^{-3}] {2.6}
Poro_tot =	total porosity of soil [m^3 air · m^{-3} soil]

The overall thermal conductivity of the soil KT can be calculated according to the procedure described in Hillel (1998, pp. 309ff.):

$$K_T = \frac{\begin{pmatrix} \text{VWC} \cdot \text{KT\_water} + \text{f\_clay} \cdot \text{KT\_clay} \cdot \text{Ks\_clay} + \\ \text{f\_sand} \cdot \text{KT\_quartz} \cdot \text{Ks\_quartz} + \text{f\_om} \cdot \text{KT\_om} \cdot \text{Ks\_om} + \\ \text{Poro\_air} \cdot \text{KT\_air} \cdot \text{Ka\_air} \end{pmatrix}}{\begin{pmatrix} \text{VWC} + \text{f\_clay} \cdot \text{Ks\_clay} + \text{f\_sand} \cdot \text{Ks\_quartz} + \\ \text{f\_om} \cdot \text{Ks\_om} + \text{Poro\_air} \cdot \text{Ka\_air} \end{pmatrix}} \tag{3.3/3}$$

Table 3.1. Thermal properties of various soil materials (see Hillel, 1982a; Campbell, 1985, p.32).

Materials	Density [g · cm^{-3}]	Specific heat [J · g^{-3} · K^{-1}]	Volumetric heat capacity [J · m^{-3} · K^{-1}]	Thermal conductivity [W · m^{-1} · K^{-1}]
Quartz†	2.66	0.80	2.128E+6	8.80
Clay minerals	2.65	0.90	2.385E+6	2.92
Organic matter	1.30	1.92	2.496E+6	0.25
Water	1.00	4.18	4.180E+6	0.57
Air (20°C)	0.0012	1.01	0.001212E+6	0.025
Ice	0.92	1.88	1.7296E+6	2.18

*Note the sign 'E+6' is another notation for 10^6. The value of the volumetric heat capacity has to be multiplied by this number.
†The mineralogy of sand and silt particles is dominated by quartz.

where:

KT_values = thermal conductivities for the various constituents
KT_water = thermal conductivity of water $[W \cdot m^{-1} \cdot K^{-1}]$ {see Table 3.1}
KT_clay = thermal conductivity of clay $[W \cdot m^{-1} \cdot K^{-1}]$ {see Table 3.1}
KT_quartz = thermal conductivity of quartz or sand $[W \cdot m^{-1} \cdot K^{-1}]$ {see Table 3.1}
KT_om = thermal conductivity of organic matter $[W \cdot m^{-1} \cdot K^{-1}]$ {see Table 3.1}
KT_air = thermal conductivity of air $[W \cdot m^{-1} \cdot K^{-1}]$ {see Table 3.1}

Ks_values are ratios of the average temperature gradient in the solids relative to the corresponding gradients in the soil water phase. These factors depend on the particle size, particle shape and mode of packing, as well as on the composition of mineral and organic material (Hillel, 1998, p. 318). (The Ka_value is the corresponding ratio for air relative to the water phase.)

Ks_clay = ratio of clay relative to water phase [–] {0.4}
Ks_quartz = ratio of quartz relative to water phase [–] {0.4}
Ks_om = ratio of organic matter relative to water phase [–] {0.4}
Ka_air = ratio of air relative to water phase [–]{1.4}

The influence of heat transfer via water vapour movement in the soil air phase can also be taken into account. This involves the calculation of an additional conductivity associated with water vapour movement (KT_vapour) which is added to the thermal conductivity of air (KT_air). Equation 3.3/8 should therefore be expanded to:

$$KT = \frac{\begin{pmatrix} VWC \cdot KT\_water + f\_clay \cdot KT\_clay \cdot Ks\_clay + \\ f\_sand \cdot KT\_quartz \cdot Ks\_quartz + f\_om \cdot KT\_om \cdot Ks\_om + \\ Poro\_air \cdot (KT\_air + KT\_vapour) \cdot Ka\_air \end{pmatrix}}{\begin{pmatrix} VWC + f\_clay \cdot Ks\_clay + f\_sand \cdot Ks\_quartz + \\ f\_om \cdot Ks\_om + Poro\_air \cdot Ka\_air \end{pmatrix}} \tag{3.3/4}$$

The thermal conductivity associated with water vapour movement (KT_vapour) can be calculated with the procedure described by Campbell (1985, eqn 9.11). However, this derivation assumes that the reader already has an advanced understanding of the physical relationships related to the soil energy balance. It will therefore be introduced in Chapter 5, after we have covered these important aspects. However, the reader should keep in mind that the relationship describing the water vapour movement in soil is mainly governed by soil temperature gradients (see Sections 5.8 and 5.9 for details).

Instead of calculating the KT_vapour values, Hillel (1977) used a look-up procedure to determine these values. A non-linear curve fit to values reported by Hillel (1977) resulted in the relationship:

$$\text{KT\_vapour} = \text{KT\_vapour\_A} \cdot \exp(\text{KT\_vapour\_B} \cdot \text{T\_soil}) \quad (3.3/5)$$

with:

KT_vapour_A = empirical constant $[\mathrm{W} \cdot \mathrm{m}^{-1} \cdot \mathrm{K}^{-1}]$ {0.024519238}
KT_vapour_B = empirical constant $[°\mathrm{C}^{-1}]$ {0.055441093}
T_soil = soil temperature[°C] (from input file)

This relationship will be used until we have developed the more process-orientated method of calculating KT_vapour values (see Chapter 5).

Before we develop the solution of eqns 3.2/6 and 3.2/3, we shall have a closer look at the thermal soil properties: volumetric heat capacity (Ch, eqn 3.3/1), thermal conductivity (KT, eqns 3.3/3 and 3.3/4) and thermal diffusivity (DT, eqn 3.2/4).

Application in ModelMaker

For the calculation, the <variables> presented in Fig. 3.3 are inserted into model *Mod3–1a.mod*. The two variables Ch_adj (units: MJ $\mathrm{m}^{-3}\ \mathrm{K}^{-1}$) and DT_adj (units: $\mathrm{mm}^2\ \mathrm{s}^{-1}$) are additionally defined to scale the Ch and DT values so that they can be easily compared with the values of thermal conductivity (KT, units: $\mathrm{J\ m}^{-1}\ \mathrm{K}^{-1}$). Soil temperature and soil volumetric water content, which are needed for the calculations, are introduced to the model via an <input file> ‘Input’. The model is currently set up for 100 time steps (in seconds) and the input parameters are set up in such a way that both the soil temperature T_soil and the soil volumetric water content VWC increase linearly from −5 to 40°C and 0 to 0.5 $\mathrm{m}^3\ \mathrm{m}^{-3}$, respectively. To provide the user with a range of options, an <independent event> ‘User_choice’ has been inserted, which allows the calculation of both variable and constant soil temperature and volumetric water content, as well as the calculation with or without thermal conductivity related to water vapour in soil. The notation in <actions> of the <independent event> ‘User_choice’ is:

```
Temp_choice=GetChoice("Use variable (Y) or constant (N)
        temperature? (Y/N)","Temperature choice",Temp_choice);
if(Temp_choice=0){
 GetValue("Input a temperature value (-5 to 50):","Temperature
        input (constant)",T_const, T_const > −5.1 and T_const <
        50.1);
}
VWC_choice=GetChoice("Use variable (Y) or constant (N) vol.
        water content? (Y/N)","Vol. Water Content
        choice",VWC_choice);
if(VWC_choice=0){
 GetValue("Input a VWC value (0.1 to 0.7):","VWC input
        (constant)",VWC_const, VWC_const > 0.09 and VWC_const <
        0.71);
}
KTvap_choice=GetChoice("Calculate thermal conductivity due to
        water vapour (KT_vapour) ? (Y/N)","Water Vapour
        movement",KTvap_choice);
```

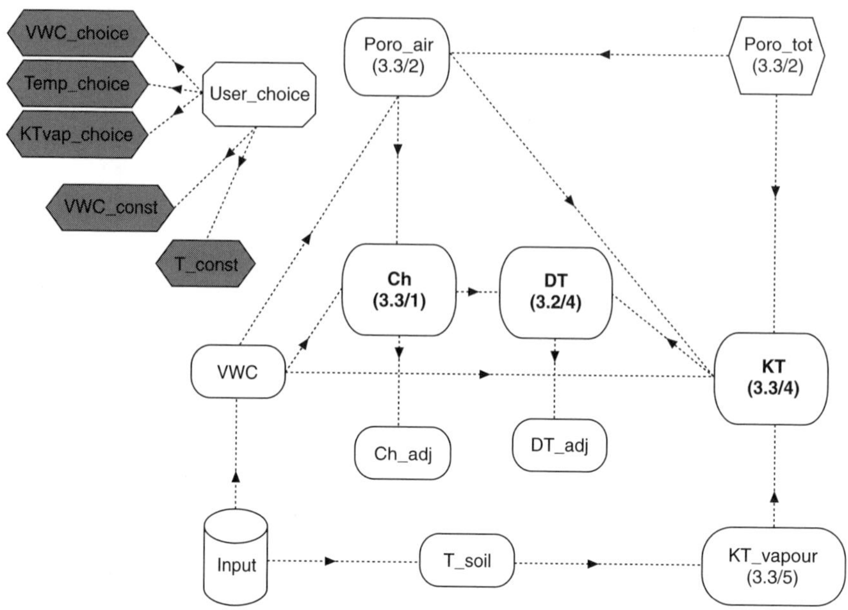

Fig. 3.3. Notation for the calculation of thermal conductivity (*KT*), volumetric heat capacity (*Ch*) and thermal diffusivity (*DT*) in soil, including the notation for the <independent event> 'User_choice' (*Mod3-1a.mod*).

The 'if-statements' are needed to allow the input of constant values for T_soil and VWC.

The following <variables> are defined conditionally in order to allow the execution of the various user options:

Variable	In response to <define>
VWC	VWC_choice
T_soil	Temp_choice
KT_vapour	Ktvap_choice

After all the components and the definitions (eqn 3.3/1 to 3.3/4) have been inserted, the model should be similar to the notation displayed in Fig. 3.3.

Volumetric heat capacity and thermal conductivity are presented in Fig. 3.4 for a clay soil (f_clay = 0.4; f_sand = 0.05) and a sandy soil (f_clay = 0; f_sand = 0.3) (all other parameters are the same as given above), with the user option:

- variable VWC, variable temperature and calculation of KT_vapour.

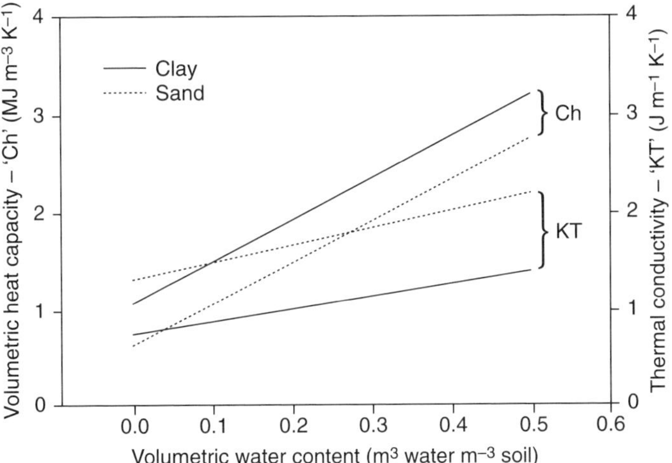

Fig. 3.4. Volumetric heat capacity (Ch) and thermal conductivity (KT) for a clay and a sand soil (*Mod3-1a.mod*).

Before we continue with the development of the soil heat flow models, it is a good idea to try several other options in order to become familiar with the dynamics of the various soil thermal properties in response to environmental conditions. The following general points can be made (see also Fig. 3.4):

- Both Ch and KT values increase with soil moisture.
- Over the whole range of soil moisture, sandy soil has a higher thermal conductivity than clay soil.
- The volumetric heat capacities of both soils are very similar (sandy soil slightly lower than clay soil).

3.4 Heat Flow Model with Constant Thermal Properties

The first model we develop is the solution to eqn 3.2/6, which assumes constant soil thermal properties and is therefore a simplified version of eqn 3.2/3. To solve the equation, we have to define a multilayered soil system. Provided we work with *n* layers where each layer is specified by an index *i*, and assuming that the thermal properties are constant for the whole profile, the notation of eqn 3.2/1 in difference form is (see also Section 1.3):

$$F_i = -KT \cdot \frac{T_i - T_{i-1}}{z_i - z_{i-1}} \qquad (3.4/1)$$

In the case where all layers are equally thick, we can also write eqn 3.4/1 as:

$$F_i = -KT \cdot \frac{T_i - T_{i-1}}{\Delta z} \qquad (3.4/2)$$

To derive the continuity equation, we have to apply eqn 3.2/2 (see also Fig. 3.1):

$$Ch \cdot \frac{\partial T_i}{\partial t} = \frac{\Delta F_i}{\Delta z}$$

and after combining it with eqn 3.4/2 (see also Fig. 3.1):

$$Ch \cdot \frac{\partial T_i}{\partial t} = \frac{F_i - F_{i+1}}{\Delta z} \qquad (3.4/3)$$

with:

$$F_i = -KT \cdot \frac{T_i - T_{i-1}}{\Delta z} \quad \text{or} \quad F_i = KT \cdot \frac{T_{i-1} - T_i}{\Delta z} \qquad (3.4/4)$$

[3]

and:

$$F_{i+1} = -KT \cdot \frac{T_{i+1} - T_i}{\Delta z} \quad \text{or} \quad F_{i+1} = KT \cdot \frac{T_i - T_{i+1}}{\Delta z} \qquad (3.4/5)$$

where:

KT = thermal conductivity $[W \cdot m^{-1} \cdot K^{-1}]$ {1.5}
F = steady-state heat flow $[W \cdot m^{-2}]$
T = temperature [K] or [°C][4]
z = layer thickness [m]

Combining eqns 3.4/3 to 3.4/5, we obtain:

$$Ch \cdot \frac{\partial T_i}{\partial t} = \frac{\left(KT \cdot \frac{T_{i-1} - T_i}{\Delta z}\right) - \left(KT \cdot \frac{T_i - T_{i+1}}{\Delta z}\right)}{\Delta z} \qquad (3.4/6)$$

Recall that our assumption was that the thermal properties of the soil would be constant, therefore, it is possible to rewrite eqn 3.4/6 as:

$$Ch \cdot \frac{\partial T_i}{\partial t} = KT \frac{\left(\frac{T_{i-1} - T_i}{\Delta z}\right) - \left(\frac{T_i - T_{i+1}}{\Delta z}\right)}{\Delta z}$$

[3] The alternative expressions are given because they cause less confusion with the minus sign and are therefore used later on in the actual ModelMaker notation.
[4] Note: both units (K and °C) can be used, because they are subtracted from each other.

After combining the Δz terms, we obtain:

$$Ch \cdot \frac{\partial T_i}{\partial t} = KT \frac{(T_{i-1} - T_i) - (T_i - T_{i+1})}{\Delta z^2}$$

Finally, applying eqn 3.2/4 and simplifying the numerator on the right-hand side, we obtain the final difference equation which we have to solve:

$$\frac{\partial T_i}{\partial t} = DT \frac{T_{i-1} - 2 \cdot T_i + T_{i+1}}{\Delta z^2} \qquad (3.4/7)$$

with:

Ch = volumetric heat capacity [$J \cdot m^{-3} \cdot K^{-1}$] {$3 \times 10^6$}
DT = diffusivity (eqn 3.2/4) [$m^2 \cdot s^{-1}$] {$1.5/3 \times 10^6 = 5 \times 10^{-5}$}
T_init = initial soil temperature [°C] {15}

Recall (Section 1.3) that, in order to solve a finite-difference equation for a soil system of *n* layers, we need to specify boundary conditions. For the first layer the equation is defined by:

$$\frac{\partial T_1}{\partial t} = DT \frac{Ts - 2 \cdot T_1 + T_2}{\Delta z^2} \qquad (3.4/8)$$

and for the *n*th layer it is defined by:

$$\frac{\partial T_n}{\partial t} = DT \frac{T_{n-1} - 2 \cdot T_n + T_{n+1}}{\Delta z^2} \qquad (3.4/9)$$

We have to specify the unknown values for *Ts* and T_{n+1}.

For the upper boundary generally the soil surface temperature *Ts* is used. However, values for *Ts* are not usually available. Relationships in environmental physics can be used to derive the soil surface temperature, but this requires the calculation of an entire soil energy balance and an understanding of the combined heat-water flow dynamics and related energy exchange processes in soil. A separate chapter (Chapter 5) will present the derivation of soil surface temperature based on energy balance considerations.

In a simplified approach *Ts* may be supplied to the model in two ways:

1. Input of a constant value for *Ts*.
2. Via a harmonic (sinusoidal) function to calculate *Ts*.

The second option resembles better the naturally occurring periodic succession of changing temperatures in response to meteorological conditions. The simplest approach is to assume that the soil surface temperature oscillates around an average temperature in a harmonic (sinusoidal) pattern. In this simplified approach, the soil surface temperature can be calculated by:

$$\mathrm{Ts} = \mathrm{Tavg} + \mathrm{A0} \cdot \sin(\mathrm{w} \cdot \mathrm{t}) \tag{3.4/10}$$

with:

$$\mathrm{w} = \frac{2 \cdot \pi}{86400}$$

where:

Tavg = average soil surface temperature [°C] {15}
A0 = amplitude of the surface temperature fluctuation (difference between the average temperature 'Tavg' and minimum 'Tmin' or maximum 'Tmax' temperature) [°C] {10}
w = radial frequency of the oscillation which is $2 \cdot \pi$ times the actual frequency [s^{-1}] {on a daily cycle: $^{2\cdot\pi}/_{86400}$ }
t = time [s]

The unknown temperature T_{n+1} at the bottom boundary is usually supplied as a fixed temperature. The oscillation of the temperature will decrease with depth, as a result of absorbance and release of heat by the conducting medium (soil). Temperature at a depth of 100 cm and deeper stays more or less constant (at least over a season). We shall assume that the temperature outside the bottom boundary stays at the temperature $T_{n+1} = Tavg$ (15°C).

Considering a soil profile divided into *n* layers of equal thickness, each layer of thickness *z* (units: m) can be calculated by:

$$\mathrm{z} = \frac{\mathrm{L}}{\mathrm{layers}}$$

where:

L = thickness of the entire soil profile[m] {1}
layers = number of layers [–] {10}

The distance over which the heat flow occurs (Dist = Δz) for layers with the same thickness equals *z*. The distance for heat flow at the first and last layer (layer 1 and 10) is $^{\mathrm{Dist}}/_{2}$.

Analytical solution

Assuming that the soil surface temperature *Ts* is defined by eqn 3.4/10 and the average temperature of the entire soil profile at any depth (*z*) and time (*t*) is defined by *T*avg, the soil temperature with depth can be calculated by the harmonic function:

$$\mathrm{Tz} = \mathrm{Tavg} + \mathrm{Az} \cdot \sin\left[\mathrm{w} \cdot \mathrm{t} - {}^{\mathrm{Depth}}/_{\mathrm{z\_d}}\right] \tag{3.4/11}$$

Substituting the solution of eqn 3.4/10 into eqn 3.2/6 an analytical solution can be calculated (for more details, see Hillel, 1982a, p.168; Campbell, 1985, pp. 27–28; Monteith and Unsworth, 1990, p. 227):

$$T\_anal[i] = Tavg + A0 \cdot \exp\left(-{Depth[i]}/{z\_d}\right) \cdot \sin\left(w \cdot t - {Depth[i]}/{z\_d}\right)$$

(3.4/12)

with:

$$z\_d = \sqrt{\frac{2 \cdot DT}{w}} \qquad (3.4/13)$$

where:

z_d = the so-called 'damping depth' which defines the soil depth at which the temperature amplitude has decreased to $\frac{1}{e} = \frac{1}{2.718} \approx 0.37$ times the amplitude of the soil surface (the amplitude is attenuated exponentially with depth).

The distance from the soil surface (Depth) is calculated by:

$$Depth[i] = i * Dist - {Dist}/{2}$$

where:

i = index of soil layer (1 to 10)
Dist = layer thickness [m]

We shall not consider the analytical solution in detail, because it is only valid under particular conditions, which are not realistic enough to model heat flow in natural soils. Therefore, a numerical solution to eqn 3.2/10 is preferred, because it is a lot more flexible. However, if the reader is interested in a more detailed analysis, the excellent overview of the formal analysis of heat flow presented by Monteith and Unsworth (1990, pp. 225–229) is recommended.

Application in ModelMaker

We shall now create our first multilayered model with ModelMaker. The aim is to calculate the numerical solution to eqn 3.2/6 under the two boundary conditions (constant, harmonic soil surface temperature change) (eqn 3.4/7). The numerical solution calculated with the harmonic *Ts* function is then compared with the analytical solution (eqns 3.4/10 to 3.4/13). ModelMaker provides the facility to define indexed components (see Section 1.5), which are needed in this situation. An indexed <compartment> with ten subcomponents is, for instance, defined by:

C1[1...10] The square bracket [1…10] defines the indexed system for 'C1'.

In order to solve eqn 3.4/7 for a soil divided into ten soil layers and to apply the boundary conditions, we define layers 1 and 10 and layers 2–9 separately (once we are familiar with the new notation, we shall use one conditional notation for the entire system).

The necessary components for the numerical solution are:

<compartments>: Temp[1], Temp[2..9], Temp[10] (the symbol Temp has to be used instead of *T* because *t* is the notation for time)

The analytical solution is calculated in the <variable>: T_anal [1..10] (the symbol for the index, *i*, is defined in <Models> <Run Options> <Symbols>).

Via the <independent event> 'User_choice', the constant or harmonic soil surface option can be chosen. In response to this choice, the user is prompted for an input of the constant soil surface temperature 'Ts_const' and the soil surface temperature amplitude 'A0'. The following statements are inserted:

```
GetValue("Use constant (1) or periodically changing soil
        surface temperature (2)","Soil surface
        temperature",Ts_calc, Ts_calc = 1 or Ts_calc = 2);
if(Ts_calc=1){
 GetValue("Input value for the constant soil surface
        temperature Ts (-20 - 50°C):","Constant soil surface
        temperature",Ts_const, Ts_const >-20.1 and Ts_const <
        50.1);
}
if(Ts_calc=2){
 GetValue("Input value for the temperature amplitude A0 (0-
        20°C):","Temperature amplitude",A0, A0 >-0.1 and A0 <
        20.1);
}
```

The Ts component is made conditional to allow for calculations either with the constant or the harmonically changing soil surface temperature.

The model set up for the numerical (eqn 3.4/7) and analytical (eqn 3.4/12) solution of eqn 3.2/6 is presented in Fig. 3.5.

The final model is presented in *Mod3–1b.mod*. The model is executed on a per second time notation and is set up in its current configuration for 5 days with outputs calculated every hour (<stop value> 432000, <output steps> 120).

The parameters are set up in such a way that the analytical and the numerical solutions can be directly compared (e.g. Tinit is equal to Tavg; recall that one requirement for the analytical solution is that Tavg is representative for the entire soil profile). The analytical solution is only valid for the harmonically changing soil surface temperature. The analytical solution is set to zero when the 'constant temperature' option is chosen (T_anal [1..10] is made conditional according to <define> Ts_calc).

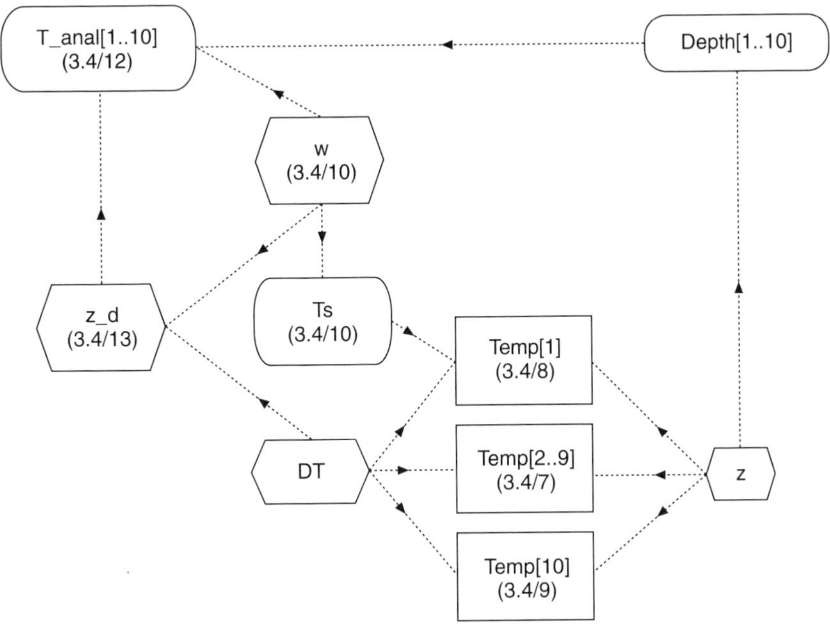

Fig. 3.5. Model for the numerical and analytical solution of the heat flow equation with constant thermal properties (*Mod3-1b.mod*).

Typical values for the diurnal damping depth (z_d) range between 10 and 15 cm (Hillel, 1982b). In the current example, the calculated damping depth (z_d) is 11.7 cm ($K_T = 1.5$ W m^{-1} K^{-1}, Ch = 3 × 10^6 J m^{-3} K^{-1}).

The model (*Mod3–1b.mod*) is able to simulate the typical heat flow pattern in soil, which is characterized by:

- attenuation of the temperature amplitude with depth (exponential attenuation, eqn 3.4/13);
- a shift of the temperature phase with depth.

The graphical output in ModelMaker (e.g. Soil temp in *Mod3–1b.mod*) allows only the separate presentation of the temperature at various depths with time. A graphical representation of temperature with depth (e.g. Fig. 3.2) has to be created in a separate graphics package (e.g. SigmaPlot: SPSS, 1997). This can be achieved by creating a table of the desired outputs with time (e.g. in Temp table of *Mod3–1b.mod* the outputs are: Ts = soil surface temperature, and Temp[1..10] for all depths). The entire output should be copied (highlight all values and <Edit> <copy>) to a spreadsheet (e.g. Excel, Microsoft Inc.). Before creating a graph, it might be necessary to transpose the values so that each column represents a time rather than a depth output (e.g. needed for SigmaPlot graphics). A typical graph of temperature (*x*-axis) versus depth (*y*-axis) is presented for several output times within the first simulated day, for both diurnally changing and constant soil surface temperature, in Fig. 3.6.

The output shows that the greatest effect of harmonically changing soil surface temperature occurs within the top 40 cm of soil, which is consistent with the theory outlined above. In the constant temperature simulation where the upper boundary is set to 25°C and the lower to 15°C, it can be seen that the soil profile is warming up from above.

The current model *Mod3–1b.mod* is only able to calculate heat flow in soil for simple sets of thermal properties and boundary conditions. However, in Section 3.3, we have seen that the thermal properties change with the soil moisture status. Therefore a more realistic simulation of heat flow in soil requires the solution of eqn 3.2/3.

Before we move on to the next section, an alteration to the graphical appearance of the model is introduced. Currently, the numerical solution is calculated in three separate compartments, where the first and last compartments (Temp[1] and Temp[10]) contain the solution for the boundary layers. However, it is also possible to carry out the calculations of all three compartments with one conditional compartment Temp[1..10]), where calculations are performed according to the index *i*, thus reducing the number of compartments. The updated code is presented in *Mod3–1c.mod*.

3.5 Heat Flow Model with Dynamically Changing Thermal Properties

A heat flow model with dynamically changing thermal properties (eqn 3.2/3) can be created by combining the notations derived in Sections 3.3 and 3.4. The thermal properties K_T and Ch for each soil layer change mainly in response to changing volumetric water contents. Soil volumetric water content is usually not measured in the required time and space resolution, and is generally supplied to the heat flow models by a separate soil volumetric water model (Chapters 4 and 5). Since we have not yet considered the simulation of soil water content, we supply this model with simulated outputs of volumetric water content provided in an input file (modelled with the energy balance (Chapter 5)).

Specific thermal conductivities (K_T[i]) and volumetric heat capacities (Ch[i]) for each soil layer are calculated with eqn 3.3/3 and 3.3/1 respectively. Volumetric water content (VWC) and air-filled porosity values (Poro_air) have to be exchanged with layer specific notations (i.e. VWC[i] and Poro_air[i]). It is assumed that the texture for each soil layer is the same (i.e. *f*_clay, *f*_sand and *f*_om are not defined in an index notation).

In the heat flow model derived in the last section, we considered a 1-m-deep profile divided into ten equally thick soil layers. Such a notation might not be realistic enough if we need to work with variable soil layer thicknesses. In accordance

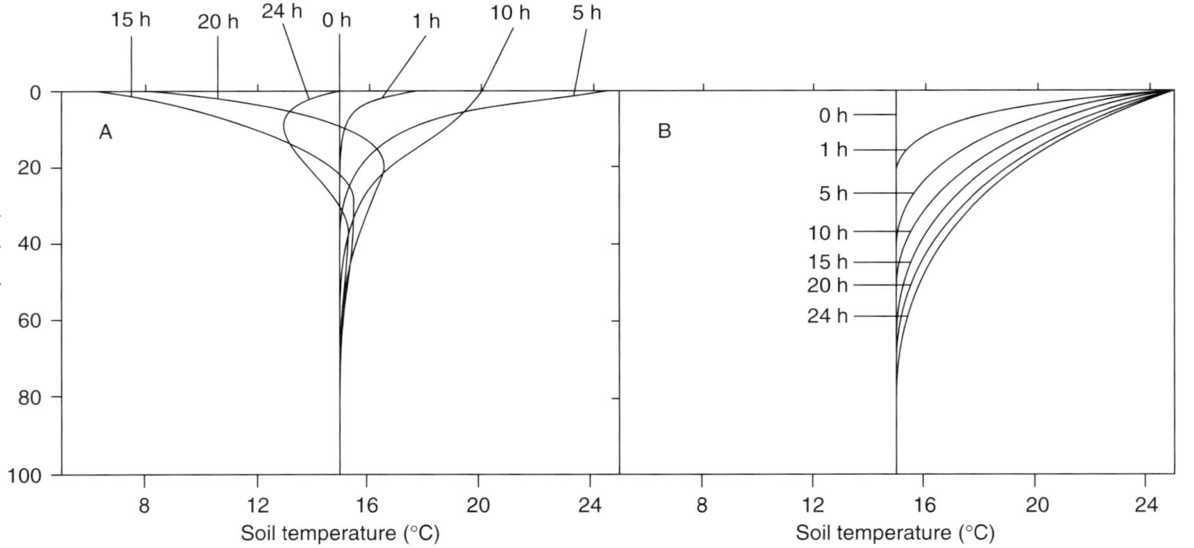

Fig. 3.6. Temperature profiles after 0, 1, 5, 10, 15, 20 and 24 h calculated with soil surface temperatures (*Ts*) either harmonically changing (*A*) (parameters: A0 = 10, Tavg = 15, Temp_init = 15) or constant (*B*) (parameters: Ts_const = 25, Tavg = 15, Temp_init = 15) (*Mod3-1b.mod*).

with Hillel (1977), a soil profile with 14 soil layers, each with variable thickness, is defined. The flow distance between adjacent layers (Dist[i]) is calculated by:

$$\mathrm{Dist}[\mathrm{i}] = \frac{\mathrm{z}[\mathrm{i}-1]+\mathrm{z}[\mathrm{i}]}{2} \tag{3.5/1}$$

for the boundary layers:

$i = 1$:

$$\mathrm{Dist}[\mathrm{i}] = \frac{\mathrm{z}[\mathrm{i}]}{2}$$

and $i = 15$:

$$\mathrm{Dist}[\mathrm{i}] = \frac{\mathrm{z}[\mathrm{i}-1]}{2}$$

The depth from the soil surface to the mid-point of each soil layer is calculated by:

$$\mathrm{Depth}[\mathrm{i}] = \mathrm{Depth}[\mathrm{i}-1] + \frac{\mathrm{z}[\mathrm{i}-1]+\mathrm{z}[\mathrm{i}]}{2} \tag{3.5/2}$$

for the boundary layers:

$i = 1$:

$$\mathrm{Depth}[\mathrm{i}] = \frac{\mathrm{z}[\mathrm{i}]}{2}$$

and $i = 15$:

$$\mathrm{Depth}[\mathrm{i}] = \mathrm{Depth}[\mathrm{i}-1] + \frac{\mathrm{z}[\mathrm{i}-1]}{2}$$

Since the flow of heat occurs between adjacent soil layers, we have to calculate appropriate thermal conductivities which are representative for the K_T values in the intervening soil layers. An average conductivity has to take into account the possibility of variable soil layer thickness and is therefore calculated by (e.g. Hillel, 1977):

$$\mathrm{AvgKT}[\mathrm{i}] = \frac{(\mathrm{z}[\mathrm{i}-1]+\mathrm{z}[\mathrm{i}])}{\left(\frac{\mathrm{z}[\mathrm{i}-1]}{\mathrm{KT}[\mathrm{i}-1]}+\frac{\mathrm{z}[\mathrm{i}]}{\mathrm{KT}[\mathrm{i}]}\right)} \tag{3.5/3}$$

At the upper boundary ($i = 1$), it is defined by:

$$\text{AvgKT[i]} = \text{KT[i]}$$

The notation for the upper boundary is a simplification, since the thermal conductivity directly at the soil surface is likely to be somewhat different from the thermal conductivity at the midpoint of the first layer ($K_T[i]$).

The heat flow between adjacent soil layers is calculated according to eqn 3.4/4, including eqn 3.5/3, as:

$$\text{FT[i]} = \text{AvgKT[i]} \cdot \frac{\text{T[i}-1] - \text{T[i]}}{\text{Dist[i]}} \tag{3.5/4}$$

For the upper boundary ($i = 1$), the flow is defined by:

$$\text{FT[i]} = \text{AvgKT[i]} \cdot \frac{\text{Ts} - \text{T[i]}}{\text{Dist[i]}}$$

where:

Ts = soil surface temperature [°C]

For the bottom layer ($i = 15$), we assume that the temperature stays constant, which leads to a zero heat flow, i.e.

$$\text{FT[i]} = 0$$

The soil surface temperature (Ts) will be taken as an input variable (from the soil energy balance, Chapter 5). The option is provided to use either a diurnally changing or a constant soil surface temperature (Ts).

The change of heat flow for each soil layer, which defines the volumetric heat content (Section 3.2), is calculated by:

$$\text{NetFT[i]} = \text{FT[i]\_FT[i + 1]} \tag{3.5/5}$$

The units of Net FT[i] are in W m^{-2}; therefore, integrating eqn 3.5/5 will lead to the volumetric heat content of the soil layer (VolH[i], units: J m^{-2}; integration of W m^{-2} or J s^{-1} m^{-2}) (Section 3.2). Therefore the differential equation which has to be solved is:

$$\frac{d\text{VolH[i]}}{dt} = \text{NetFT[i]} \tag{3.5/6}$$

with the initial volumetric heat content of each soil layer calculated by:

$$\text{VolH\_init[i]} = \text{Ch\_init[i]} \cdot \text{z[i]} \cdot \text{Temp\_init[i]}$$

where (see eqn 3.3/1):

$$\text{Ch\_init}[i] = \begin{pmatrix} \text{Dens\_water} \cdot \text{C\_water} \cdot \text{VWC\_init}[i] + \\ \text{Dens\_air} \cdot \text{C\_air} \cdot (\text{Poro\_tot} - \text{VWC\_init}[i]) + \\ \text{Dens\_clay} \cdot \text{C\_clay} \cdot \text{f\_clay} + \\ \text{Dens\_quartz} \cdot \text{C\_quartz} \cdot \text{f\_sand} + \\ \text{Dens\_om} \cdot \text{C\_om} \cdot \text{f\_om} \end{pmatrix} \cdot 10^6 \tag{3.5/7}$$

(Note: the only change to eqn 3.3/1 for volumetric heat capacity is the introduction of layer-specific initial volumetric water contents and air-filled porosities.)

VWC_init[i] = initial soil volumetric water content [m^3 water • m^{-3} soil] {0.45 for all layers}

The temperature of each soil layer can now be calculated (combining eqns 3.3/1, 3.3/2 and 3.5/6) by:

$$\text{Temp}[i] = \frac{\text{VolH}[i]}{z[i] \cdot \text{Ch}[i]} \tag{3.5/8}$$

where:

Temp_init[i] = initial soil temperature [°C] {15 for all layers}

Application in ModelMaker

The final model for heat flow with dynamic thermal properties is presented in *Mod3–1d.mod* (Fig. 3.7). To allow for the two different simulation scenarios (diurnally changing and constant soil surface temperature), an <independent event> 'User_choice' with the statement for 'Ts_calc' and 'Ts_const', as described in Section 3.4, has been inserted.

Running the model

The model is set up to simulate a 1-day period with outputs every 5 mins. In <model> <run options> <integration>, 'fixed step length' is selected with five steps between outputs. Performing the calculations with fixed step in comparison to variable (default) step length will speed up the simulations considerably. On a PC equipped with a 100 MHz Pentium, calculation with the 'variable step length' takes approximately 15 min, while with the 'fixed step length' it only takes about 40 s.

However, it should be noted that the combination of <stop value> <output steps> and 'fixed steps' in between outputs also determines whether or not the numerical solution is stable. For instance, if one step is used instead of five steps, the numerical solution would become erratic and breaks down after some time. It is important to find an appropriate combination of the above-mentioned <run

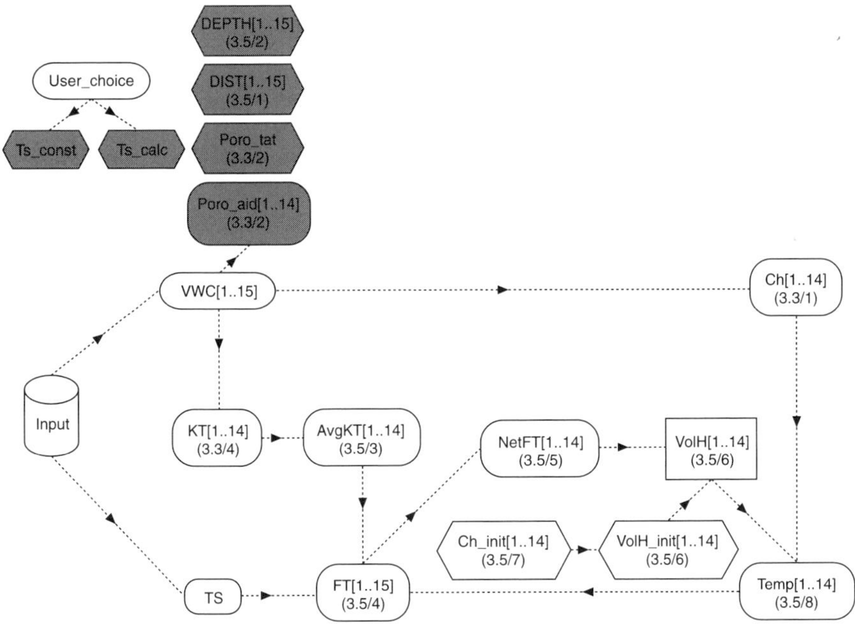

Fig. 3.7. Model for soil heat flow with dynamically changing thermal properties (*Mod3-1d.mod*).

options> in order to optimize simulation speed and still carry out the calculations correctly.

The output of the model *Mod3–1d.mod* is similar to the output of the model with constant thermal properties (*Mod3–1b.mod*). However, there is a slight influence of soil volumetric water content on the soil temperature. The effect of variable soil moisture on the thermal soil properties and the subsequent influence on soil temperature can best be seen by selecting the ‘constant soil surface temperature’ option and running the model for several days. Figure 3.8 presents simulated soil temperature at 2 cm depth for a period of 5 days.

The last scenario, where the soil surface temperature stays constant while volumetric water content undergoes diurnal fluctuation, does not occur in nature. It is included to demonstrate the effect of VWC on heat flow in soil. The input file to the current model (*Inp3–1d.txt*) contains approximately 5 days of input data, which precludes longer simulation periods. However, the input variables supplied to the heat flow model (VWC and *Ts*) are usually simulated themselves and therefore do not represent any restriction to the simulation length. It is therefore necessary to create models for soil volumetric water content (Chapter 4) and soil surface temperature (Chapter 5).

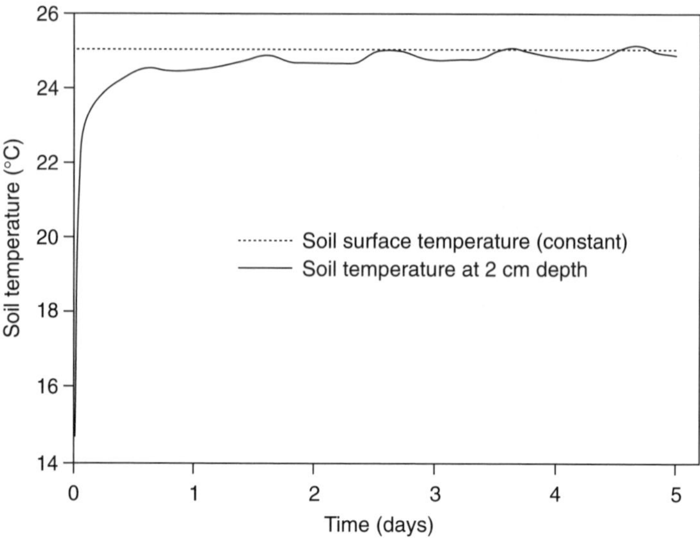

Fig. 3.8. Soil temperature at the surface and at 2 cm depth simulated with the heat flow model including dynamic thermal properties (*Mod3-1d.mod*).

Soil Water 4

4.1 Introduction

The importance of soil moisture and its relationship with the various microbiological transformations (Chapter 2), as well as its influence on thermal properties and heat flow in soil (Chapter 3), has already been presented. Furthermore, soil water content is inversely related to the soil air content and therefore determines the rate of gas exchange between the soil and the atmosphere (see Sections 2.3.3 and 2.4.4). In addition, the availability of soil water to roots influences plant growth (Chapter 6).

In Section 2.3.4, we have already introduced expressions for soil moisture. The most common ones are:

- soil gravimetric water content (θ_g), which refers to the mass of water per mass of soil (units: g water g^{-1} soil);
- soil volumetric water content (θ_v), which refers to the volume of soil occupied by water (units: m^3 water m^{-3} soil) θ_g is converted to θ_v via multiplication with the soil bulk density (units: g cm^{-3} soil, multiply by 10^6 to convert from cm^{-3} to m^{-3}; the underlying assumption is that the density of water equals 1 g cm^{-3});
- water-filled porosity (WFP) refers to the per cent saturation of the soil and is calculated with eqn 2.3/39.

In the following section we introduce another way of expressing soil moisture, in terms of its energy content or so-called 'water potential'.

4.2 Potential Concept

Soil may be regarded as a porous medium containing a range of different pore sizes. At saturation, all soil pores are filled with water. During subsequent drying, first the largest pores and then successively smaller soil pores are emptied. The reason for this is the capillarity forces (adhesion and adsorption of water), which increase with surface area (i.e. water is held more tightly in smaller pores). Therefore, to extract water from dry soil requires more energy than from wet soil. Classical thermodynamics, as outlined in many soil physics textbooks (e.g. Hillel, 1998, pp. 129ff.), states that the overall energy content of an entire system does not change (energy conservation law). However, energy can be converted from one form into another. The energy content of soil water is related to the force with which it is held in the soil. Therefore, if the forces which hold the water in the soil are high, the actual energy content of the water is low, and vice versa.

The soil water potential at a particular point in the soil profile is expressed as the difference in 'free energy' per unit volume or mass of water in comparison with water held in a reference pool of free water (set to zero) (Marshall and Holmes, 1998, p. 35). The energy content of the soil water is usually below that of the reference pool, which means the soil water potential is negative.

Soil water potential, usually given in units of pressure (e.g. bar, Pascal), may be regarded as the difference in pressure of an open-ended water column and the pressure at some point at depth *h*, where the pressure reading occurs (Fig. 4.1).

In soil we deal most of the time with 'negative pressure' (potential), indicated by negative values. However, since it is more convenient to work with positive numbers, the soil water potential (negative value) is sometimes expressed as soil water suction (positive value). In the case presented in Fig. 4.1, the water potential Ψ (units: Pa) at the height of pressure reading can be calculated as:

$$\Psi = \rho \cdot g \cdot h$$

where:

$\rho =$ density of water $[\text{kg} \cdot \text{m}^{-3}]$ {1000}
$g =$ acceleration due to gravity $[\text{m} \cdot \text{s}^{-2}]$ {9.81}
$h =$ height [m]

The units of the potential (Ψ) are $\left[\frac{\text{kg}}{\text{m}^3} \cdot \frac{\text{m}}{\text{s}^2} \cdot \text{m}\right]$, which is equal to pressure in units of [Pa] (Pascal) (see Table 4.1 for details; $[\text{Pa}] = \left[\frac{\text{J}}{\text{m}^3}\right] = \left[\frac{\text{N} \cdot \text{m}}{\text{m}^3}\right]$; $[\text{N}] = \left[\frac{\text{kg} \cdot \text{m}}{\text{s}^2}\right]$).

The most commonly used dimensions for potential, together with the appropriate transformations, based on eqn 4.2/1, are summarized in Table 4.1. Throughout the text, the unit metre (m) is used. This is a unit which can better be pictured (e.g. a column of water of a height in m is easier to picture than the unit

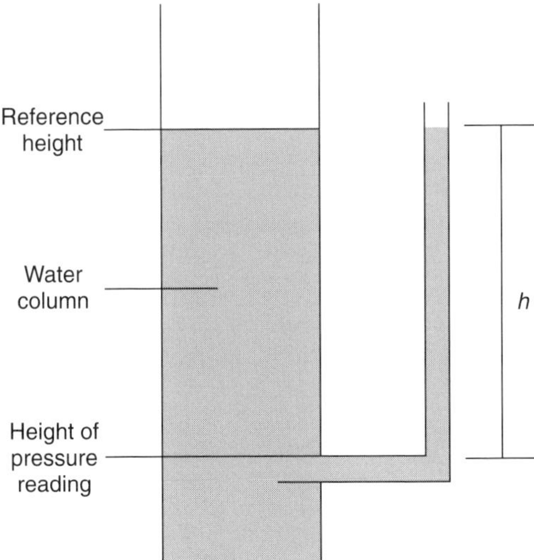

Fig. 4.1. Graphical representation of pressure (potential) as a water column.

J kg^{-1}), which is an advantage for the beginner. Furthermore, it is often the preferred dimension for water potential in soil water models (e.g. Hillel, 1977).

The total soil water potential is the result of various forces which influence the energy status of the soil water. The potentials resulting from those forces are as follows:

- *Gravitational potential* (Ψ_g) is related to the gravitational force and is determined by the difference in height between the reference height and the height of measurement (e.g. in Fig. 4.1 the height difference (h) is a measure for Ψ_g).
- *Matric potential* (Ψ_m) is the result of the adsorptive forces between soil water and the soil matrix (as already described above). Compared with free water (Ψ_m set to zero), the matric potential declines with drying of the soil and generally has negative values.
- *Osmotic potential* (Ψ_o) is related to the effects of solutes, which decrease the energy content of the soil water. This potential is particularly important in the interactions of soil and plants. Plants can maintain water uptake by lowering the plant water potential, which is mainly governed by the osmotic potential (see Chapter 6, Section 6.5.3).
- *Overburden potential* (Ψ_p) refers to an actual pressure (positive potential values) of the soil water due to the pressure of the overlying soil matrix. This potential can be substantial in soils where an overburden pressure exists, such as in saturated clay soils.

Sometimes the term hydraulic potential (Ψ_h) is used, which refers to the sum of the gravitational potential (Ψ_g) and matric potential (Ψ_m). The hydraulic potential at the measuring height in Fig. 4.1 is the sum of $\Psi_g = h$ plus a pressure resulting

Table 4.1. Units used for soil water potential and their transformations.

Units	Transformations	Example	Remarks
Pressure notation (P): [Pa] or $[J \cdot m^{-3}]$	$P = \rho \cdot g \cdot h$	For h = 1 m $P = 100 \cdot 9.81 \cdot 1 = 9810$ Pa or 9.81 kPa (≈ 10 kPa)	ρ = density of water $[kg \cdot m^{-3}]$ {1000} g = gravity $[m \cdot s^{-2}]$ {9.81} h = height [m]
Energy per unit mass: $[J \cdot kg^{-1}]$ or [kPa] (ρ =1000 kg m^{-3})	$\frac{P}{\rho} = g \cdot h$	For P = 10 kPa = 10,000 Pa $\frac{10{,}000}{1000} = 10\left[J \cdot kg^{-1}\right];$ $\left[\frac{Pa}{\frac{kg}{m^3}}\right] = \left[\frac{\frac{J}{m^3}}{\frac{kg}{m^3}}\right] = \left[\frac{J}{kg}\right]$	$[Pa] = \left[\frac{J}{m^3}\right]$
Height: [m]	$\frac{P}{\rho \cdot g} = h$	For P = 10 kPa = 10,000 Pa $\frac{10{,}000}{1000 \cdot 9.81} = 1.019\,[m]$ $\left[\frac{Pa}{\frac{kg}{m^3} \cdot \frac{m}{s^2}}\right] = \left[\frac{\frac{J}{m^3}}{\frac{kg}{m^3} \cdot \frac{m}{s^2}}\right] = \left[\frac{\frac{N \cdot m}{m^3}}{\frac{N}{m^3}}\right] = [m]$	$[N] = \left[\frac{kg \cdot m}{s^2}\right]$
Old pressure unit: [bar]	1 bar = 100 kPa	P = 10 kPa = 0.1 bar	
pF (logarithmic notation): log (base 10) of h in cm	$pF = 100 \cdot \lvert h \rvert$	For h = 1 m = 100 cm pF = 2	

from the water column, $\Psi_p = \rho \cdot g \cdot h$ (assuming that no solutes are present in the water which would give rise to an osmotic potential, i.e. assuming $\Psi_o = 0$).

For further details of the various soil water potentials and their relationships in soil, the following chapters in textbooks are recommended: Campbell (1985), Chapter 5; Marshall and Holmes (1988), Chapter 2; Hillel (1998), Chapter 6.

4.3 Soil Moisture Characteristic

Soil volumetric water content and matric potential are related to each other by what is called the 'soil moisture characteristic curve'. This relationship, which is unique for each soil type, can be measured and determines the energy content of a certain soil volumetric water content. For modelling purposes, it is convenient to express these relationships in the form of a function. In textbooks, we find both matric potential (or suction) as a function of volumetric water content ($\Psi_m(\theta)$ or, in the notation here, MPot(VWC)) (e.g. Hillel, 1998, pp. 155ff.) and volumetric water content (VWC) as a function of matric potential (suction) (VWC(MPot)) (e.g. Marshall and Holmes, 1988, Chapter 2).

Various empirical equations have been developed to describe mathematically the soil moisture characteristic curves (see Marshall and Holmes, 1988, Chapter 8; Hillel, 1988, pp. 155ff.).

One of the simplest expressions relating soil volumetric water content to soil matric potential is the power function (Campbell, 1985, p. 43; Marshall and Holmes, 1988, p. 34). (Note: this function calculates a positive value for MPot but can be converted to the correct value by multiplication with −1. The value obtained by this equation is also referred to as suction.)

$$-\text{MPot} = e^{a} \cdot \left({}^{\text{VWC}}\!/\!_{\text{VWC\_sat}} \right)^{b} \tag{4.3/1}$$

where:

−MPot = soil matric potential [m]
VWC = soil volumetric water content [$m^3 \cdot m^{-3}$]
VWC_sat = saturated soil volumetric water content (also 'Poro_tot') [$m^3 \cdot m^{-3}$]
a and b = empirical constants

The constants *a* and *b* can be determined for a particular soil by linear regression of $\ln\left({}^{\text{VWC}}\!/\!_{\text{VWC\_sat}}\right)$ vs. ln(−MPot). To illustrate this equation, measured points of volumetric water content (*x*-axis) versus soil matric potential (*y*-axis) (0–10 m soil matric suction) are plotted in Fig. 4.2A.

The first step is to replot the data so that both *x*- and *y*-axes represent natural log scales. Linear regression on the log-transformed values yields the slope (*a*) and the intercept (*b*) of this line (Fig. 4.2B). For the data in Fig. 4.2B, the parameter values are: $a = -2.90$ (= ln(a)) and $b = -9.37$. The linear function in Fig. 4.2B is described by:

$$\ln(-\text{MPot}) = \text{a} + \text{b} \cdot \ln\left(\text{VWC}/\text{VWC\_sat}\right) \qquad (4.3/2)$$

which can be transformed into eqn 4.3/1 (applying some exponential laws):

$$-\text{MPot} = \text{e}^{\text{a}+\text{b}\cdot\ln\left(\text{VWC}/\text{VWC\_sat}\right)} \text{ and } -\text{MPot} = \text{e}^{\text{a}} \cdot \left(\text{VWC}/\text{VWC\_sat}\right)^{\text{b}}$$

(See Section 1.2.3.)

The constant e^a is also referred to as the air-entry suction for the soil (Campbell, 1985, eqn 5–9). The air-entry value determines the potential when the largest pores begin to drain. Thus, for all volumetric water contents higher than the volumetric water content at the air-entry suction (VWC_e), the soil is saturated and therefore at its maximum matric potential (MPot = 0).

The measured and modelled relationships are plotted together in Fig. 4.2C. The largest potential (MPot = 0) at the air entry point is at volumetric water contents slightly less than saturation.

It should be noted that the soil moisture characteristic curves are different if they are determined during the wetting or drying process of soil. This is due to the hysteresis effect in soil. A detailed description of this phenomenon is given by Hillel (1998, Chapter 6).

Application in ModelMaker

Data from Hillel (1977) are used to illustrate the determination of the moisture characteristic relationship. Parameters a and b are determined according to the procedure described above ($a = -2.5$, $b = -3$, VWC_sat = 0.5, VWC_e = 0.49). The soil moisture release curve (eqn 4.3/1) can be calculated with *Mod4–1a.mod* (Fig. 4.3). The <lookup table> 'T1' contains volumetric water content data, which are converted to soil water suction (S) via the <variable> 'MPot'. To illustrate the soil moisture characteristic curve (SMC), the graph of 'VWC' vs. 'MPot' (log scale) is presented.

4.4 Hydraulic Conductivity

The flow of water in the soil (F_w) occurs in response to soil water potential gradients. The 'proportionality constant' (Section 1.3) in the soil water flow derivation is the soil hydraulic conductivity (K_w). The basic equation was derived by a French engineer in the last century and is named after him as 'Darcy's law':

$$\text{Fw} = -\text{Kw} \cdot \frac{d\text{HPot}}{dz} \qquad (4.4/1)$$

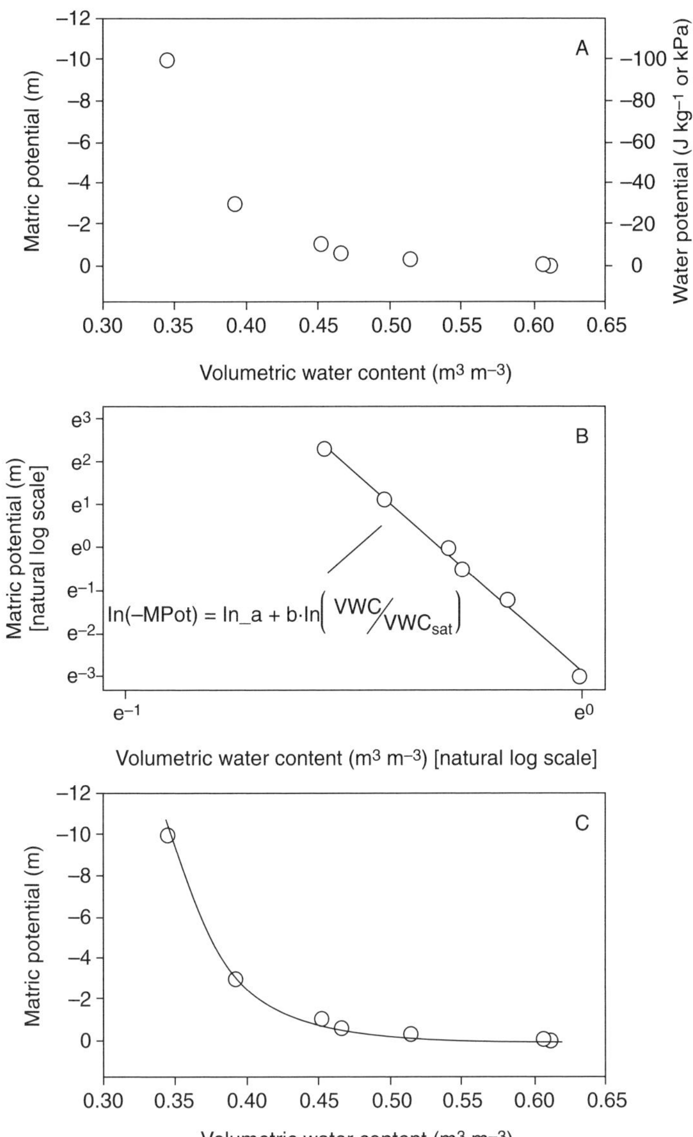

Fig. 4.2. Measured points on the soil moisture characteristic curve (A), the linear regression of log-normal transformed volumetric water contents versus soil matric suction (negative matric potentials) (B) and the calculated soil moisture characteristic (C).

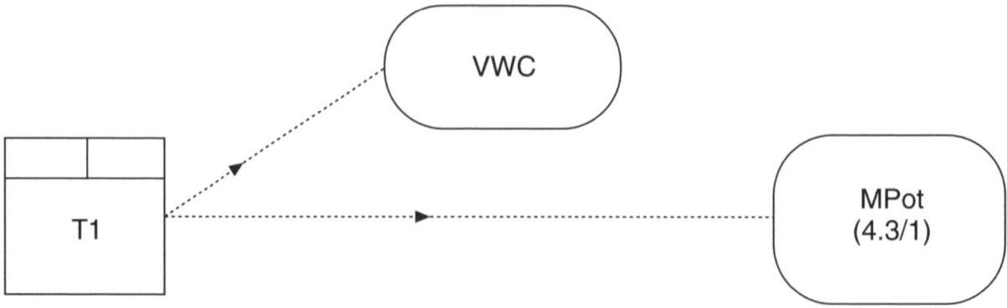

Fig. 4.3. Calculation of the soil moisture characteristic (*Mod4-1a.mod*) (T1 = input table of volumetric water contents; VWC = volumetric water content, MPot = matric potential, negative value).

where:

Fw = flow of water $[m \cdot s^{-1}]$
Kw = hydraulic conductivity $[m \cdot s^{-1}]$
HPot = sum of the matric potential and gravitational potential (= hydraulic potential) [m]
z = flow distance [m] (see Section 3.5, eqn 3.5/1 and 3.5/2)

The most common units for hydraulic conductivity are m s^{-1}. Other units in terms of length per time can be derived from this unit by simple transformation. Appropriate units for hydraulic conductivity and potential have to be chosen which result in the desired unit for flow, F_w (e.g. K_w in m s^{-1}, HPot in m, results for F_w in m s^{-1}). Campbell (1985), for instance uses kg s m^{-3} for K_w and J kg^{-1} for water potential. Hydraulic conductivity in units of kg s m^{-3} can be converted into units of m by multiplication with gravity (g, units: m s^{-2}) and division by density (ρ, units: kg m^{-3}).

The units for water flow F_w calculated with the units given by Campbell (1985) are:

$$[Fw] = \left[\frac{kg \cdot s}{m^3}\right] \cdot \left[\frac{J \cdot m}{kg}\right] = \left[\frac{kg \cdot s}{m^3}\right] \cdot \left[\frac{N \cdot m}{kg \cdot m}\right] = \left[\frac{kg \cdot s}{m^3}\right] \cdot \left[\frac{kg \cdot m \cdot m}{s^2 \cdot kg \cdot m}\right] = \left[\frac{kg}{m^2 \cdot s}\right]$$

The units $\left[\frac{kg}{m^2 \cdot s}\right]$ can be converted to $\left[\frac{m}{s}\right]$ by division by the water density, ρ (units: $\left[\frac{kg}{m^3}\right]$ {1000}).

The values for hydraulic conductivity vary approximately between 10^{-4} and 10^{-6} m s^{-1} in sand and between 10^{-6} and 10^{-9} m s^{-1} in clay soil (Hillel, 1998, pp. 203ff.). The variation is mainly due to the effect of soil water content on this property. Analogous to the equation for the soil moisture characteristics, it is also possible to calculate hydraulic conductivity as a function of volumetric water content. The hydraulic conductivity function described by Campbell (1985, Chapter 6) will be used to model the 'K_w − VWC' relationship:

$$Kw = Kw\_sat \cdot \left(VWC/VWC\_sat\right)^m \quad (4.4/2)$$

with:

$$m = 2 \cdot b + 3$$

where:

Kw = hydraulic conductivity [$m \cdot s^{-1}$]
Kw_sat = hydraulic conductivity under saturated conditions [$m \cdot s^{-1}$] (usually a measured value)
VWC_sat = saturated volumetric water content (equal to the total porosity of soil [$m^3 \cdot m^{-3}$] (determined according to eqn 2.3/40)
b = constant [–]

Saturated hydraulic conductivity (Kw_sat) may be either measured or derived from textural data (see Campbell, 1985, Chapter 6). The constant *b* changes with soil texture and ranges approximately between 2 (clay soil) and 24 (sandy soil).

An example for hydraulic conductivity calculated with eqn 4.4/2 is presented in Fig. 4.4 for data presented by Hillel (1977).

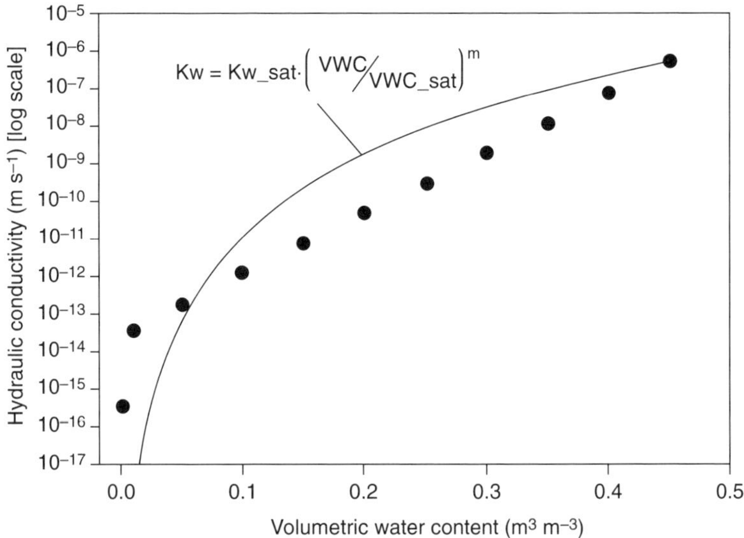

Fig. 4.4. Hydraulic conductivity (data points) and modelled K_w (parameter: Kw_sat = 5e-7, VWC_sat = 0.45, b = 1.5) (data taken from Hillel, 1977) (*Mod4-1b.mod*).

So far we have considered only the hydraulic conductivity for liquid water. In addition, some water may flow as a result of water vapour movement in the soil profile, which becomes important in very dry soil. The contribution of water vapour to hydraulic conductivity is added to the hydraulic conductivity of liquid water. This additional hydraulic conductivity is a function of vapour pressure, temperature and water vapour diffusivity in soil. The calculation procedure is described in connection with soil energy balance considerations in Chapter 5.

Application in ModelMaker

The calculation of hydraulic conductivity is presented in *Mod4–1b.mod*. The calculation is carried out with the additional <variable> 'K_w' added to the previous model. The K_w–VWC relationship is presented in a separate graph. The parameters 'b_Kw', 'VWC_sat' and 'Kw_sat' (for values see Fig. 4.4) are added to the model. While 'VWC_sat' (total porosity) and 'Kw_sat' are usually derived from measurements, the parameter 'b_Kw' is chosen so that observations are described well (in this case, the data by Hillel, 1977) (Fig. 4.4).

4.5 Basic Water Flow Model

Water flows in response to water potential gradients ($\frac{dHPot}{dz}$) with soil hydraulic conductivity (K_w) as the proportionality constant (Darcy's law, eqn 4.4/1). The continuity equation for water flow is developed in a similar manner to that already described for heat flow (Chapter 3). Water always flows from a region of high water potential (less negative) to low water potential (more negative). Considering one-dimensional flow in and out of a soil layer (in analogy to Fig. 3.1) and applying the law of mass conservation (i.e. flow in and out of a layer equals the change of water in that layer), we can represent this relationship in the mathematical notation (see also the derivation in Chapter 3):

$$\frac{\partial VWC}{\partial t} = \frac{\partial Fw}{\partial z} \qquad (4.5/1)$$

Combining eqns 4.5/1 and 4.4/1, we obtain the common form of the continuity equation for water flow:

$$\frac{\partial VWC}{\partial t} = \frac{\partial}{\partial z}\left(-K_w \cdot \frac{\partial HPot}{\partial z}\right) \qquad (4.5/2)$$

The equation is named 'Richard's equation', after its developer in the early 1930s. The independent variables are time (t) and depth (z) and the dependent variables are volumetric water content (VWC) and hydraulic potential (HPot).

The movement of water in a soil profile can also be illustrated by a graphical analysis of soil water potential profiles. The soil profile in Fig. 4.5 shows the grav-

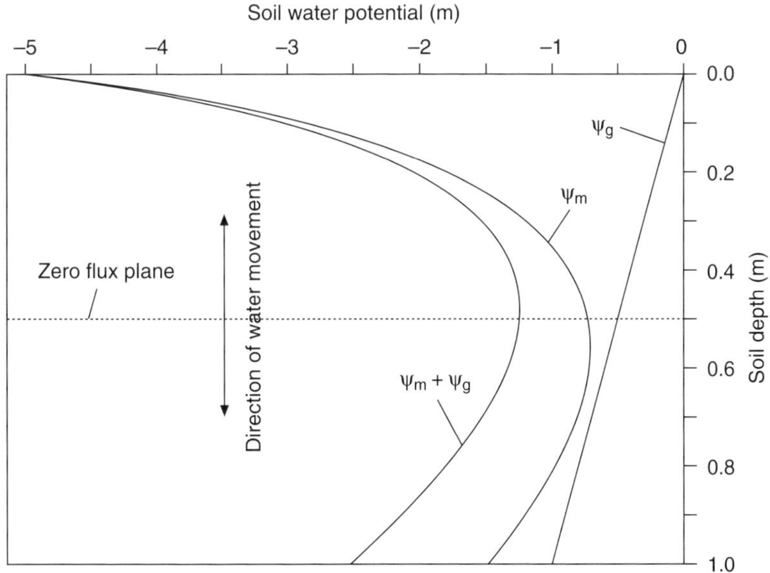

Fig. 4.5. Soil water potential (gravitational potential, matric potential and hydraulic potential) and corresponding water movement within a 1-m-deep soil profile.

itational soil water potential (Ψ_g), the matrix soil water potential (Ψ_m) and the combined effect of both (hydraulic potential).

The graphical analysis of soil water potential profiles can be carried out in a similar way to that already presented for the soil heat flow analysis (see Section 3.2). Soil water moves from high to low potential (or low to high suction). Therefore, the direction of water movement can be determined (Fig. 4.5). In a region where the flow of water changes direction, we find a 'zero-flux plane', which is characterized by a more or less constant water potential gradient. In Fig. 4.5, such a 'zero-flux plane' exists at a soil depth of 0.5 m.

To solve the water flow equation numerically, we first rewrite eqn 4.5/2 in a layered notation in the same way as for the heat flow analysis (Section 3.5):

$$\frac{\partial \mathrm{VWC}[i]}{\partial t} = \frac{\mathrm{Fw}[i] - \mathrm{Fw}[i+1]}{\mathrm{Dist}[i]} \tag{4.5/3}$$

with:

$$\mathrm{Fw}[i] = -\mathrm{Kw}[i] \cdot \frac{\mathrm{HPot}[i] - \mathrm{HPot}[i-1]}{\mathrm{Dist}[i]} \quad \text{or} \quad \mathrm{Fw}[i] = \mathrm{Kw}[i] \cdot \frac{\mathrm{HPot}[i-1] - \mathrm{HPot}[i}{\mathrm{Dist}[i]} \tag{4.5/4}$$

$$\mathrm{Fw[i+1]} = -\mathrm{Kw[i+1]} \cdot \frac{\mathrm{HPot[i+1]} - \mathrm{HPot[i]}}{\mathrm{Dist[i+1]}} \quad \text{or}$$

$$\mathrm{Fw[i+1]} = \mathrm{Kw[i+1]} \cdot \frac{\mathrm{HPot[i]} - \mathrm{HPot[i+1]}}{\mathrm{Dist[i+1]}} \tag{4.5/5}$$

where:

VWC[i] = volumetric water content of layer i [m^3 water · m^{-3} soil]
Kw[i] = hydraulic conductivity of layer i [m · s^{-1}] (according to eqn 4.4/2)
HPot[i] = hydraulic potential of layer i (sum of the matrix potential determined with eqn 4.3/1 and the gravitational potential, Depth[i], eqn 3.5/2) [m]
Dist[i] = Distance for water flow measured from the midpoint of one layer to the midpoint of the adjacent layer [m]

Boundary conditions

For the upper boundary (layer $i = 1$), we solve the equation:

$$\frac{\partial \mathrm{VWC[1]}}{\partial t} = \frac{\mathrm{Fw[1]} - \mathrm{Fw[2]}}{\mathrm{Dist[1]}} \tag{4.5/6}$$

with the flows:

$$\mathrm{Fw[1]} = \mathrm{Kw[1]} \cdot \frac{\mathrm{HPot[0]} - \mathrm{HPot[1]}}{\mathrm{Dist[1]}}$$

and:

$$\mathrm{Fw[2]} = \mathrm{Kw[2]} \cdot \frac{\mathrm{HPot[1]} - \mathrm{HPot[2]}}{\mathrm{Dist[2]}}$$

For the lower boundary (layer n) we solve the equation:

$$\frac{\partial \mathrm{VWC[n]}}{\partial t} = \frac{\mathrm{Fw[n]} - \mathrm{Fw[n+1]}}{\mathrm{Dist[n]}} \tag{4.5/7}$$

with the flows:

$$\mathrm{Fw[n]} = \mathrm{Kw[n]} \cdot \frac{\mathrm{HPot[n-1]} - \mathrm{HPot[n]}}{\mathrm{Dist[n]}}$$

and:

$$Fw[n+1] = Kw[n+1] \cdot \frac{HPot[n] - HPot[n+1]}{Dist[n+1]}$$

Both the flows (Fw[1]) (mainly due to the potential HPot[0]) and Fw[n + 1]) are unknown.

Under field conditions, soil water exchange on the upper boundary of bare soil (soil surface) arises mainly from:

- water inputs into the soil due to rainfall;
- water loss due to evaporation driven by environmental conditions.

In both cases, the rate of water extracted or added to the soil is dependent on the rate of the evaporation or rainfall passing through the soil surface. The flow and redistribution of water into deeper soil layers is then dependent on the hydraulic conductivity and the soil water potential gradients within the soil profile.

At the bottom of the soil profile, the conditions might be defined by:

- an impermeable layer (e.g. rock layer), which prevents any water movement into deeper soil regions;
- free drainage into deeper soil regions;
- a water-table, which might lead to upward movement of water.

In the basic water flow model, we consider a steady extraction of water out of the soil due to an evaporational demand of the atmosphere (also described by the term 'potential evaporation'). Evaporation can be regarded as a transport process between the soil atmosphere and the air. Evaporation continues at its potential rate (according to atmospheric demand) as long as the relative humidity of the soil air is close to saturation. While evaporation continues to dry out the soil, the soil water potential gradients are adjusted to keep up with the decreasing hydraulic conductivity so that the evaporational demand can be met. This drying process is also referred to as first-stage drying (e.g. Hillel, 1982a). If the soil profile dries out further, the relative humidity of the soil atmosphere may drop below saturation. Under these conditions, the rate of water transfer from the soil atmosphere to the air is restricted by the rate of soil water flow to the site of evaporation (also called second-stage drying). Consequently, the actual evaporation rate from the soil may drop below the potential evaporation rate. There exists a relationship between relative humidity and soil water potential. It is therefore possible to define a soil water potential at which this change from first- to second-stage drying occurs. In the next chapter, we calculate the relative humidity of the soil atmosphere, which enables us to link evaporation with relative soil humidity. At the end of first-stage drying, the soil moisture of the soil surface is more or less at its final moisture content (Hillel, 1977).

To include the above considerations into our model, the flux of water across the soil surface Fw[1] (eqn 4.5/5) is set to the potential evaporation when the hydraulic potential of the first soil layer (HPot[1]) is above a defined soil potential

MinPot. If HPot[1] is lower than MinPot, the rate of evaporation is determined by the rate of water movement into the first layer Fw[2], i.e.:

$$\text{Fw}[1] = -\text{Evap} \tag{4.5/8}$$

with:

if HPot[1]>MinPot then Evap = PET
if HPot[1]<MinPot and PET<Fw[2] then Evap = PET
if HPot[1]<MinPot and PET>Fw[2] then Evap = Fw[2]

where:

PET = evaporational demand of the atmosphere (potential evaporation) $[\text{m} \cdot \text{s}^{-1}]$ { set to: 30 mm day^{-1} = 30/1000/86400 m s^{-1}}
MinPot = hydraulic potential when the actual drops below the potential evaporation [m] {−1000}
Evap = actual evaporation rate $[\text{m} \cdot \text{s}^{-1}]$
HPot[1] = hydraulic potential of layer 1 [m]

The evaporational demand of the atmosphere is set to a fixed value (Evaporation, units: mm day^{-1}). The conversion into units of m s^{-1} is facilitated in PetRate.

$$\text{PetRate} = \frac{\text{Evaporation}}{1000 \cdot 86400} \tag{4.5/9}$$

The condition for the lower boundary layer (impermeable layer) is incorporated into the model by setting the flux

$$\text{Fw}[n+1] = 0 \tag{4.5/10}$$

The soil water model is constructed similarly to the heat flow model (Section 3.5). The depth of the soil profile and the distances between the soil layers are defined by eqns 3.5/1 and 3.5/2. Average hydraulic conductivities (AvgKw[i]) are calculated for the water flows between layers by (e.g. Hillel, 1977):

$$\text{AvgKw}[i] = \frac{\text{Kw}[i-1] \cdot z[i-1] + \text{Kw}[i] \cdot z[i]}{2 \cdot \text{Dist}[i]} \tag{4.5/11}$$

for $i = 1$:

$$\text{AvgKw}[i] = \text{Kw}[i]$$

for $i = 15$:

$$\text{AvgKw}[i] = \frac{\text{Kw}[i-1] + \text{Kw}[i]}{2}$$

The matric potentials (MPot[i]) are calculated with eqn 4.3/1 and they are added to Depth[i] (eqn 3.5/2) to obtain the hydraulic potential (HPot[i]).

The flows Fw[i] are calculated by eqns 4.5/4 and 4.5/5. For the first layer ($i = 1$) the flow is defined by the evaporation rate calculated with eqn 4.5/8 and for the bottom layer by eqn 4.5/10.

The change in water flow (net flow) is calculated for each layer by:

$$\text{NetFw[i]} = \text{Fw[i]} - \text{Fw[i + 1]} \tag{4.5/12}$$

The volume of water held in each layer (units: m or $m^3\ m^{-2}$) can now be calculated by integrating eqn 4.5/12:

$$\frac{d\text{VolVWC[i]}}{dt} = \text{NetFw[i]} \tag{4.5/13}$$

with the initial volume of water in each layer (units: m) calculated by:

$$\text{VolVWC\_init[i]} = \text{VWC\_init[i]} \cdot \text{z[i]} \tag{4.5/14}$$

where:

VWC_init[i] = initial volumetric water content $[m^3 \cdot m^{-3}]$ {0.35}

The volumetric water content is calculated according to eqn 4.5/3 by:

$$\text{VWC[i]} = \frac{\text{VolVWC[i]}}{\text{z[i]}} \tag{4.5/15}$$

The volumetric water content just outside the soil profile is set to a defined value VWC_bottom {0.01} and the conditional notation:

$$\text{if} \quad \frac{\text{VolVWC[i]}}{\text{z[i]}} > \text{VWC\_sat[i]} = \text{VWC\_sat[i]}$$

is included in the <variable> 'VWC[i]' to set the upper limit of the volumetric water content to saturation. This last addition is important once we have introduced the notation for 'Rainfall' in the next section.

Application into ModelMaker

The basic model for water flow in soil, considering only steady evaporation at the soil surface and an impermeable layer at the bottom, is presented in *Mod4–1c.mod* (Fig. 4.6).

In addition to the components described above, the cumulative potential and actual evaporation (CumPet and CumEvap, units: mm) are added to the model and the actual evaporation rate EvapRate (units: mm $hour^{-1}$). However, these components are only for output consideration, so that potential and actual evaporation rates can be compared. The current set-up of the model is for a simulation period of 10 days (<Stop value> = 864000) with hourly outputs (<output steps> = 240). The output of evaporation with this model shows the typical dynamics for drying, including first- and second-stage drying (Hillel, 1998, pp. 520–527).

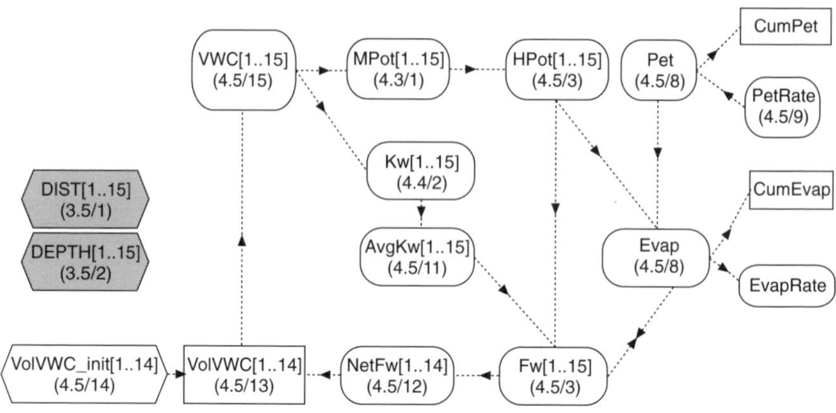

Fig. 4.6. Basic water flow model (*Mod4-1c.mod*).

Typical outputs of the model are considered in more detail at the end of the next section. In the next section, the effect of various boundary conditions typically found in nature are considered.

4.6 Effect of Other Boundary Conditions

In this section we shall consider the effect of several other surface and bottom boundary conditions and their combinations on the water flow in soil.

Diurnally changing evaporation rate

Instead of a steady evaporation rate, it is more realistic to consider a diurnally changing evaporation rate. In the simplest case, this can be achieved by replacing the current steady evaporation rate (PetRate) with a harmonic function, i.e.:

$$\text{PetRate} = \text{pi} \cdot \left(\frac{\text{Evaporation}}{1000 \cdot 86400} \right) \cdot \sin\left(\frac{2 \cdot \text{pi} \cdot \text{t}}{86400} \right) \qquad (4.6/1)$$

where:

pi = π {3.14159}
t = time [s]
Evaporation = daily evaporation rate [$\text{mm} \cdot \text{day}^{-1}$]{30}

To avoid negative values the function is made conditional so that all negative values are set to zero (if PetRate < 0, PetRate = 0).

Steady and diurnally changing rainfall

Evaporation is the main pathway of water loss through the soil surface. Precipitation (or irrigation) is the main pathway of water addition via the soil surface to the soil profile.

Rainfall is treated in the same way as evaporation except that the rates are not subtracted from but are added to the soil surface layer. The assumption is made that all rainfall, regardless of its rate, will enter the top soil layer.

If the precipitation rate (Rain) is given in units of mm day^{-1}, then the rate added for each time unit (Rain_inp in m s^{-1}) is:

$$\text{Rain\_inp} = \frac{\text{Rain}}{1000 \cdot 86400} \tag{4.6/2}$$

Distribution of the total rainfall (Rain) over an entire day with a harmonic function can be achieved by setting Rain_inp to:

$$\text{Rain\_inp} = \text{pi} \cdot \left(\frac{\text{Rain}}{1000 \cdot 86400} \right) \cdot \sin\left(\frac{2 \cdot \text{pi} \cdot t}{86400} \right) \tag{4.6/3}$$

where:

Rain = rate of precipitation [$\text{mm} \cdot \text{day}^{-1}$] {10, 20}

Negative values (output of eqn 4.6/3) are excluded via the conditional statement ‘Rainfall’: if Rain_inp < 0 then Rain_inp = 0.

The flow of water down the soil profile will distribute the added water to the entire profile. However, once the soil is saturated, any additional water will pond on the soil surface and eventually lead to runoff. We can monitor the runoff in the model with an additional notation (Runoff). Runoff occurs once the soil is saturated and the rate of rainfall into the soil is higher than the drainage of water down the profile. In mathematical terms we can define this relationship by:

if VWC[1] ≥ VWC_sat and Rainfall > Fw[2]:
Runoff = Rainfall – Fw[2] (4.6/4)

where:

Rainfall = rate of rainfall [$\text{m} \cdot \text{s}^{-1}$] (eqn 4.6/3)
Fw[2] = flow of water from the first to the second soil layer [$\text{m} \cdot \text{s}^{-1}$] (eqn 4.5/6)
VWC[1] = volumetric water content of the first soil layer [$\text{m}^3 \cdot \text{m}^{-3}$] (eqn 4.5/15)

To allow for maximum model flexibility, additional boundary conditions are included where either evaporation or rainfall or both are set to zero. This option becomes interesting when the effects of specific boundary conditions are studied without interference from each other. In this case, eqns 4.5/8 and 4.6/1 (PetRate) and eqns 4.6/2 and 4.6/3 (Rain_Inp) can optionally be set to zero, i.e.:

$$\text{PetRate} = 0 \quad (4.6/5)$$

$$\text{Rain\_inp} = 0 \quad (4.6/6)$$

The flow of water across the soil surface, previously defined by eqn 4.5/8, is now, with the additional rainfall notation, given by:

$$\text{Fw}[1] = -\text{Evap} + \text{Rainfall} \quad (4.6/7)$$

where:

Evap = evaporation rate (eqn 4.5/8) $[\text{m} \cdot \text{s}^{-1}]$
Rainfall = conditional statement for eqn 4.6/3 $[\text{m} \cdot \text{s}^{-1}]$

Bottom boundary conditions

Apart from an impermeable layer at the bottom of the soil profile as used in the basic water flow model, we shall consider two other boundary conditions:

- Free drainage.
- A water-table.

Free drainage occurs in soil if the soil water is moving downwards as a result of the gravitational force. The matrix potential gradient under these conditions is zero. Considering Darcy's law (eqn 4.4/1), the potential gradient under free drainage ($\frac{d\Psi}{dz}$) is due only to gravity (i.e. $\frac{d\Psi_g}{dz} = 1$) and the water flow equation reduces to Fw = Kw. Therefore, free drainage out of the bottom layer (Fw[n + 1], eqn 4.5/10) can be incorporated into the model with the notation (see also Hillel, 1977):

$$\text{Fw}[n+1] = \text{Kw}[n] \quad (4.6/8)$$

where:

Kw[n] = hydraulic conductivity of the bottom layer (*n*) $[\text{m} \cdot \text{s}^{-1}]$

To model the effect of a water-table at the bottom of the soil profile, we can define the flow at the bottom boundary (Fw[n + 1], eqn 4.5/7) by setting the matrix potential outside the bottom layer to zero. This means that the hydraulic potential for the bottom layer HPot[n + 1] in the presence of a water-table is defined by the gravitational potential, i.e.:

$$\text{HPot}[n+1] = -\text{Depth}[n+1] \quad (4.6/9)$$

where:

Depth[n + 1] = gravitational potential at the bottom of the soil profile (negative depth) [m]

The hydraulic potential calculated with eqn 4.6/9 will be used in eqn 4.5/7 to calculate the flow of water at the bottom of the soil profile.

The value of the volumetric water content at the bottom just outside the soil profile VWC[15] is set to VWC_sat[i] if the bottom boundary option: 'impermeable layer' or 'water table' is chosen. For the option 'free drainage' the value of VWC[15] is set to VWC[14].

Application in ModelMaker

To investigate the additional boundary conditions we have to add the notations for 'Evap' and include the new notation for 'Rainfall' and 'Runoff' (Fig. 4.7, *Mod4–1d.mod*).

To allow for maximum user flexibility, an <independent event> 'Choice' is added to the model, which prompts the user for the input of specific boundary conditions:

```
GetValue("Evaporation: diurnal=1; linear=2; no=3","Evaporation
        Choice",EVAPchoice,EVAPchoice=1 or EVAPchoice=2 or
        EVAPchoice=3);
GetValue("Rainfall: diurnal=1; steady=2;
        no=3","RainfallChoice",RainfallChoice,RainfallChoice=1
        or RainfallChoice=2 or RainfallChoice=3);
GetValue("Bottom layer: impermeable=1; water table=2, free
        drainage = 3","Bottom Boundary
        Choice",Bottomchoice,Bottomchoice=1 or Bottomchoice=2
        or Bottomchoice=3);
```

It is now possible to model a range of combinations of surface and bottom boundary conditions. To execute the calculation, the following components had to be made conditional:

> PetRate (according to <define> EvapChoice), Rain_Inp (according to <define> RainfallChoice) and F_w[1..15] (according to <define> BottomChoice).

The notation in eqn 4.6/9 is added to the <variable> 'HPot[1..15]'. This can safely be done because the water flows at the bottom boundary for the other bottom boundary conditions are defined in Fw[1..15] and do not use HPot[15]. Furthermore, the components RainfallRate, CumRainfall, RunoffRate and CumRunoff are added to the model to monitor model outputs in a suitable notation.

Figure 4.8 presents model outputs under consideration of an impermeable bottom layer and either evaporation (Fig. 4.8A) or rainfall (Fig. 4.8B). Typical soil moisture profiles under the two regimes, evaporation and rainfall, are observed (see Hillel, 1982a, pp. 278ff., for evaporation; Hillel, 1982a, pp. 238ff., or Hillel, 1998, pp. 589ff. for rainfall). Outputs during evaporation closely follow the hypothetical drying of soil as described by Hillel (1998, pp. 520ff.). During

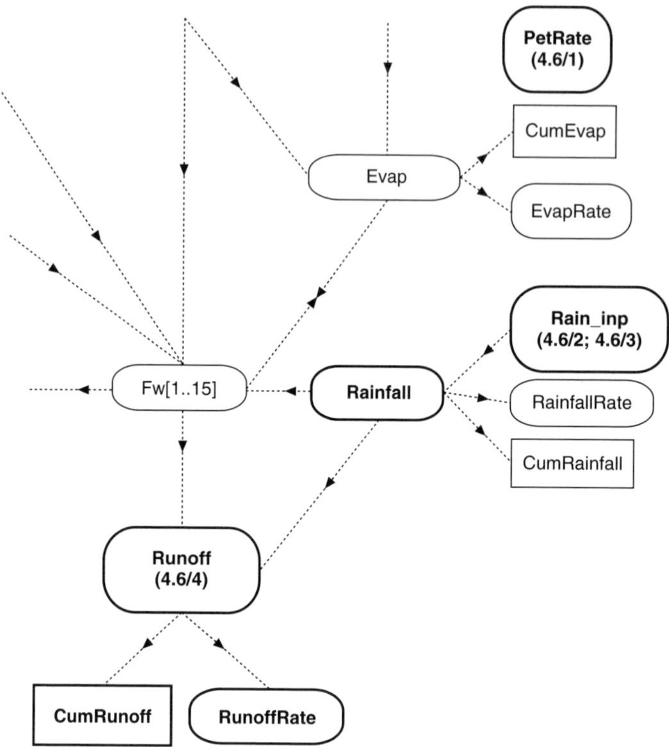

Fig. 4.7. Additon to the basic water flow model (*Mod4-1c.mod*) to include the notation for 'Rainfall' and 'Runoff' (*Mod4-1d.mod*).

infiltration of water into soil (Fig. 4.8B), a typical wetting-front is observed (Hillel, 1998, pp. 385–395).

The accumulation of water under the evaporation regime at the bottom of the profile (Fig. 4.8A) is due to the impermeable layer option at the bottom, which does not allow for drainage (see Hillel, 1977, p. 50).

4.7 Infiltrability

In the last section, we assumed that all rainfall, irrespective of its rate, would enter the soil surface layer. However, a more mechanistic model of infiltration would take account of a variable infiltration. Hillel (1977) calculated an 'infiltrability' term in his model, which is defined as the downward flux of water when the soil surface is covered with a thin layer of water (during rainfall events). The infiltrability of the soil surface depends mainly on the hydraulic gradients in the soil.

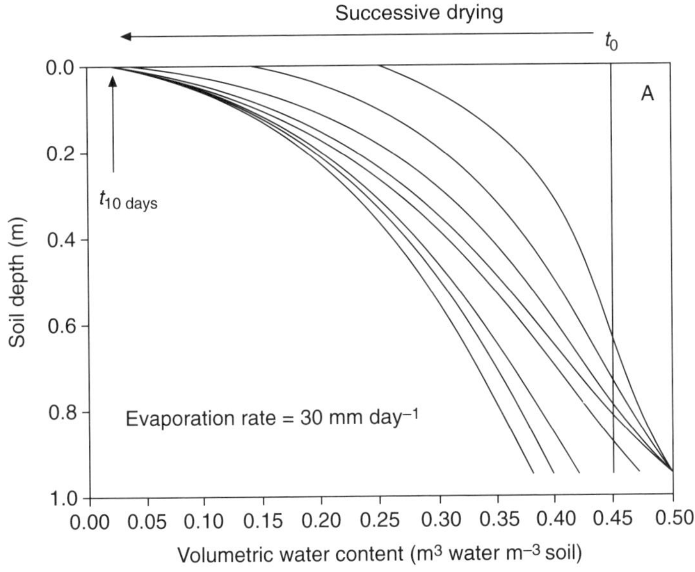

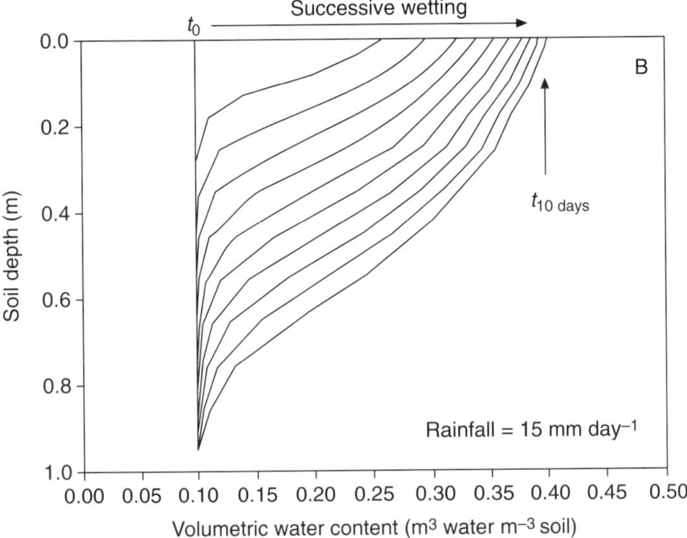

Fig. 4.8. Soil water profiles under successive drying (A) and wetting (B) (*Mod4-1d.mod*).

During infiltration, the soil surface is saturated and the hydraulic potential is therefore zero (matrix and gravitational potential are zero at the soil surface). The infiltration capacity (InfCap) can be calculated by:

$$\text{InfCap} = \frac{\text{Kws} + \text{Kw}[1]}{2} \cdot \frac{0 - \text{HPot}[1]}{\text{Dist}[1]} \qquad (4.7/1)$$

where:

Kws = saturated hydraulic conductivity of the soil $[\text{m} \cdot \text{s}^{-1}]$ $\{5 \times 10^{-7}\}$
Kw[1] = hydraulic conductivity of the first soil layer $[\text{m} \cdot \text{s}^{-1}]$
HPot[1] = hydraulic potential of the first soil layer [m]

The infiltration rate into the soil surface can be calculated by comparing the rainfall rate (Rainfall) and infiltration capacity (InfCap). The entire rainfall will enter the soil if the infiltration capacity is at least as high as the rainfall rate. In cases where the infiltration capacity is lower than the rainfall rate, water at a rate set by InfCap enters the soil and the rest runs off. According to these considerations, it is now possible to formulate revised notations for infiltration and runoff at the soil surface.

If Rainfall < InfCap:
InfRate = Rainfall (4.7/2)

If Rainfall > InfCap
InfRate = InfCap
Runoff = Rainfall – InfCap

The water flow in the top soil layer is now defined by:

Fw[1] = -Evap + InfRate (4.7/3)

where:

Evap = evaporation rate $[\text{m} \cdot \text{s}^{-1}]$ (eqn 4.5/8)

Application in ModelMaker

To implement the infiltration notation into ModelMaker, the new variables InfCap (eqn 4.7/1) and InfRate (eqn 4.7/2) have to be added to the model. In addition, the notation in Runoff has to be adjusted according to eqn 4.7/2. The additions and changes to the model are presented in *Mod4–1e.mod* (Fig. 4.9).

In addition, all parameters are set up in a layer-specific notation. For instance, the 'A' and 'B' parameters for calculating the moisture characteristic are defined for each layer separately, which allows the definition of layer-specific inputs, which is needed if water flow in a multilayered soil profile is to be investigated.

The two components <variable> 'InfiltRate' (unit: mm hour^{-1}) and <compartment> 'CumInfilt' (unit: mm) are added to the notation for output considerations, so that they can be compared with the respective notations for rainfall, evaporation and runoff.

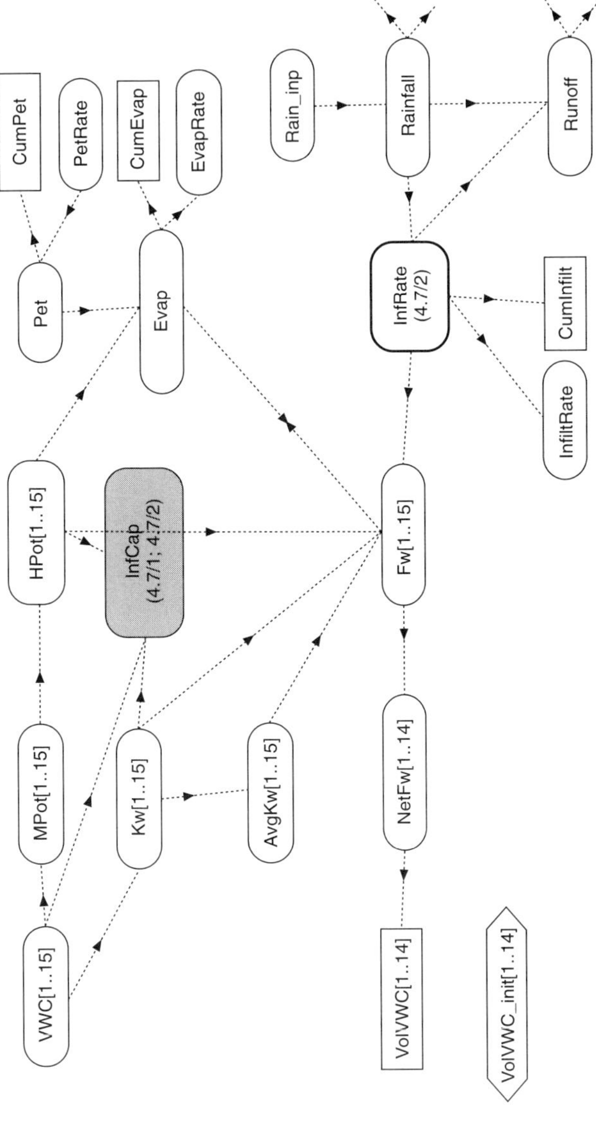

Fig. 4.9. Notation for the infiltrability calculation in the soil water model (*Mod4-1e.mod*).

4.8 Future Development

Further additions to the water model may include the effect of a rough soil surface, which may retain some water on the soil surface (ponding) before actual runoff occurs. Furthermore, a more realistic scenario would preclude evaporation during rainfall events. The above-mentioned considerations are included in the model by Hillel (1977, pp. 79ff.) and it is not too difficult to incorporate them into the current model.

However, the main objective of the modelling exercise in this chapter is to present a basic water flow model and show the effect of some characteristic boundary conditions. With the final models (*Mod4–1d.mod* and *Mod4–1e.mod*), it is possible to simulate the effect of a range of typical scenarios for evaporation and rainfall or the combination of both on water flow in soil. In addition, a range of typical bottom boundary conditions on the dynamics of water flow in soil can be investigated. The water model can serve as an investigation tool to better understand the movement of water in soil. It is impossible to illustrate the outputs for all possible user choices here. However, before going ahead and combining the heat and water flow model in the next chapter, the reader is advised to become familiar with the mathematics and the dynamics of the current model and to try out as many user options as possible.

In the next chapter, the heat and the water flow models will be combined and a soil energy balance will be calculated, so that generally available climate data (radiation, rainfall, air temperature, etc.) can be used to drive the energy relationships (heat plus water) in soil.

Soil Energy Balance 5

5.1 Introduction

The understanding of soil temperature and moisture relationships in soil is a prerequisite for modelling various processes within the soil–biosphere. Furthermore, each variable is influenced by the other. The thermal properties thermal conductivity K_T and volumetric heat capacities Ch are strongly influenced by soil moisture (Chapter 3). Water vapour movement in soil, which might contribute to water movement in dry soil, is enhanced by soil temperature gradients. In Chapter 3, a heat flow model was introduced, which was driven by the soil heat flow occurring at the soil surface. The soil surface temperature (*TS*), which is needed in these calculations, was supplied to the model as an input parameter. However, soil temperature is a result of the combined effects of energy relations on the soil surface, such as incoming solar radiation and processes such as conduction, absorption and reflection of heat (Hillel, 1977, pp. 61ff.; 1998, pp. 309ff.).

The model for water flow in soil (Chapter 4) was driven by the evaporative demand of the atmosphere, which was supplied to the model as an input parameter. However, this variable is dependent on radiation, air temperature, humidity and wind speed. For both the heat flux at the soil surface and the evaporative demand of the atmosphere, it is possible to define relationships with climatic conditions to derive these quantities.

In this chapter, we shall introduce the essential elements which are needed to calculate soil surface temperature and evaporative demand, based on an energy balance for bare soil. Relationships in environmental physics are defined which use available weather data.

5.2 Combined Soil Temperature–Moisture Model

Before we introduce the soil energy relations, the first step is to combine the soil temperature (Chapter 3) and soil water model (Chapter 4) into one model. The essential step is to add the temperature flow model to the water flow model and exchange the notation for volumetric water content (VWC[1..15]) (supplied via an input table in the temperature model) with the calculated volumetric water content (VWC[1..15] eqn 4.5/15).

In addition, the saturated volumetric water content (supplied as a parameter in the water model) is set equal to the total porosity (Poro_tot[i], eqn 3.3/2). Furthermore, the minimum volumetric water content is calculated from the minimum potential (gravitational + matric potential) the soil will obtain (parameter MinPot eqn 4.5/8) by rearranging eqn 4.3/1:

$$\mathrm{VWC\_min[i]} = \left\{ \frac{(\mathrm{MinPot[i]} + \mathrm{Depth[i]})}{-e^{A[i]}} \right\}^{\left(1/_{B(i)}\right)} \cdot \mathrm{Poro\_tot[i]} \qquad (5.2/1)$$

where:

(MinPot[i] + Depth[i]) = MPot_min[i] = minimum matric potential for each soil layer

VWC_sat[i] = Poro_tot[i]

For all other parameters and variables, see eqn 4.3/1 (note: instead of VWC and MPot, the minimum values VWC_min and MPot_min are used).

Application in ModelMaker

The addition of the temperature model (*Mod3–1d.mod*) to the water model (*Mod4–1e.mod*) can be carried out by copying the necessary notations and parameters to the water model. Both models were already set up with an identical layered system so that no procedural adjustments are needed to combine both models. All components except the VWC[1..15] component are copied. The statements of the <independent event> 'User_choice' are combined into one <independent event> 'Choice'. The parameters needed for the calculation of the thermal properties have to be copied to the combined model as well. The parameter VWC_sat[i] is deleted and replaced by the <define> 'Poro_tot[i]' (*Mod3–1d.mod*). The notations in some components where VWC_sat[i] was used must be replaced with Poro_tot[i]. All layer-specific parameters are set up in an array system (e.g. f_clay, f_sand…) to ease the simulation of flows in layered soil profiles. An additional <define> 'VWC_min[1..14]' (made globally available) has been inserted.

The final model, which includes both the temperature model (*Mod3–1d.mod*) and the water model (*Mod4–1e.mod*), is presented in *Mod5–1a.mod* (Fig. 5.1). The model is set up for a total simulation time of five days with outputs every 10 min (<stop value>: 432000, <output steps>: 720). The integration options are set to <fixed step length> using 5 steps between outputs (= calculation every 2 min). Soil surface temperature is supplied to the model via the <input file> *Inp5–1a.txt*.

5.3 Radiation Balance of Bare Soil

The main modes of energy transfer are radiation, convection and conduction. Radiation refers to the emission of energy in the form of photons or electromagnetic waves. Solids, liquids and gases have the intrinsic property to emit and absorb radiation, which is associated with energy changes in atoms and molecules. The energy emitted by molecules, for instance, is related to the movement (vibration and rotation) of individual atoms within the molecules (Monteith and Unsworth, 1990, pp. 24ff.). The total energy emitted by a body (units: W m^{-2}) is proportional to the fourth power of its absolute surface temperature (temperature in Kelvin) and can be determined by the Stefan–Boltzmann law (Monteith and Unsworth, 1990, eqn 3.17):

$$St = \varepsilon \cdot \sigma \cdot T^4 \tag{5.3/1}$$

where:

$\sigma =$ Stefan–Boltzmann constant $[W \cdot m^{-2} \cdot K^{-1}]$ $\{5.67 \times 10^{-8}\}$
$\epsilon =$ emissivity constant [–] {1 for perfect emitter = black body}
$T =$ surface temperature of the emitter [K]

The wavelength (λ, units: μm) where maximum energy is emitted ($E(\lambda_m)$) is inversely proportional to the surface temperature of the emitter (Wien's law) (Monteith and Unsworth, 1990, eqn 3.16):

$$\lambda_m = \frac{C_w}{T} \tag{5.3/2}$$

where:

$C_w =$ Wien's constant [μm · K] {2897}
$T =$ surface temperature of the emitter [K]
$\lambda_m =$ wavelength where emitted energy reaches a maximum [μm]

Considering the sun with a surface temperature of approximately 6000 K and soil at approximately 300 K (27°C), the wavelength where the emitted energy is at its maximum is for the solar spectrum 0.48 μm and for the soil 9.7 μm (Monteith and Unsworth, 1990, p. 26).

However, bodies will emit a whole range of wavelengths, each associated with an energy flux. The energy flux for each wavelength over the whole wavelength distribution can be calculated by Planck's law (Monteith and Unsworth, 1990, eqn 3.19):

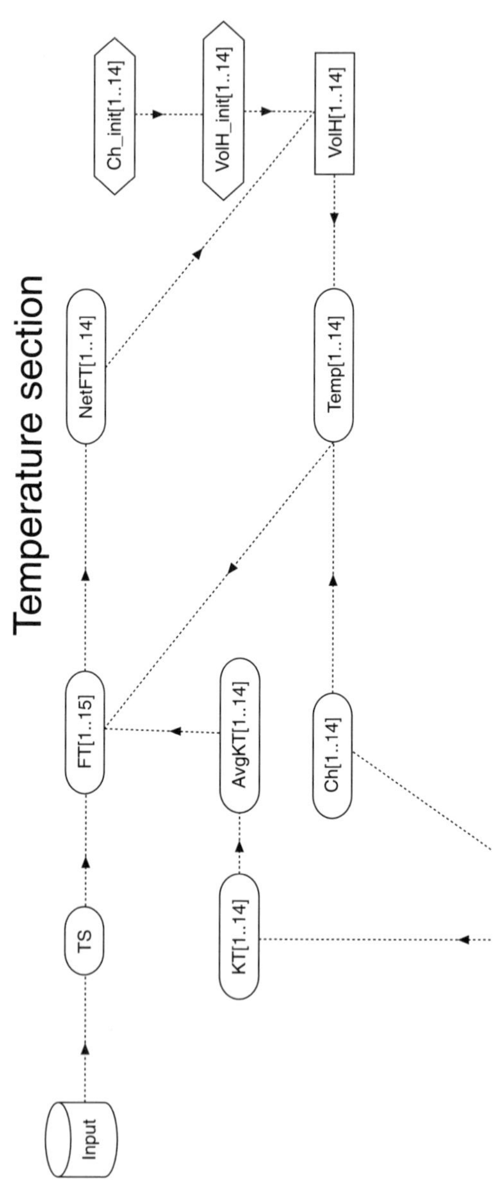
Temperature section
Ch_init[1..14]
VolH_init[1..14]
VolH[1..14]
NetFT[1..14]
Temp[1..14]
FT[1..15]
AvgKT[1..14]
Ch[1..14]
TS
KT[1..14]
Input

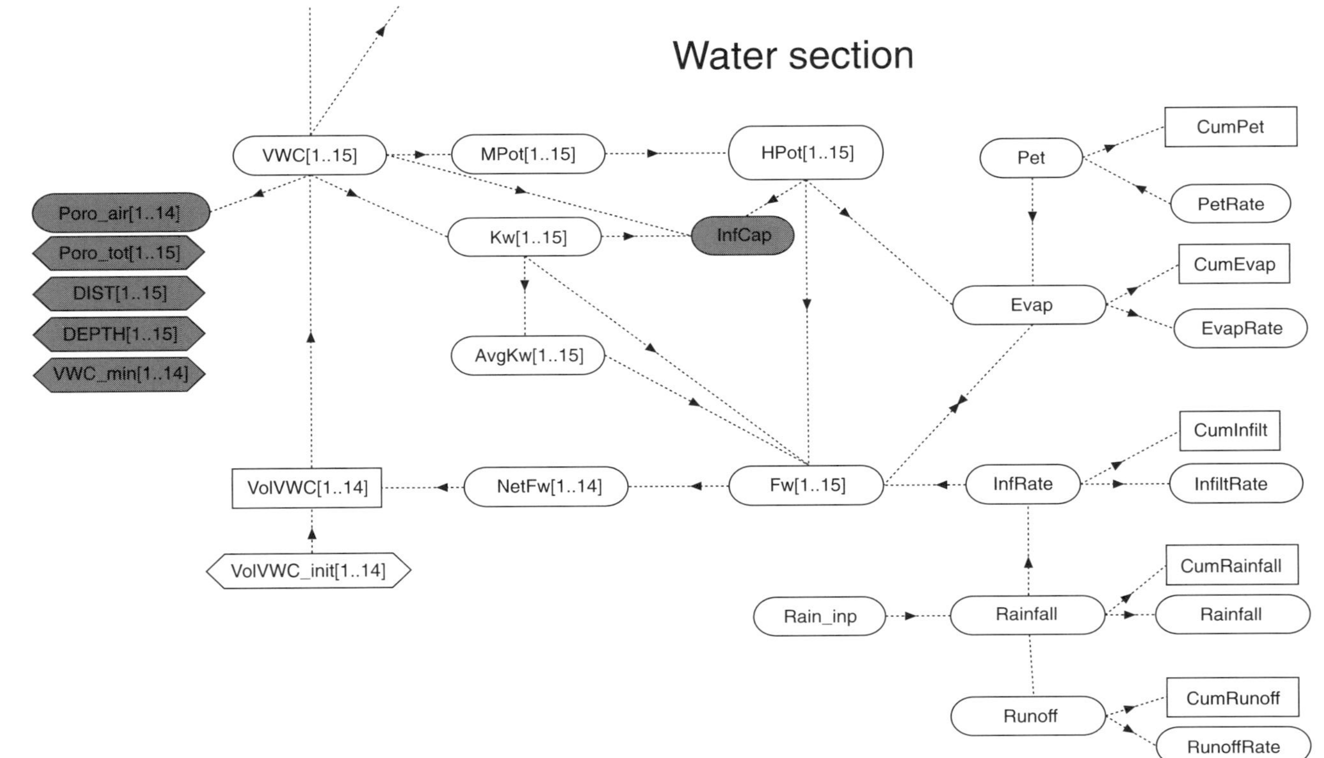

Fig. 5.1. Combined temperature (*Mod3-1d.mod*) – water flow (*Mod4-1e.mod*) model (*Mod5-1a.mod*) (see also Figs 3.7, 4.6, 4.7 and 4.9).

$$E(\lambda) = \frac{8 \cdot \pi \cdot h \cdot c}{\lambda^5 \cdot e^{\left\{\frac{h \cdot c}{\sigma \cdot \lambda \cdot T} - 1\right\}}} \qquad (5.3/3)$$

where:

h = Planck's constant $[J \cdot s]$ $\{6.63 \times 10^{-34}\}$
c = velocity of light $[m \cdot s^{-1}]$ $\{3 \times 10^{8}\}$
λ = wavelength [m]
T = surface temperature [K]
σ = Stefan–Boltzmann constant $[W \cdot m^{-2} \cdot K^{-1}]$ $\{5.67 \times 10^{-8}\}$

The total energy over the entire spectrum (all wavelengths) is given by eqn 5.3/1.

The wavelength spectrum of the sun is approximately between 0.15 and 3 μm (including the visible light, from 0.3 to 0.7 μm) and the spectrum of terrestrial systems ranges approximately from 3 to 50 μm. Since the two spectra are different, with very little overlap, they are also referred to as short-wave radiation (from the sun, *S*) and long-wave radiation (from the earth or sky, *L*).

Convection is another mode of energy transport, related to the transfer of energy with a moving mass, e.g. infiltration of warm water into cold soil (Hillel, 1998, p. 311).

The third mode of energy transport, conduction, has already been introduced in connection with the heat flow in soil (Chapter 3). It is related to the transfer of kinetic energy from a warmer to a colder part via collisions and rapid movements of molecules. On the soil surface, heat flow occurs into the soil (ground heat flux, *G*) and into the air (sensible heat flux, *C*).

Another important process of energy transfer is the composite phenomenon of latent heat transfer (*LE*). The process of evaporation transforms liquid water into water vapour. Through this process energy is absorbed and it is released again when the vaporized water condenses (transformed to liquid water). Such a process also occurs during the transition from ice to liquid water (Hillel, 1998, p. 309–340).

Radiation and energy balance for bare soil

Radiation received and emitted by a system (e.g. soil) is transferred to flows of energy. It is therefore possible to formulate both a radiation balance, which considers all incoming and outgoing radiative fluxes, and a corresponding energy balance for a system. The net radiation (sum of incoming minus sum of outgoing radiation) received by the system (e.g. soil surface) is transformed into energy and determines whether the soil is actually warming up or cooling down. Since energy relations in soil are governed by the conditions at the soil surface (see Chapter 3), we can concentrate on formulating a radiation and energy balance for that part of the system. The other modes of energy transport will transmit the conditions of the soil surface to the rest of the system. The various radiative and energy flows on the soil surface of bare soil are illustrated in Fig. 5.2.

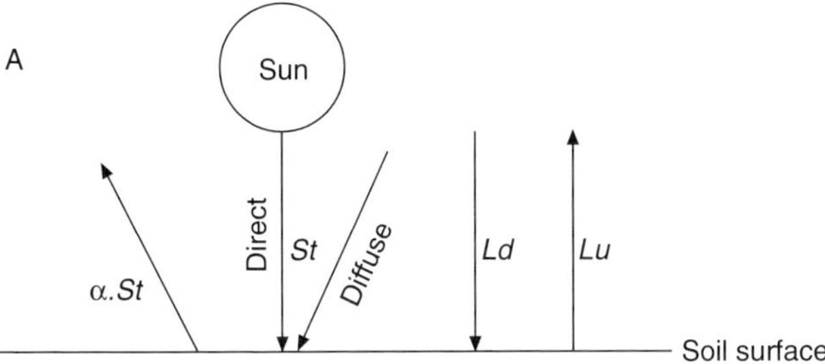

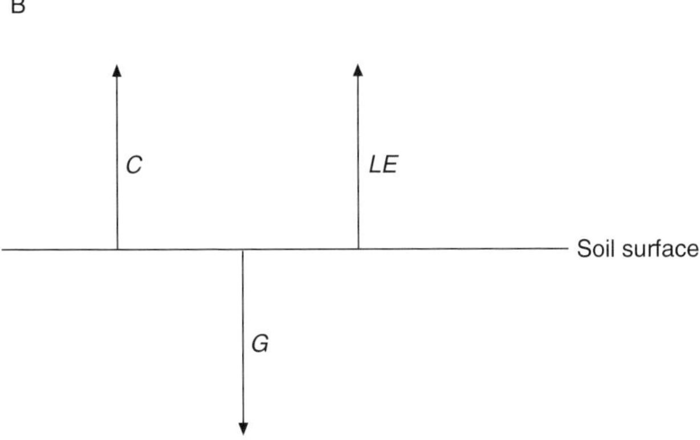

Fig. 5.2. Radiation balance (A) and energy balance (B) for the soil surface of bare soil (see text for abbreviations).

The radiation balance for the soil surface is illustrated in Fig. 5.2A. Incoming short-wave radiation, *St*, is the sum of direct and diffuse radiation (from the atmosphere). Depending on the colour and wetness of the soil surface, a part of this short-wave radiation is reflected, $\alpha \cdot St$ (α = reflectivity coefficient or albedo) (Hillel, 1998, pp. 589ff.).

Apart from short-wave radiation, there is also some long-wave radiation emitted from the atmosphere (*Ld*, long-wave radiation down) which will reach the soil surface. From the soil surface some energy is lost in the form of long-wave radiation (*Lu*, long-wave radiation up). The net radiation, *Rn* (incoming minus outgoing radiation), can therefore be calculated by:

$$Rn = St \cdot (1 - \alpha) + Ld - Lu \qquad (5.3/4)$$

where:

Rn = net radiation $[W \cdot m^{-2}]$
St = short-wave radiation (from the sun) $[W \cdot m^{-2}]$
α = reflectivity coefficient (albedo) [–] {0.1–0.4}
Ld = long-wave radiation from the atmosphere $[W \cdot m^{-2}]$
Lu = long-wave radiation from the soil $[W \cdot m^{-2}]$

Net radiation received by the soil surface is transformed into flows of energy. The energy balance of bare soil is illustrated in Fig. 5.2B and the net radiation from the soil energy balance perspective is defined as:

$$Rn = C + LE + G \qquad (5.3/5)$$

where:

C = heat flux (sensible heat flux) from the soil surface to the air above $[W \cdot m^{-2}]$
G = heat flux from the soil surface into the soil $[W \cdot m^{-2}]$
LE = evaporative heat flux or latent heat flux $[W \cdot m^{-2}]$

The latent heat flux (*LE*) is the product of the evaporation rate, *E*, and the latent heat transferred per unit quantity of water evaporated, *L* (tabulated, e.g. Monteith and Unsworth, 1990, Table A.3, p. 268). The units of the evaporation rate (*E*) are generally given in $m\ s^{-1}$ and the latent heat of vaporization (*L*) in units of $J\ g^{-1}$. To express the latent heat flux (*LE*) in units of $W\ m^{-2}$, the evaporation rate has to be multiplied by the density of water (ρ_w 'dens_water', units: $g\ m^{-3}$). The resulting units are:

$$LE\left[\frac{W}{m^2}\right] = L\left[\frac{J}{g}\right] \cdot E\left[\frac{m}{s}\right] \cdot \rho_w\left[\frac{g}{m^3}\right]$$

(note: units $[J \cdot s^{-1}] = [W]$)

where:

L = latent heat of vaporization (in Monteith and Unsworth, 1990, Table A.3, sign: λ) $[J \cdot g^{-1}]$
E = evaporation rate $[m \cdot s^{-1}]$
ρ_w = density of water $[g \cdot m^{-3}]$

The entire soil surface energy balance can be calculated by combining eqns 5.3/4 and 5.3/5:

$$St \cdot (1 - \alpha) + Ld - Lu = C + LE + G$$

or:

$$St \cdot (1 - \alpha) + Ld - Lu - C - LE - G = 0 \qquad (5.3/6)$$

Components directed towards the soil surface are conventionally taken as positive and components directed away from the soil surface as negative. As we shall see later on, some components change signs throughout the day (e.g. the ground heat flux G is negative during the night and positive during the day).

5.4 Calculation of the Various Terms of the Radiation Balance

In this section, we shall derive relationships which enable us to calculate the terms of the radiation balance (eqn 5.3/4). The available input variables for the calculations are the following climatological data:

- Short-wave radiation (*St*).
- Windspeed (*U*).
- Precipitation (*Rain*).
- Air temperature (*TA*).
- Dew-point temperature (*DP*).

The dew-point temperature (*DP*) is the temperature to which the air temperature has to be cooled down in order to become saturated with water molecules (i.e. the drier the air, the higher the difference between the actual measured air temperature (*TA*) and the dew point temperature (*DP*)). In weather stations, an approximate value for *DP* is the 'wet-bulb' temperature, which is measured in a psychrometer. All the variables are measured at a standard height of 2 m above ground.

Short-wave radiation (*St*) is readily measured in the field and is therefore directly supplied to the radiation calculations.

Long-wave radiation from the sky (Ld)

This term is calculated according to eqn 5.3/1 (in ModelMaker notation) as (see also Monteith and Unsworth, 1990, pp. 50–52):

$$\mathrm{Ld} = \mathrm{ea} \cdot \mathrm{sigma} \cdot \mathrm{TA}^4 \qquad (5.4/1)$$

where:

ea = emissivity of the atmosphere [–] (see below)
sigma = Stefan–Boltzmann constant [$\mathrm{W \cdot m^{-2} \cdot K^{-1}}$] {$5.67 \times 10^{-8}$}
TA = air temperature [K] (at standard height, 2 m)

The emissivity from the atmosphere (ea) is strongly dependent on air humidity (HA) and can be calculated with the 'Brunt formula' (Hillel, 1977, p. 62):

$$\mathrm{ea} = \mathrm{Brunt\_a} + \mathrm{Brunt\_b} \cdot \sqrt{1.41 \cdot \mathrm{HA}} \qquad (5.4/2)$$

where:

Brunt_a, Brunt_b = parameters for estimating the emissivity [–] {0.605, 0.039}

The humidity of air or absolute humidity (HA) defines the mass of water per unit volume of moist air (units: g m^{-3}). This quantity is independent of temperature. However, a functional relationship exists between vapour pressure (e, units: kPa) and absolute humidity (HA). To calculate the amount of moisture in air, it is convenient to work with the saturation vapour pressure (SVP, units: kPa), which is the partial pressure of water vapour in air when no further water molecules can be absorbed by the air. Any further increase in water content of the air would lead to condensation (e.g. fog). The saturation vapour pressure is strongly temperature-dependent and can be approximated with empirical equations (e.g. Monteith and Unsworth, 1990, pp. 8–10), such as:

$$SVP = SVP\_0 \cdot e^{\left\{\frac{A \cdot ((DP+273.16)-273.16)}{((DP+273.16)-T_{SVP})}\right\}} \tag{5.4/3}$$

where:

SVP_0 = saturation vapour pressure at 0°C (273 K) [kPa] {0.611}
A = empirical constant [–] {17.27}
DP = dew point temperature [K] or [°C + 273.16] {from weather data}
T_{SVP} = empirical temperature constant [K] {36}

The saturation vapour pressure (SVP) can be converted into absolute humidity (HA, units: g m^{-3}) with the equation (see Monteith and Unsworth, 1990, p. 11):

$$HA = \frac{Molwt\_H2O \cdot SVP}{R \cdot DP} \cdot 1000 \tag{5.4/4}$$

where:

Molwt_H2O = mol weight of water [g • mol^{-1}] {18}
R = gas constant [J • mol^{-1} • K^{-1}] {8.314}
SVP = saturation vapour pressure (eqn 5.4/3) [kPa]
DP = dew point temperature [K] (input variable)

Equation 5.4/4 has to be multiplied by 1000 so that the resulting units are in g m^{-3}.

$$[HA] = \frac{[g \cdot mol^{-1}] \cdot [kPa]}{[J \cdot mol^{-1} \cdot K^{-1}] \cdot [K]} = \frac{[g] \cdot [kPa]}{[J]}$$

Since $[Pa] = \left[\frac{J}{m^3}\right]$ and $[kPa] = \left(\left[\frac{J}{m^3}\right] \cdot 1000\right)$, we can also write the above equation as: $[HA] = \frac{[g] \cdot [J \cdot m^{-3}]}{[J]} = \left[\frac{g}{m^3}\right]$.

Long-wave radiation from the soil (Lu)
The equation for long-wave radiation emitted by the soil surface is also in the form of eqn 5.3/1 and is calculated by:

$$Lu = es \cdot \sigma \cdot TS^4 \qquad (5.4/5)$$

where:

es = emissivity from soil (see below) [–] {between 0.9 and 1.0}
σ = Stefan–Boltzmann constant [$W \cdot m^{-2} \cdot K^{-1}$] {$5.67 \times 10^{-8}$}
TS = soil surface temperature (calculated with energy balance relationships derived in the next section) [K] or [°C + 273.16]

The soil emissivity (es) is dependent on an emissivity factor (em) and the soil surface water content and can be calculated by (Hillel, 1977, pp.62–69):

$$es = em + \frac{VWC[1] \cdot VWC\_min[1]}{Poro\_tot[1]} \qquad (5.4/6)$$

where:

em = soil emissivity factor [–] {0.9}
VWC[1] = volumetric water content of the soil surface layer [$m^3 \cdot m^{-3}$]
Poro_tot[1] = saturated volumetric water content of the soil surface layer [$m^3 \cdot m^{-3}$]
VWC_min[1] = minimum volumetric water content of the soil surface layer (eqn 5.2/1) [$m^3 \cdot m^{-3}$]

While the terms for short-wave radiation (St) are taken from weather data and sky long-wave radiation (Ld) can easily be calculated from available data, it is a lot more complicated to calculate long-wave radiation emitted by the soil (Lu). This is mainly because of the unknown soil surface temperature (TS), which is needed for the calculation of Lu. Both long-wave radiations (Ld and Lu) stay more or less constant throughout the day (e.g. Monteith and Unsworth, 1990, p. 55, Fig. 4.12). The next two sections are dedicated to the derivation of relationships based on the soil energy balance (eqn 5.3/5) to calculate the soil surface temperature.

Before we can attempt to calculate the terms in the energy balance (eqn 5.3/5) it is essential to understand how the transfer of heat and water vapour takes place from the soil surface to the atmosphere. Therefore, in the next section, we shall revisit some essentials regarding these soil–atmosphere exchange processes.

5.5 Exchange Between Soil and Atmosphere

The soil surface is in direct contact with air. Considering the soil energy balance (eqn 5.3/5) there are two flows which transfer sensible heat (C) and latent heat (LE, transfer of heat via water vapour) into the air above. This layer of air is also called the boundary layer. Adjacent to the soil surface might be a very thin layer (maybe only 1 mm thick) of still air, where transport occurs via diffusion. This part of the boundary layer is called the laminar boundary layer. However, depending on the roughness of the soil surface and the wind speed, the air above this laminar boundary layer is usually well mixed. The current understanding is that transport processes within this turbulent boundary layer carry heat and water vapour via little microcurrents of air which swirl around in a more or less irregular pattern (called eddies) (Hillel, 1998, pp. 589ff.). All our measurements are usually performed within this turbulent boundary layer.

Fluxes of heat and water vapour within the turbulent boundary layer are considered to be constant, which is, for example, an essential prerequisite for the successful application of micrometeorological techniques to determine carbon dioxide or water vapour fluxes. The thickness of the boundary layer depends primarily on the length of the uniform rough surface over which air is moving (the length is called fetch). The ratio of fetch length (x) to boundary layer thickness (h) (x/h) varies between 20 and 70, depending on the roughness of the surface (Monteith and Unsworth, 1990, pp. 231ff.). Considering a 100-m-long fetch at an x/h ratio of 20, the boundary layer height would be approximately 5 m. Therefore, all micrometeorological measurements which are related to fluxes from the soil surface should be performed within this turbulent boundary layer.

The transition from the laminar to the turbulent boundary layer is related to forces associated with horizontal movement (inertial forces) and to forces related to vertical exchange and attraction of molecules between adjacent horizontal layers (Fig. 5.3). The combination of both forces creates a shearing stress or friction, which is also called momentum transfer (Monteith and Unsworth, 1990, p. 18, pp. 101ff.). Under natural conditions, the horizontal component is wind speed, which increases with distance from the soil surface. Assuming that the air above the soil surface is in layers the forces between adjacent layers will lead to a transfer of momentum (Fig. 5.3). This momentum is transferred from layer to layer until it reaches the soil surface, where it is absorbed and creates a frictional force in the direction of the wind movement. The frictional drag on the air leads to a force opposite to the direction of wind movement (Fig. 5.3). According to the environmental conditions, the transfer of momentum can be directed towards (positive) or away from the soil surface (negative).

Usually, distinct wind profiles develop over a soil surface which are related to those frictional forces. Close to the soil surface, the wind velocity is zero and it generally increases logarithmically with distance from the soil surface. If there are elements on the soil surface (called roughness elements, e.g. plants with height z), an 'equivalent soil surface' or what is called a zero plane displacement (d) can be defined. In this case the height above the soil surface where the wind speed is zero is given by $z_0 + d$, where z_0 is a parameter known as the roughness length. Empirical relationships are developed to derive these parameters (see below, eqn 5.5/3).

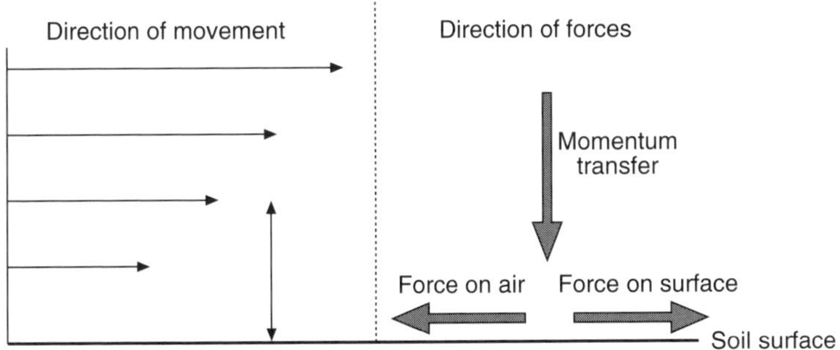

Fig. 5.3. Direction of movement over a soil surface, the related forces and the transfer of momentum (after Monteith and Unsworth, 1990, p. 19) (see text for further explanations).

The level of turbulence which exerts a transfer of momentum in the boundary layer is dependent on the roughness of the surface (e.g. bare soil, short grass, tall crops), the distance from the surface and the thermal stratification of the boundary layer (Campbell, 1985, p. 138). Depending on the conditions, three stability states in air flow can be distinguished: unstable, neutral and stable (see also Monteith and Unsworth, 1990, pp. 234ff.). These stability states can be distinguished according to the shape of the wind-speed profile. Under neutral conditions, the shearing stress is independent of height and the wind-speed profile is represented by a perfect logarithmic shape. The transport mechanism in a turbulent boundary layer under neutral conditions can be characterized by circular microcurrents (eddies) (Fig. 5.4).

If the soil surface is strongly heated (e.g. during the day), the vertical movement via eddy velocities may be enhanced by strong thermal stratification (buoyancy effect), creating unstable conditions and a wind profile which is stretched beyond the perfect logarithmic shape (Monteith and Unsworth, 1990, pp. 235ff.). These conditions can also be illustrated by vertically elongated eddy structures (Fig. 5.4). The other extremes are stable conditions, such as during clear nights with light winds. Stable conditions are characterized by wind profiles reduced below the logarithmic shape and eddy structures which are damped (Fig. 5.4).

The transfer of heat and momentum is clearly dependent on the atmospheric stability state. Environmental physicists have developed relationships based on weather data which characterize the system appropriately (Campbell, 1985, pp. 138ff.).

The flows of heat (C) and latent heat (LE) from the soil to the air above have to take the stability conditions into account. The basic calculation is related to the principles already described in Chapter 1, where the flow of a quantity is proportional to the difference in quantity ($C1 - C2$). A proportionality factor (f) is

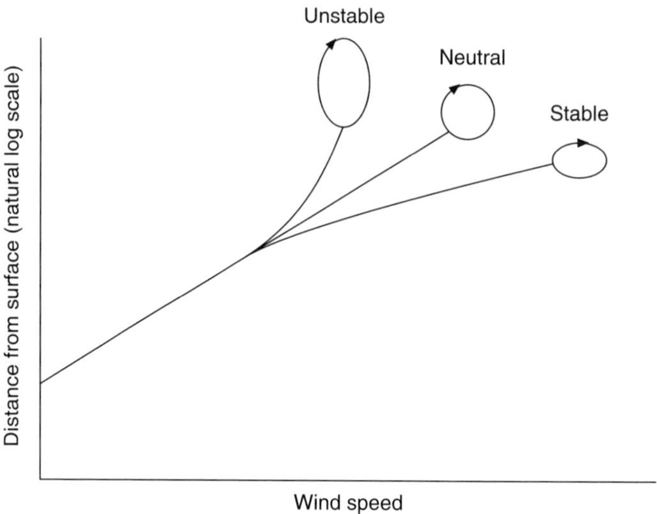

Fig. 5.4. Wind-speed profiles and eddy structures for neutral, stable and unstable stability states in air (modified according to Monteith and Unsworth, 1990, p. 235).

introduced in order to create an appropriate equation (($C1 - C2$) · f). For describing the transport in the turbulent boundary layer, this proportionality factor is referred to as boundary layer conductance for momentum transfer, *Kh* (e.g. Campbell, 1985, p. 138). Most texts use an Ohm's law analogy for the calculation, where, instead of multiplying by a conductance, the difference in quantity ($C1 - C2$) is divided by a boundary layer resistance, *Ra* (e.g. Monteith and Unsworth, 1990, pp. 117ff.) ($\frac{(C1 - C2)}{Ra}$). However, this will lead to the same result, because the relationship between boundary layer conductance and boundary layer resistance is:

$$\mathrm{Kh} = \frac{1}{\mathrm{Ra}} \qquad (5.5/1)$$

where:

Kh = boundary layer conductance [$\mathrm{m \cdot s^{-1}}$]
Ra = boundary layer resistance [$\mathrm{s \cdot m^{-1}}$]

Boundary-layer conductance

The following text focuses on the derivation of an appropriate notation for the boundary layer conductance (*Kh*) including consideration of stability. The derivation follows the notation by Campbell (1985, pp. 138–140).

A stability parameter (*SP*) is calculated, which defines whether the stability states in the air flow near the ground are neutral, stable or unstable. The two best-established methods for calculating the stability state are the Richard's number, which is calculated from temperature gradients and windspeed (e.g. Hillel, pp. 61–70), and the Monin–Obukov length, which is a function of fluxes from heat and momentum (e.g. Campbell, 1985, pp. 138–140). For the stability parameter defined in this section, a modified Monin–Obukov length (see Monteith and Unsworth, 1990, p. 237) is used. The stability parameter (*SP*), which we shall use in this chapter, is defined by (Campbell, 1985, pp. 138–140):

$$SP = \frac{-\text{kaman} \cdot \text{Zmeas} \cdot \text{gravity} \cdot C}{\text{Ch\_air} \cdot (TA + 273.16) \cdot \text{U\_frict}^3} \qquad (5.5/2)$$

with:

Ch_air = C_air • Dens_air • 1000000 (see eqn 3.3/1)

where:

kaman = von Karman's constant [–] {0.41}
Zmeas = height where weather data are measured [m] {2}
gravity = acceleration due to gravity [m • s^{-2}] {9.81}
C = heat flux (sensible heat flux) from the soil surface to the air above [W • m^{-2}] (eqn 5.3/5, calculation see next section)
Ch_air = volumetric specific heat of air (see eqn 3.3/1) [J • m^{-3} • K^{-1}]
C_air = specific heat of air [J • g^{-1} • K^{-1}] {1.01}
Dens_air = density of air [g • cm^{-3}] {0.0012} (multiply by 1,000,000 to convert to units: g m^{-3})
TA = air temperature measured at height Zmeas [°C]
U_frict = friction velocity [m • s^{-1}] (eqn 5.5/3)

The friction velocity (U_frict) is a measure of the turbulent velocity fluctuations in the air and is defined by:

$$\text{U\_frict} = \frac{U \cdot \text{kaman}}{\ln\left\{\frac{(\text{Zmeas} - d + \text{Zm})}{\text{Zm}}\right\}} + \text{Fm} \qquad (5.5/3)$$

with:

d = 0.77 • h
Zm = 0.13 • h

where:

U = wind speed (from input variables) [m • s^{-1}]
d = zero plane displacement [m]
Zm = roughness length for momentum [m]

h = height of the roughness element [m] {0.01}
Fm = stability correction for momentum [–]

The stability correction for momentum (*Fm*) is defined differently, depending on whether the stability conditions are neutral, stable or unstable. The stability correction is calculated in response to the stability parameter (*SP*, eqn 5.5/2):

if $-0.001 < SP < 0.001$ we have *neutral* conditions:
$Fm = 0$
if $SP > 0.001$ we have *stable* conditions:
$Fm = 4.7 \cdot SP$ (5.5/4)
if $SP < -0.001$ we have *unstable* conditions:
$Fm = 0.6 \cdot Fh$

where:

Fh = stability correction for heat (see below, eqn 5.5/6).

Finally, we are able to define the boundary layer conductance (*Kh*) (combining eqn 5.5/2 to 5.5/4):

$$Kh = \frac{kaman \cdot Ch\_air \cdot U\_frict}{\ln\left[\frac{Zmeas - d + Zh}{Zh}\right]} + Fh \qquad (5.5/5)$$

with:

$Zh = Zm \cdot 0.2$

where:

Zh = roughness length for heat flow [m]
Zm = roughness length for momentum [m]

The stability correction for heat, *Fh* (units: m), is calculated conditionally in response to the stability state (see also eqn 5.5/4) as:

if $-0.001 < SP < 0.001$ we have *neutral* conditions:
$Fh = 0$
if $SP > 0.001$ we have *stable* conditions: (5.5/6)
$Fh = 4.7 \cdot SP$
if SP < -0.001 we have *unstable* conditions:

$$Fh = -2 \cdot \ln\left[\frac{1 + \sqrt{1 - 16 \cdot SP}}{2}\right]$$

The boundary layer resistance (*RAC*) can now be calculated as:

$$\text{RAC} = \frac{1}{\text{Kh}} \quad (5.5/7)$$

At this stage, it might be noticed that the various notations describing the stability functions and the boundary layer conductance are actually dependent on each other. However, the actual calculation procedure will be dealt with in a separate section after we have defined the notations to calculate the terms in the soil energy balance (eqn 5.3/5) in the next section.

5.6 Calculation of the Various Terms of the Energy Balance

Recall that the terms of the soil energy balance (eqn 5.3/5) which still have to be calculated are: *C* the sensible heat flow, *LE*, the latent heat flow and *G*, the heat flow into the soil.

Sensible heat flow (C)

The sensible heat flow (*C*, units W m^{-2}) can be calculated, according to eqn 13.1 in Monteith and Unsworth (1990) or eqn 2.15 in Hillel (1977), as:

$$\text{C} = \frac{\text{Ch\_air} \cdot (\text{TS} - \text{TA})}{\text{RAC}} \quad (5.6/1)$$

where:

Ch_air = volumetric heat capacity of air [$J \cdot m^{-3} \cdot K^{-1}$] (see eqn 5.5/2)
TS = soil surface temperature [°C]
TA = air temperature (input variable) [°C]
RAC = boundary layer resistance [$s \cdot m^{-1}$]

The units for this equation are as follows:

$$[\text{C}] = \frac{[\text{J} \cdot \text{m}^{-3} \cdot \text{K}^{-1}] \cdot [\text{K}]}{[\text{s} \cdot \text{m}^{-1}]} = \left[\frac{\text{J} \cdot \text{m} \cdot \text{K}}{\text{m}^3 \cdot \text{s} \cdot \text{K}}\right] = \left[\frac{\text{J}}{\text{s} \cdot \text{m}^2}\right] = \left[\frac{\text{W}}{\text{m}^2}\right]$$

Latent heat flux (LE)

The latent heat flux (LE) is the product of the latent heat of vaporization (L) and evaporation rate (E) (Hillel, 1977, pp. 61ff.).

$$\text{LE} = \text{L} \cdot (\text{Dens\_water} \cdot 1000000) \cdot \text{E} \quad (5.6/2)$$

where:

L = latent heat of vaporization (see below) [$J \cdot g^{-1}$]
Dens_water = density of water [$g \cdot cm^{-3}$] {1} (to obtain units: g m^{-3}, it is necessary to multiply by 1,000,000)
E = rate of evaporation (see below) [$m \cdot s^{-1}$]

The units for the latent flux (LE, units: W m^{-2}) are derived in the following way:

$$[LE] = [J \cdot g^{-1}] \cdot [g \cdot m^{-3}] \cdot [m \cdot s^{-1}] = \left[\frac{J \cdot g \cdot m}{g \cdot m^3 \cdot s}\right] = \left[\frac{J}{s \cdot m^2}\right] = \left[\frac{W}{m^2}\right]$$

The latent heat of vaporization (L) is a property of water vapour which changes with temperature, and it can be looked up in tables (e.g. Monteith and Unsworth, 1990, Table A.3, p. 268). However, a linear regression of temperature versus tabulated L was performed, so that this property can be readily calculated from available temperature data:

$$L = A\_L - B\_L \cdot TS \quad (5.6/3)$$

where:

TS = soil surface temperature [°C]
A_L = intercept of the relationship temperature vs. L (value of L at TS = 0°C) [$J \cdot g^{-1}$] {2501}
B_L = slope of the relationship temperature'vs. L [$J \cdot g^{-1} \cdot °C^{-1}$] {2.37272}

The evaporation rate (E, units: m s^{-1}) can be calculated from the difference in water vapour concentration between the soil surface and the measuring height divided by the boundary layer resistance (RAC) (Hillel, 1977, pp. 61ff.):

$$E = \frac{HS - HA}{RAC \cdot (Dens\_water \cdot 1000000)} \quad (5.6/4)$$

where:

HS = water vapour concentration at the soil surface [$g \cdot m^{-3}$]
HA = water vapour concentration at the measuring height [$g \cdot m^{-3}$] (eqn 5.4/4)
Dens_water = density of water [$g \cdot m^{-3}$] {1} (to obtain units: g m^{-3} it is necessary to multiply by 1,000,000)

Note that in the calculation of the latent heat (LE) we have basically divided (eqn 5.6/4) and multiplied by the density of water (eqn 5.6/2). This is done to have dimensional consistency, i.e. expressing evaporation rate (E) in units of m s^{-1}. The evaporation rate (E) can be used directly as the upper boundary expression to drive the water flow model by setting (see eqn 4.6/1).

$$Pet = E \quad (5.6/5)$$

The difference between the water vapour concentration at the soil surface and the air (HS−HA) is sometimes referred to as the 'drying power' of the atmosphere, which determines the rate at which water evaporates from the soil. To calculate the water vapour concentration at the soil surface, we make use of a relationship between the relative humidity of air and soil water potential. Relative humidity of air (Hrel) is defined as the ratio of actual vapour pressure to saturation vapour pressure at the same temperature (Monteith and Unsworth, 1990, p. 12). The water vapour concentration at the soil surface (*HS*, units: g m^{-3}) can be calculated from the relationship (Campbell, 1985, p. 99):

$$HS = Hrel \cdot HS\_sat \tag{5.6/6}$$

where:

HS_sat = saturation water vapour concentration of the soil surface [g · m^{-3}]
Hrel = relative humidity of the soil surface [–]

The saturation water vapour concentration of the soil surface (HS_sat) is calculated in the same way as the humidity of air, using eqs 5.4/3 and 5.4/4. The only difference is that, instead of the dew-point temperature (DP), the soil surface temperature (TS) is used for the calculations, i.e.

$$HS\_sat = \frac{Molwt\_H2O \cdot SVP \cdot 1000}{R \cdot TS} \tag{5.6/7}$$

with:

$$SVP = SVP\_0 \cdot e^{\left\{\frac{A \cdot ((TS+273.16)-273.16)}{((TS+273.16)-T_{SVP})}\right\}}$$

where:

SVP = saturation vapour pressure [kPa]
SVP_0 = saturation vapour pressure at 0°C (273 K) [kPa] {0.611}
A = empirical constant [–] {17.27}
TS = soil surface temperature (see next section) [°C + 273.16]
T_{SVP} = empirical temperature constant [K] {36}
Molwt_H2O = mol weight of water [g · mol^{-1}] {18}
R = gas constant [J · mol^{-1} · K^{-1}] {8.314}

If liquid water is in thermodynamic equilibrium with water vapour, the water potentials of both phases are the same. The relative humidity at the soil surface (Hrel) can then be calculated as a function of soil water matric potential of the top soil layer (MPot[1]) (Campbell, 1985, pp. 47, 99–100; Marshall and Holmes, 1988, pp. 72–75; Monteith and Unsworth, 1990, pp. 12–13):

$$Hrel = e^{\left\{\frac{Molwt\_H2O \cdot MPot[1] \cdot gravity}{R \cdot (TS+273.16) \cdot 1000}\right\}} \tag{5.6/8}$$

The notation which we shall use for water vapour concentration at the soil surface (HS) can now be formulated by combining eqns 5.6/6 to 5.6/8.

$$HS = HS\_sat \cdot e^{\left\{\frac{MPot[1] \cdot Molwt\_H2O}{R \cdot (TS+273.16)} \cdot \frac{gravity}{1000}\right\}} \qquad (5.6/9)$$

where:

Molwt_H2O =	mol weight of water $[g \cdot mol^{-1}]\{18\}$
Mpot[1] =	soil water matric potential at the top soil layer (see eqn 4.3/1) [m]
R =	gas constant $[J \cdot mol^{-1} \cdot K^{-1}]$ {8.314}
TS =	soil surface temperature [°C]
gravity =	acceleration due to gravity $[m \cdot s^{-2}]$ {9.81}

The multiplication of ($\frac{gravity}{1000}$) is needed so that the dimensions are formulated correctly, i.e.:

$$[HS] = \left[\frac{g}{m^3}\right] \cdot \exp\left\{\frac{[m] \cdot [g \cdot mol^{-1}]}{[J \cdot mol^{-1} \cdot K^{-1}] \cdot [K]} \cdot \frac{m}{s^2} \cdot \frac{kg}{1000 \cdot g}\right\} = \left[\frac{g}{m^3}\right] \cdot \exp\{-\}$$

since $[J] = [N] \cdot [m] = \left[\frac{kg \cdot m}{s^2}\right] \cdot [m]$ (see Table 4.1). The important transformation is the dimensional change of the matric potential (MPot) from units in m to units in Pa.

The only missing component of the soil energy balance (eqn 5.3/5) is now the ground heat flux (G).

Ground heat flux (G)

The ground heat flux (G, eqn 5.3/5) defines the flow of energy which occurs across the soil surface. This flux is dependent on the energy relations on top of the soil surface, as well as on conditions in the soil. During the night, when the soil temperature is higher than the air temperature, the direction of G is from the soil to the atmosphere, resulting in a 'negative' ground heat flux. In Chapter 3, we have already established the notation which describes the flow from the soil surface into the soil (eqn 3.5/4) as the product of the temperature gradient on the soil surface and thermal conductivity, i.e.:

$$G = AvgKT[1] \cdot \frac{TS - Temp[1]}{Dist[1]} \qquad (5.6/10)$$

where:

AvgKT[1] = average thermal conductivity in the soil surface layer [$W \cdot m^{-1} \cdot K^{-1}$] (eqn 3.5/3)
TS = soil surface temperature [°C]
Temp[1] = temperature at the midpoint of the soil surface layer [°C]
Dist[1]= distance from the soil surface to the midpoint of the soil surface layer [m]

However, we are now in a position to calculate G in a different way, according to the radiation–energy balance (eqn 5.3/6). After we have defined the various radiation–energy terms, we are able to calculate the missing ground heat flux term (G) by rearrangement of eqn 5.3/6:

$$G = St \cdot (1 - albedo) + Ld - Lu - C - LE \qquad (5.6/11)$$

The ground heat flux (G) calculated via the energy–radiation balance is exchanged with the required heat flux on top of the soil surface (eqn 5.6/10). Thus the flow of heat into the soil on the soil surface (eqn 3.5/4) is defined by:

$$FT[1] = G \qquad (5.6/12)$$

We are also in a position to calculate the soil surface temperature (TS) by rearrangement of eqn 5.6/10:

$$TS = Temp[1] + \frac{G \cdot Dist[1]}{AvgKT[1]} \qquad (5.6/13)$$

The interesting thing is that the soil surface temperature which we can finally calculate was already required for the calculation of some of the radiation and energy terms (e.g. Lu, C, LE). How it is possible to perform this seemingly impossible calculation is illustrated in the next section.

5.7 Calculation Procedure for Soil Surface Temperature and Evaporation

We have derived equations which enable us to calculate soil surface temperature (*TS*) from a soil energy balance perspective. However, the problem is that the value of *TS* is already required in calculations of some radiation (*Lu*) and energy terms (*C*, *LE*). Ideally, we would like to simultaneously solve all equations which are dependent on each other. Such equations are referred to as simultaneous equations.

The mathematical procedure used in this section to solve the simultaneous equations works in the following way. Initially, we shall supply an arbitrary value for *TS* which we call a 'first guess' for *TS*. With this value and some measured inputs (e.g. *St*, *TA*) it is possible to solve all the equations, including the ground heat flux (*G*). As a final step we are also able to calculate a *TS* value, using eqn

5.6/13. This newly calculated *TS* is constrained through the calculation procedure and will approximate the 'real' *TS* a bit better than the initial guess. The trick is to repeat the calculations with the calculated *TS* value and again obtain a new value for *TS*, which is even closer to the exact *TS* value. If this procedure is repeated many times, the value for *TS* comes closer and closer to the exact value. The principle of this procedure is to repeat the calculations in a loop and reuse the final result to obtain an even closer approximation (Fig. 5.5A). These loop calculations (or iterations) are repeated until we are happy with the calculated *TS* value. Such a solution is also referred to as an iterating solution (see also Newton–Raphson algorithm, Appendix). The difference between the current and the previously calculated 'TS' values becomes smaller as the exact solution is better approached (Fig. 5.5B).

Despite the fact that we are able to come arbitrarily close to the exact solution, we are never able to reach the exact solution (Fig. 5.5B). Therefore, if the number of calculations in the loop are not restricted, the iterations would carry on for ever. A convenient way of defining a 'stop condition' for the iteration loop is to check the difference between the calculated *TS* value from the current iteration with the value from the previous iteration. This difference provides an indication of how close the calculated *TS* value is to the exact *TS* value. The condition for the calculations in the iteration loop to continue is set to:

$$\text{if abs}\left[\text{TS}_{\text{current}} - \text{TS}_{\text{previous}}\right] > \text{Iter\_stop} \tag{5.7/1}$$

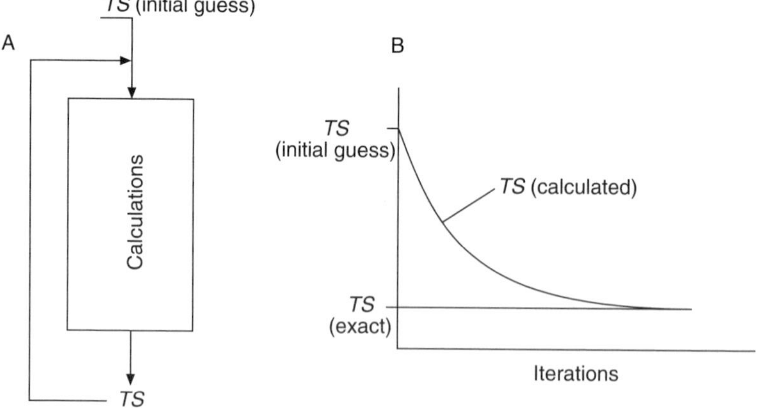

Fig. 5.5. Schematic iteration procedure for *TS* (A) and calculated *TS* with iterations (B).

where:

$TS_{current}$ = soil surface temperature calculated within the current iteration [°C]

$TS_{previous}$ = soil surface temperature calculated within the previous iteration [°C]

Iter_stop = temperature difference between current and previously calculated *TS* value which should be reached before the iterations will stop [°C] {0.001}

A value for Iter_stop of 0.001°C means that the calculated *TS* is iterated to within 0.001°C of the exact *TS* value.

All equations which are influenced by the soil surface temperature have to be included in the iteration loop. These are the boundary layer considerations (Section 5.5), as well as radiation and energy terms (Sections 5.4 and 5.6):

$$U\_frict = \frac{U \cdot kaman}{\ln\left\{\frac{(Zmeas - d + Zm)}{Zm}\right\}} + Fm \qquad \text{(eqn 5.5/3)}$$

$$SP = \frac{-kaman \cdot Zmeas \cdot gravity \cdot C}{Ch\_air \cdot (TA + 273.16) \cdot U\_frict^3} \qquad \text{(eqn 5.5/2)}$$

$$Kh = \frac{kaman \cdot Ch\_air \cdot U\_frict}{\ln\left[\frac{Zmeas - d + Zh}{Zh}\right]} + Fh \qquad \text{(eqn 5.5/5)}$$

$$RAC = \frac{1}{Kh} \qquad \text{(eqn 5.5/7)}$$

$$Lu = es \cdot sigma \cdot TS^4 \qquad \text{(eqn 5.4/5)}$$

$$C = \frac{Ch\_air \cdot (TS - TA)}{RAC} \qquad \text{(eqn 5.6/1)}$$

$$SVP = SVP\_0 \cdot e^{\left\{\frac{A \cdot ((TS+273.16)-273.16)}{(TS+273.16)-T_{SVP}}\right\}} \qquad \text{(see eqn 5.6/7)}$$

$$HS\_sat = \frac{Molwt\_H2O \cdot SVP}{R \cdot TS} \cdot 1000 \qquad \text{(eqn 5.6/7)}$$

$$HS = HS\_sat \cdot e^{\left\{\frac{MPot[1] \cdot Molwt\_H2O}{R \cdot (TS+273.16)} \cdot \frac{gravity}{1000}\right\}} \qquad \text{(eqn 5.6/9)}$$

$$E = \frac{HS - HA}{RAC \cdot (Dens\_water \cdot 1000000)} \quad \text{(eqn 5.6/4)}$$

$$L = A\_L - B\_L \cdot TS \qquad \text{(eqn 5.6/3)}$$

$$LE = L \cdot (Dens\_water \cdot 1000000) \cdot E \quad \text{(eqn 5.6/2)}$$

$$G = St \cdot (1 - albedo) + Ld - Lu - C - LE \quad \text{(eqn 5.6/11)}$$

$$TS = Temp[1] + \frac{G \cdot Dist[1]}{AvgKT[1]} \quad \text{(eqn 5.6/13)}$$

Application in ModelMaker

In ModelMaker, calculations in iteration loops can be performed in <independent events>. Several control statements are available in ModelMaker for loop calculations: 'while', 'for' and 'do-while' loops (see Walker, 1997, pp. 120–126). In our case, we use the 'do-while' loop procedure. Initially the loop will be calculated with a 'guess value' ('do') and then the calculations are calculated 'while' the condition defined by eqn 5.7/1 is still correct.

The <independent event> 'TS_calc' is defined in ModelMaker with the following statement in <actions>:

```
TS=TA;
Do {
  FTS=TS;
  U_frict=U*kaman/(ln((Zmeas-d+zm)/zm)+Fm);
  Sp=-(kaman*Zmeas*gravity*C)/((Dens_air*C_air*1000000)
     *(TA+273.16)*U_frict^3);
  Kh=kaman*U_frict/(ln((Zmeas-d+zh)/(zh))+Fh);
  RAC=1/Kh;
  Lu=(es*Sigma*(FTS+273.16)^4);
  C=(FTS-TA)*(Dens_air*C_air*1000000)/RAC;
  SVP=((SVP_0*exp(A*((273.16+FTS)-273.16)/((273.16+FTS)
       -Tsvp)));
  HS_sat=(SVP*Molwt_H2O*1000)/(R*(273.16+FTS));
  HS=HS_sat
       *exp((MPOT[1]*Molwt_H2O*gravity)/(R*1000*(FTS+273.16)));
  E=(HS-HA)/(RAC*Dens_water*1000000);
  L = (A_L+B_L*FTS);
  LE=L*1000000*Dens_water*E;
  G=St*(1-albedo)+Ld-Lu-C-LE;
  TS=Temp[1]+(G*DEPTH[1]/AvgKT[1]);
}
while(abs(TS-FTS)>Iter_stop);
```

Within the loop, the notation for *TS* is changed to *FTS* which is a so-called 'dummy' variable for *TS* during the calculations. A dummy variable is only needed for certain operations in this case for the calculation within the loop. The change from *TS* to *FTS* (*FTS* = *TS*) occurs at the beginning of the loop. This is needed because the current and the previous *TS* value have to be compared at the end of the iteration loop, where a new *TS* value is calculated. The initial calculations are performed with a 'guess value' for *TS* set to air temperature 'TA'. The comparison at the end of the loop between *FTS* (previous value for *TS*) and *TS* (new calculated) will provide the stop conditions for the loop iteration. The calculations in the loop are performed every 120 time units (seconds) and starting at $t = 0$. These values are inserted into the 'triggers' definition of the <independent event> 'TS_calc' and are held in the parameters TriggerPer = 120 and TriggerStart = 0.

After exiting the loop, the various components still have to be calculated with the latest *TS* value. Therefore the entire notation within the loop is inserted below the 'do-while' loop and *TS* is exchanged for *FTS*. The notations below the loop definition are:

```
U_frict=U*kaman/(ln((Zmeas-d+zm)/zm)+Fm);
Sp=-(kaman*Zmeas*gravity*C)/((Dens_air*C_air*1000000)
     *(TA+273.16)*U_frict^3);
Kh=kaman*U_frict/(ln((Zmeas-d+zh)/(zh))+Fh);
RAC=1/Kh;
Lu=(es*Sigma*(TS+273.16)^4);
C=(TS-TA)*(Dens_air*C_air*1000000)/RAC;
SVP=((SVP_0*exp(A*((273.16+FTS)-273.16)/((273.16+FTS)
     -Tsvp)));
HS_sat=(SVP*Molwt_H2O*1000)/(R*(273.16+FTS));
HS=HS_sat
      *exp((MPOT[1]*Molwt_H2O*gravity)/(R*1000*(TS+273.16)));
E=(HS-HA)/(RAC*Dens_water*1000000);
L =(A_L+B_L*TS);
LE=L*1000000*Dens_water*E;
G=St*(1-albedo)+Ld-Lu-C-LE;
Rn_soil=C+LE+G;
Rn_rad=St*(1-albedo)+Ld-Lu;
```

All the components which are calculated within the loop are held in <define> values, which are linked to the <independent event> 'TS_calc'. In addition, variables *Fm* (eqn 5.5/4), *Fh* (eqn 5.5/6), *HA* (eqn 5.4/4), *ea* (5.4/2), *es* (eqn 5.4/6) and *Ld* (eqn 5.4/1) have to be inserted and linked to the components where they are needed. Furthermore, input variables held in an <input file> 'Inputs' are read into variables *TA* (air temperature), *DP* (dew point temperature), *U* (wind speed) *Prec* (precipitation) and *St* (short-wave radiation), which are made globally available to the calculations.

For an additional check that the energy balance is calculated correctly, both the net radiation using the radiation terms (eqn 5.3/4) and the net radiation using the soil energy balance (eqn 5.3/5) are calculated at the end of the definition of the

<independent event> 'TS_calc'. They are held in the <defines> 'Rn_rad' and 'Rn_soil', respectively, and should give the same results if the calculations are performed correctly.

Since some of the variables are required from a soil water–heat model (e.g. AvgKT[1], Temp[1], MPot[1]), the notation for the energy balance can only be calculated if it is directly linked to the soil water–temperature model. Therefore, the notation for the energy balance is included and linked to the model *Mod5–1a.mod*. The combined model is presented in *Mod5–1b.mod* and the additional notation is presented in Fig. 5.6.

In the <independent event> 'Choice', only the user choices for the bottom boundary BottomChoice and the RainfallChoice are kept. For the rainfall choice the statement is now:

```
RainfallChoice=GetChoice("With Rainfall? (Y/N)","Rainfall
    Choice",RainfallChoice);
```

If the user chooses 'Yes', the rainfall data from the input file *Inp5–1b.txt* are used. This file contains hourly data of the input variables: air temperature (*TA*), dewpoint temperature (*DP*), wind speed (*U*), precipitation (*Prec*) and solar radiation (*St*). The variable Rain_inp is defined conditionally according to RainfallChoice (Yes(1): Rain_inp = Prec, No(0): Rain_inp = 0).

Running the model

Before running the model, the <Model> <Run options> and the values for the Trigger period (TriggerPer) and starting value (TriggerStart) of the <independent event> 'TS_calc' have to be defined appropriately. Tests showed that the model runs satisfactorily if calculations are carried out every 2 min of simulation time. Therefore 'TS' should also be iterated at the same time intervall (TriggerPer = 120). The temperature–water section uses some outputs from the iteration procedure (e.g. Pet = E, FT[1] = G). Close scrutiny of the results provided evidence that the calculations in the water–temperature section were taking *E* and *G* values not from the current calculation but from the previous calculation step (2 min earlier), which indicates the way in which ModelMaker performs calculations. To alleviate this problem, the iterations for the energy balance have to be carried out just before the other calculations. To avoid artefacts, it was decided to perform the iterations always 1 s before the rest of the calculations. To simulate for 10 days with outputs every 10 min, the parameters in <Model> <Run Options> have to be set to:

Start Value = 1
Stop Value = 864001
Output Steps = 1440

To perform calculations every 2 min, the 'fixed step length' has to be chosen with five fixed steps between the outputs.

To perform the TS_calc iterations always 1 s before the rest of the model calculations, the 'periodic trigger values' have to be set to:

TriggerPer = 120
TriggerStart = 0

There are enough data in the input file *Inp5–1b.txt* to simulate 10 days. The data are from a sunny, early summer day where there was no rainfall. The rainfall data are only added to the input file for demonstration purposes. Therefore a more realistic scenario is simulated if the RainfallChoice = No is chosen.

A typical output for 1 day of the various energy terms of the model, with the 'user Choice': RainfallChoice = No and BottomChoice = 3 (free drainage), is presented in Fig. 5.7.

At this stage, it is advisable to compare the output with some typical measured values. For example, comparisons of the simulations with the measurements presented by Monteith and Unsworth (1990, pp. 53–57) or Hillel (1998, p. 601) show that the simulations are close to reality. In order to become familiar with the model, it is advisable to run the model with different user-definable scenarios. In addition to the test data set (*Inp5–1e.txt*), a monthly data set (*Inp5–1real.txt*) is provided, which contains values logged every half-hour from an automatic weather station in Linden (Germany) for May 1997. To exchange the current input file (*Inp5–1e.txt*) with *Inp5–1real.txt*, double-click on the <Lookup file> 'Inputs', go to 'Find file…' and choose the appropriate data file.

In Chapters 3 and 4, it was mentioned that the water vapour movement in soil will contribute to the thermal conductivity (K_T) and the water flow in soil. In the next two sections, we extend the basic energy balance model to include these aspects.

5.8 Contribution of Water Vapour Movement to Soil Thermal Conductivity

In Chapter 3, we introduced a relationship which defined the contribution of water vapour to thermal conductivity in soil (KT_vapour, eqn 3.3/5). The empirical relationship in Chapter 3 was based on data by Hillel (1977, pp. 66–69). However, it is possible to define the thermally induced vapour flow and its effect on thermal conductivity with relationships based on environmental physics. The flow of water vapour in soil transfers latent heat and contributes to the overall soil heat flow. The relationship used is the one presented by Campbell (1985, p. 102, eqn 9.11), which computes the additional thermal conductivity related to the latent heat flow for each soil layer (KT_vapour[i], units: W m^{-1} K^{-1}):

$$\mathrm{KT\_vapour[i]} = \mathrm{Dv[i]} \cdot \mathrm{Hrel\_soil[i]} \cdot \mathrm{Del\_vp[i]} \cdot \mathrm{L\_soil[i]} \cdot \mathrm{F\_vp[i]} \cdot \frac{\mathrm{Molwt\_H2O}}{(\mathrm{Temp[i]} + 273.16) \cdot \mathrm{R}} \qquad (5.8/1)$$

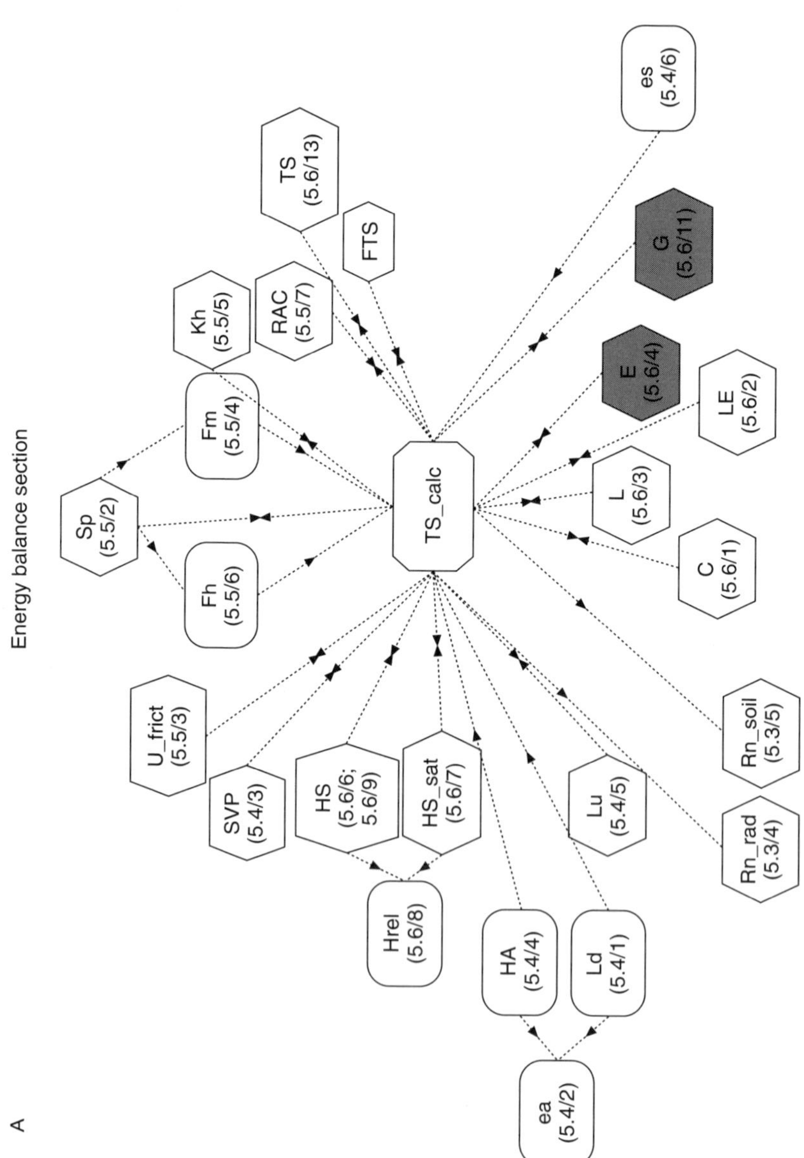
A
Energy balance section
TS_calc
es
(5.4/6)
TS
(5.6/13)
FTS
G
(5.6/11)
RAC
(5.5/7)
Kh
(5.5/5)
E
(5.6/4)
LE
(5.6/2)
Fm
(5.5/4)
Sp
(5.5/2)
L
(5.6/3)
C
(5.6/1)
Fh
(5.5/6)
U_frict
(5.5/3)
SVP
(5.4/3)
HS
(5.6/6;
5.6/9)
HS_sat
(5.6/7)
Lu
(5.4/5)
Rn_soil
(5.3/5)
Rn_rad
(5.3/4)
Hrel
(5.6/8)
HA
(5.4/4)
Ld
(5.4/1)
ea
(5.4/2)

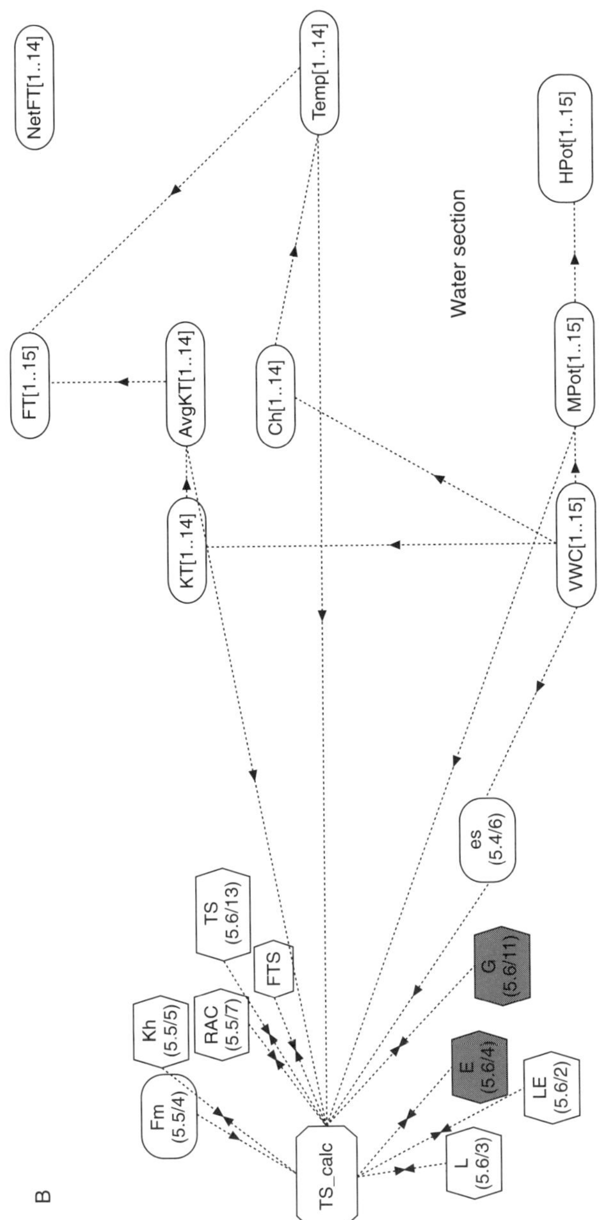

Fig. 5.6. Energy balance section (A) included and linked (B) to the existing water–temperature model (*Mod5-1b.mod*).

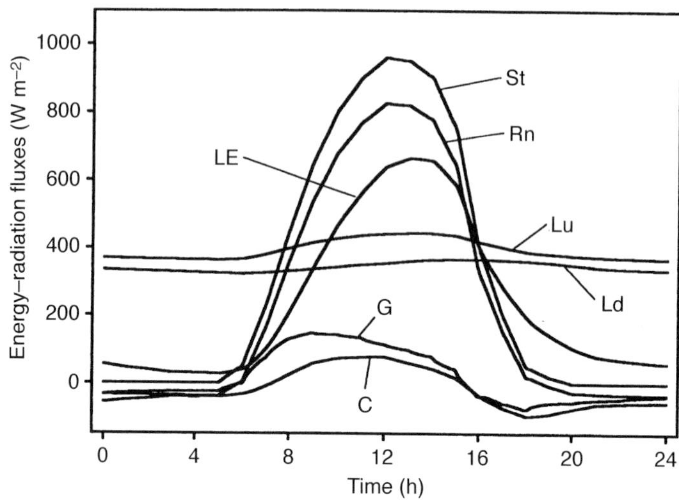

Fig. 5.7. Energy and radiation outputs of the energy balance model (*Mod5-1b.mod*) (St = solar radiation; Rn = net radiation; LE = latent heat flux; C = sensible heat flux; G = ground heat flux; Lu = long-wave radiation up; Ld = long-wave radiation down).

where:

i =	soil layer index
Dv[i]=	diffusivity for water vapour $[m^2 \cdot s^{-1}]$
Hrel_soil[i] =	relative humidity of soil [–]
Del_vp[i] =	change of saturation vapour pressure $[Pa \cdot K^{-1}]$
L_soil[i]	latent heat of vaporization in soil $[J \cdot g^{-1}]$
F_vp[i] =	enhancement function for water vapour flow in soil [–]
Molwt_H2O =	molecular weight of water $[g \cdot mol^{-1}]$ {18}
Temp[i] =	soil temperature [°C] or [K] (add 273.16)
R =	gas constant $[J \cdot mol^{-1} \cdot K^{-1}]$ {8.314}

The last term, '$\dfrac{\text{Molwt\_H2O}}{(\text{Temp[i]}+273.16)\cdot \text{R}}$', is only added to the notation in order to express KT_vapour in the same units as the thermal conductivity in soil (*KT*, units: $W\ m^{-1}\ K^{-1}$).

The units of eqn 5.8/1 are:

$$[\text{KT\_vapour}] = \left[\frac{m^2}{s}\right]\cdot\left[\frac{Pa}{K}\right]\cdot\left[\frac{J}{g}\right]\cdot\frac{\left[\dfrac{g}{mol}\right]}{[K]\cdot\left[\dfrac{J}{mol\cdot K}\right]} \quad \left(\text{recall: } [Pa]=\left[\frac{J}{m^3}\right]\right)$$

simplified:

$$= \left[\frac{m^2}{s}\right] \cdot \left[\frac{\frac{J}{m^3}}{K}\right] \cdot \left[\frac{J}{g}\right] \cdot \left[\frac{g}{J}\right]$$

$$= \left[\frac{m^2}{s}\right] \cdot \left[\frac{\frac{J}{m^3}}{K}\right] = \left[\frac{m^2}{s}\right] \cdot \left[\frac{J}{m^3 \cdot K}\right] = \left[\frac{J}{s \cdot m \cdot K}\right]$$

$$= \underline{\underline{\left[\frac{W}{m \cdot K}\right]}} \quad \left(\text{recall: } \left[\frac{J}{s}\right] = [W]\right)$$

Thermal conductivity KT [i] is adjusted according to eqn 3.3/4 (see also Hillel, 1998, p. 319) by adding the conductivity due to water vapour flow (KT_vapour[i]) to the conductivity of air (KT_air).

The calculation of the various terms in eqn 5.8/1 are explained below.

Diffusivity of water vapour (Dv[i])

This property is calculated using the gas diffusivity equation introduced in Chapter 2 (eqns 2.3/21–2.3/23). The notation used for water vapour diffusivity (Dv[i], units: $m^2\ s^{-1}$) is:

$$Dv[i] = Diff0\_H2O \cdot \left(\frac{Temp[i] + 273.16}{273.16}\right)^{Diff\_n} \cdot Diff\_b \cdot (Poro\_air[i])^{Diff\_m} \tag{5.8/2}$$

where:

Diff0_H2O =	diffusivity of water vapour through air at NTP (273.16 K, 101,300 Pa) $[m^2 \cdot s^{-1}]$ $\{2.12 \times 10^{-5}\}$
Diff_n =	parameter to adjust diffusivity for temperature [–] {1.75}
Diff_b =	parameter to adjust diffusivity for air-filled porosity [–] {0.66}
Diff_m =	parameter to adjust diffusivity for air-filled porosity [–] {1}
Poro_air[i] =	air-filled porosity (eqn 3.3/2, Poro_tot[i] – VWC[i]) $[m^3 \cdot m^{-3}]$

Relative humidity of soil (Hrel_soil[i])

The relative humidity of soil (Hrel_soil[i]) is calculated in the same way as the relative humidity for the soil surface (Hrel[i], eqns 5.6/6–5.6/9), exchanging the soil surface temperature (TS) with the soil layer temperature (Temp[i]). The equations for the saturated vapour concentration (Hso_sat[i]), the saturated vapour pressure (SVP_soil[i]), the vapour concentration of the soil (Hso[i]) and the relative humidity (Hrel_soil[i]) are:

$$\text{Hso\_sat}[i] = \frac{\text{Molwt\_H2O} \cdot \text{SVP\_soil}[i] \cdot 1000}{\text{R} \cdot \text{Temp}[i]} \tag{5.8/3}$$

$$\text{SVP\_soil}[i] = \text{SVP\_0} \cdot e^{\left\{\frac{\text{A} \cdot ((\text{Temp}[i]+273.16)-273.16)}{((\text{Temp}[i]+273.16)-\text{T}_{\text{SVP}})}\right\}} \tag{5.8/4}$$

$$\text{Hso}[i] = \text{Hso\_sat}[i] \cdot e^{\left\{\frac{\text{MPot}[1] \cdot \text{Molwt\_H2O}}{\text{R} \cdot (\text{Temp}[i]+273.16)} \cdot \frac{\text{gravity}}{1000}\right\}} \tag{5.8/5}$$

$$\text{Hrel\_soil}[i] = \frac{\text{Hso}[i]}{\text{Hso\_sat}[i]} \tag{5.8/6}$$

where:

SVP_0 =	saturation vapour pressure at 0°C (273 K) [kPa] {0.611}
A =	empirical constant [–] {17.27}
Temp[i] =	soil temperature [°C + 273.16]
T_{SVP} =	empirical temperature constant [K] {36}
Molwt_H2O =	mol weight of water [g · mol^{-1}] {18}
R =	gas constant [J · mol^{-1} · K^{-1}] {8.314}
MPot[i] =	matrix potential of top soil layer [m]

Change of saturation vapour pressure (Del_vp[i]) and latent heat of vaporization (L_soil[i])

The change of saturation vapour pressure is an important quantity in micrometeorology which defines the rate of increase of saturation vapour pressure (eqn 5.8/4) with temperature (Monteith and Unsworth, 1990, p. 10). This quantity is tabulated in meteorological tables (e.g. Monteith and Unsworth, 1990, Table A.4, D(*T*)) and is usually expressed in units: Pa K^{-1}. The 'change of saturation vapour pressure' is strongly temperature-dependent and has to be re-evaluated for each soil temperature. For temperatures up to 40°C, it is possible to calculate Del_vp[i] with the equation 2.24 given by Monteith and Unsworth (1990):

$$\text{Del\_vp}[i] = \frac{\text{L\_soil}[i] \cdot \text{Molwt\_H2O} \cdot \text{SVP\_soil}[i] \cdot 1000}{\text{R} \cdot (\text{Temp}[i] + 273.16)^2} \tag{5.8/7}$$

with the latent heat of vaporization (L_soil[i]) calculated according to eqn 5.6/3:

$$L\_soil[i] = A\_L + B\_L \cdot Temp[i] \tag{5.8/8}$$

where:

A_L =	intercept of the relationship Temp[i] vs. L_soil[i] (at Temp[i] = 0 °C) [$J \cdot g^{-1}$] {2501}
B_L =	slope of the relationship Temp[i] vs. L_soil[i] [$J \cdot g^{-1} \cdot °C^{-1}$] {2.37272}
Temp[i] =	soil temperature [°C + 273.16]
Molwt_H2O =	mol weight of water [$g \cdot mol^{-1}$] {18}
R =	gas constant [$J \cdot mol^{-1} \cdot K^{-1}$] {8.314}
SVP_soil[i] =	saturation soil vapour pressure (eqn 5.8/4) [kPa]

Enhancement factor for water vapour transport (F_vp[i])

Observations provide evidence that water vapour transport under temperature gradients is only calculated realistically by multiplication with a dimensionless water vapour enhancement factor (between 1 and 12) (see Campbell, 1985, p. 101). The flow of water vapour under temperature gradients (non-isothermal conditions) is presented in more detail in Section 5.10. However, to calculate the effect of water vapour transport on thermal conductivity, it is necessary to include this enhancement factor here. The dimensionless enhancement factor is a function of soil water content (VWC[i]) and is calculated with the notation given by Campbell (1985, p. 101):

$$F\_vp[i] = Avp + (Bvp * VWC[i]) - (Avp - Dvp) \cdot e^{\left\{-\left[(Cvp \cdot VWC[i])^{Evp}\right]\right\}} \tag{5.8/9}$$

with:

$$Cvp = Cvp1 + \frac{1}{\sqrt{Cvp2 \cdot f\_clay[i]}}$$

where:

Avp =	parameter for the enhancement factor [–] {9.5}
Bvp =	parameter for the enhancement factor [–] {6}
Dvp =	parameter for the enhancement factor [–] {1}
Evp =	parameter for the enhancement factor [–] {4}
Cvp1 =	parameter for the Cvp calculation [–] {1}
Cvp2 =	parameter for the Cvp calculation [–] {2.6}
VWC[i] =	volumetric water content of each soil layer [$m^3 \cdot m^{-3}$]
f_clay[i] =	clay fraction of each soil layer (see eqn 3.3/1) [–]

Application in ModelMaker

To add the notations described above the <variables> presented in Fig. 5.8 (bold components) have to be added to the already existing model (*Mod5–1b.mod*).

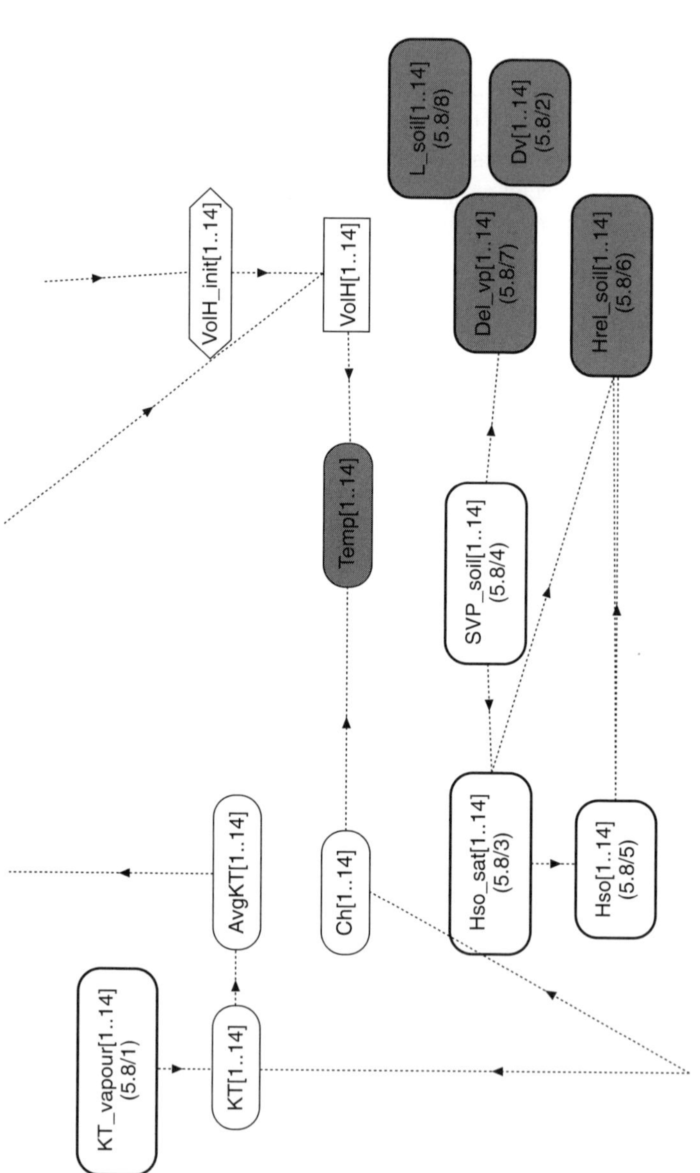

Fig. 5.8. Additions (bold components) to model *Mod5-1b.mod* to calculate the contribution of water vapour flow to soil thermal conductivity (*Mod5-1c.mod*).

They are added between the temperature and water section of the existing model because the new notations are linked to components in both sections. Since the soil temperature (Temp[i]) is used in most of the new components, it was decided to define this variable <globally> (green hatching).

To compare the output of the model with and without the contribution of water vapour flow on thermal conductivity, the following statement is added to the <independent event> 'Choice':

```
KT_vapChoice=GetChoice("Calculate water vapour flow
       contribution to KT?(Y/N)","
```

The <variable> 'KT_vapour[1..14]' is defined conditionally according to the <define> 'KT_vapChoice', which holds the user input to the above statement (*Y*=1, *N*=0). If the user desires not to calculate the water vapour contribution to the thermal conductivity, the <variable> 'KT_vapour[1..14]' is set to zero.

The effect of the water vapour flow on the thermal conductivity in the top soil layer ($i = 1$) is presented for a 10-day simulation (using data file *Inp5–1b.txt*) as a graph of thermal conductivity (KT [1]) versus time (Fig. 5.9). The user choice for this simulation run is:

RainfallChoice = 0 (no rainfall)
BottomChoice = 3 (free drainage)

The model was executed with both:

KT_vapChoice = 1 (calculate KT_vapour[i], eqn 5.8/1)

and:

KT_vapChoice = 0 (KT_vapour[i] = 0).

The thermal conductivity is slightly higher if water vapour flow is considered (Fig. 5.9). The contribution gets larger with decreasing soil volumetric water contents, which is due to the increasing contribution of the water vapour flow, compared with the liquid water flow, to the overall soil water movement (Hillel, 1998, p. 316–323). However, the effect of the additional thermal conductivity (KT_vapour[i]) on soil temperature (Temp[i],) and water content (VWC[i],) is negligible. This can be seen for the top soil layer if Temp[1] and VWC[1] with and without calculation of KT_vapour[1] are plotted in the same way as in Fig. 5.9.

The water vapour movement contributes little to the overall water flow in soil but becomes more distinct when the liquid water content is very low. To test this effect, we derive a notation for the contribution of water vapour flow to the overall water movement in soil.

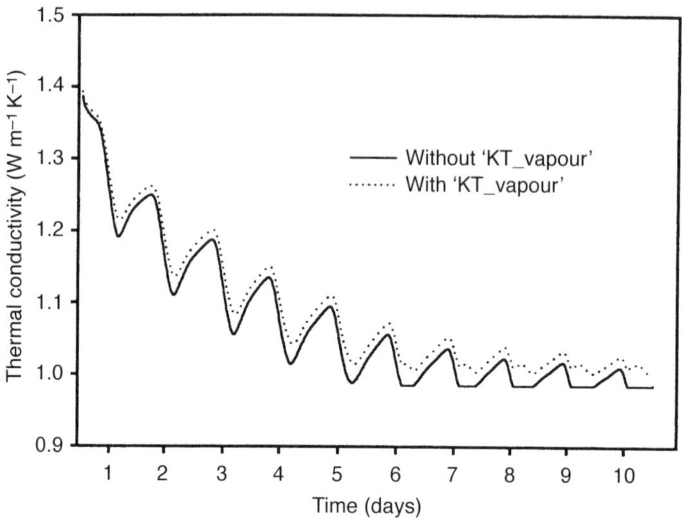

Fig. 5.9. Thermal conductivity of top soil layer ($i = 1$) calculated with and without the contribution of water vapour flow (calculated with *Mod5-1c.mod,* data: *Inp5-1b.txt*).

5.9 Contribution of Water Vapour Movement to Soil Water Flow

Similarly to the contribution of water vapour flow to thermal conductivity, it is also possible to define the contribution of water vapour flow to the overall water movement in soil. It is possible to derive a notation for water vapour flow which is similar to that for the flow of liquid water in soil (eqn 4.5/4). The flow of water vapour in soil (Fw_vapour[i]) can be calculated as (eqn 4.5/4; see also Campbell, 1985, eqn 9.6):

$$\text{Fw\_vapour}[i] = \text{Kw\_vapour}[i] \cdot \frac{\text{HPot}[i-1] - \text{HPot}[i]}{\text{Dist}[i]} \qquad (5.9/1)$$

where:

Kw_vapour[i] = hydraulic conductivity for water vapour flow [$m \cdot s^{-1}$]

Combining eqns 4.5/4 and 5.9/1, we can write the overall flow of water (liquid water + water vapour) as:

$$\mathrm{Fw[i]} = (\mathrm{Kw[i]} + \mathrm{Kw\_vapour[i]}) \cdot \frac{\mathrm{HPot[i-1]} - \mathrm{HPot[i]}}{\mathrm{Dist[i]}} \tag{5.9/2}$$

The hydraulic conductivity due to water vapour flow (Kw_vapour[i]) can be calculated with the notation presented by Campbell (1985, eqn 9.7, in conjunction with eqn 9.2):

$$\mathrm{Kw\_vapour[i]} = \frac{\mathrm{Dv[i]} \cdot \mathrm{Hso[i]} \cdot \mathrm{Molwt\_H2O}}{\mathrm{R} \cdot (\mathrm{Temp[i]} + 273.16)} \cdot \frac{\mathrm{gravity}}{1000} \cdot \frac{1}{\mathrm{Dens\_water} \cdot 1000000} \tag{5.9/3}$$

where:

Dv[i] =	diffusivity of water vapour (eqn 5.8/2) $[\mathrm{m}^2 \cdot \mathrm{s}^{-1}]$
Hso[i] =	soil vapour concentration (eqn 5.8/5) $[\mathrm{g} \cdot \mathrm{m}^{-3}]$
Temp[i] =	soil temperature[°C + 273.16]
Molwt_H2O =	mol. weight of water $[\mathrm{g} \cdot \mathrm{mol}^{-1}]$ {18}
R =	gas constant $[\mathrm{J} \cdot \mathrm{mol}^{-1} \cdot \mathrm{K}^{-1}]$ {8.314}
gravity =	acceleration due to gravity $[\mathrm{m} \cdot \mathrm{s}^{-2}]$ {9.81}
Dens_water =	density of water $[\mathrm{g} \cdot \mathrm{cm}^{-3}]$ {1}

The terms $\frac{\mathrm{gravity}}{1000}$ and $\frac{1}{\mathrm{Dens\_water} \cdot 1000000}$ have to be included in eqn 5.9/3 to obtain the desired dimensions for Kw_vapour[i] (units: m s^{-1}).

The dimensions of Kw_vapour[i] are:

$$[\mathrm{Kw\_vapour[i]}] = \frac{\left[\frac{\mathrm{m}^2}{\mathrm{s}}\right] \cdot \left[\frac{\mathrm{g}}{\mathrm{m}^3}\right] \cdot \left[\frac{\mathrm{g}}{\mathrm{mol}}\right]}{\left[\frac{\mathrm{J}}{\mathrm{mol} \cdot \mathrm{K}}\right] \cdot [\mathrm{K}]} \cdot \frac{\left[\frac{\mathrm{m}}{\mathrm{s}^2}\right]}{1000} \cdot \frac{1}{\left[\frac{\mathrm{g}}{\mathrm{m}^3}\right]}$$

$$\left(\text{recall: } [\mathrm{J}] = [\mathrm{N} \cdot \mathrm{m}] = \left[\frac{\mathrm{kg} \cdot \mathrm{m}}{\mathrm{s}^2} \cdot \mathrm{m}\right]\right)$$

(multiply 'dens_water' in: $[\mathrm{g} \cdot \mathrm{cm}^{-3}]$ by 1,000,000 to obtain $[\mathrm{g} \cdot \mathrm{cm}^{-3}]$)
(division by '1000' in the second term converts 'g' into 'kg')

We can simplify the above units to:

$$= \frac{\left[\frac{m^2}{s}\right]\cdot\left[\frac{g}{m^3}\right]}{[J]}\cdot\left[\frac{kg\cdot m}{s^2}\right]\cdot\left[\frac{m^3}{g}\right]$$

$$= \frac{\left[\frac{m^2}{s}\right]}{[J]}\cdot\left[\frac{kg\cdot m}{s^2}\right]$$

$$= \frac{\left[\frac{m}{s}\right]}{[J]}\cdot[m]\cdot\left[\frac{kg\cdot m}{s^2}\right]\frac{\left[\frac{m}{s}\right]}{[J]}\cdot[J]$$

$$= \underline{\underline{\left[\frac{m}{s}\right]}}$$

The conductivity for water vapour at the bottom just outside the soil profile (Kw_vapour[15])) is set equal to the conductivity of the previous soil layer (for $i = 15$):

$$Kw\_vapour[15] = \frac{Dv[i-1]\cdot Hso[i-1]\cdot Molwt\_H2O}{R\cdot(Temp[i-1]+273.16)} \cdot\frac{gravity}{1000}\cdot\frac{1}{Dens\_water\cdot 1000000}$$

Application in ModelMaker

The only change to the previous model (*Mod5–1c.mod*) is the addition of the <variable> 'KT_vapour[i]' (Fig. 5.10).

To allow the user to perform simulations with and without the contribution of Kw_vapour[i], an additional user-definable statement is added to the <independent event> 'Choice':

```
Kw_vapChoice=GetChoice("Calculate water vapour flow
        contribution to Kw?(Y/N)"," KW water vapour
        flow",Kw_vapChoice);
```

The variable Kw_vapour[1..15] is calculated conditionally according to the value held in the <define> 'Kw_vapChoice'. If the user answers 'No' to the above question, Kw_vapour[1..15] is set to zero; otherwise the calculation is carried out according to eqn 5.9/3. The complete model is presented *in Mod5–1d.mod.*

Before presenting some outputs, one final addition to the model will be made. In the development of the model for water flow (Chapter 4) and water vapour movement (this section), we did not consider the effect of temperature gradients on water movement in the soil. This additional effect on liquid and water vapour flow is presented in the next section.

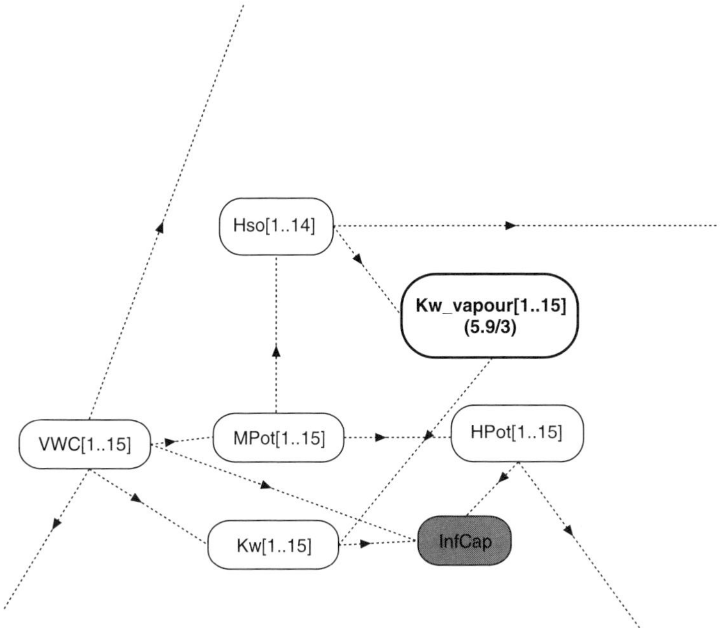

Fig. 5.10. Notation (bold component) for calculating the contribution of water vapour flow to the water movement in soil (*Mod5-1d.mod*).

5.10 Non-isothermal Water and Water Vapour Flow in Soil

If temperature rises, the pressure in both the liquid and the gaseous phase increases and water tends to move from warm to cold regions in the soil (Campbell, 1985, pp. 101–103). This phenomenon is particularly important in temperate climates during winter, when distinct freezing zones develop on the soil surface while soil temperatures deeper in the profile are considerably higher. The effect of temperature gradients on water movement in soil will be considered separately for liquid water and water vapour flow in soil.

Thermally induced liquid water flow

The notation for thermally induced water flow is given by Campbell (1985, eqn 9.8, p.101). The non-isothermal flow equation has to take into account the temperature gradient within the soil profile. By analogy to the water flow equation (eqn 4.5/4), the non-isothermal flow is defined by (see eqn 4.4/1):

$$\mathrm{Fw} = -\mathrm{Kw} \cdot \frac{d\mathrm{HPot}}{d\mathrm{T}} \cdot \frac{d\mathrm{T}}{dz} \qquad (5.10/1)$$

or in the difference notation (eqn 4.5/4):

$$\mathrm{Fw[i] = AvgKw[i] \cdot \left[\frac{HPot[i-1] - HPot[i]}{Temp[i-1] - Temp[i]} \right] \cdot \left[\frac{Temp[i-1] - Temp[i]}{Dist[i]} \right]} \tag{5.10/2}$$

or as used in the notation later on:

$$\mathrm{Fw[i] = AvgKw[i] \cdot dHPot\_dT[i] \cdot dT\_dz[i]} \tag{5.10/3}$$

with:

$$\mathrm{dHPot\_dT[i] = \left[\frac{HPot[i-1] - HPot[i]}{Temp[i-1] - Temp[i]} \right]} \tag{5.10/4}$$

$$\mathrm{dT\_dz[i] = \left[\frac{Temp[i-1] - Temp[i]}{Dist[i]} \right]} \tag{5.10/5}$$

where:

AvgKw[i] = average hydraulic conductivity between adjacent soil layers (eqn 4.5/11) [m • s^{-1}]
HPot[i] = hydraulic potential (see eqns 4.5/3–4.5/5) [m]
Temp[i] = soil temperature [°C]
Dist[i] = flow distance between adjacent layers (eqn 3.5/1) [m]

The dimensions for the non-isothermal water flow are:

$$\mathrm{[Fw] = \left[\frac{m}{s} \right] \cdot \left[\frac{m}{°C} \right] \cdot \left[\frac{°C}{m} \right] = \left[\frac{m}{s} \right]}$$

Thermally induced water vapour flow

Thermally induced water vapour flow (Fw_vapTemp[i]) can be calculated by (see Campbell, 1985, eqn 9.10):

$$\mathrm{Fw\_vapTemp[i] = Dv[i] \cdot Hrel\_soil[i] \cdot Del\_vp[i] \cdot dT\_dz[i] \cdot X} \tag{5.10/6}$$

with:

$$\mathrm{X = \frac{Molwt\_H2O}{R \cdot (Temp[i] + 273.16) \cdot Dens\_water \cdot 1000000}}$$

where:

Dv[i] =	diffusivity of water vapour (eqn 5.8/2) [$m^2 \cdot s^{-1}$]
Del_vp[i] =	change of saturation vapour pressure (eqn5.8/7) [$Pa \cdot K^{-1}$]
Hrel_soil[i] =	relative soil humidity (eqn 5.8/6) [–]
Temp[i] =	soil temperature[°C + 273.16]
Molwt_H2O =	mol weight of water [$g \cdot mol^{-1}$] {18}
R =	gas constant [$J \cdot mol^{-1} \cdot K^{-1}$] {8.314}
gravity =	acceleration due to gravity [$m \cdot s^{-2}$] {9.81}
Dens_water =	density of water [$g \cdot cm^{-3}$] {1}
dT_dz[i] =	temperature gradient in soil (eqn 5.10/5) [$°C \cdot m^{-1}$] or [$K \cdot m^{-1}$]
X =	factor to convert units: m s^{-1}

The units for the flow of 'Fw_vapTemp[i]' are:

$$[\text{Fw\_vapTemp[i]}] = \left[\frac{m^2}{s}\right] \cdot [-] \cdot \left[\frac{Pa}{K}\right] \cdot \left[\frac{K}{m}\right] \text{(X component see below)}$$

$$= \left[\frac{m^2}{s}\right]\left[\frac{\frac{J}{m^3}}{K}\right] \cdot \left[\frac{K}{m}\right]$$

$$= \left[\frac{m^2}{s}\right]\left[\frac{J}{m^3}\right] \cdot \left[\frac{1}{m}\right]$$

$$= \left[\frac{J}{s \cdot m^2}\right]$$

and for the *X* component

$$[X] = \frac{\left[\frac{g}{mol}\right]}{\left[\frac{J}{mol \cdot K}\right] \cdot [K] \cdot \left[\frac{g}{m^3}\right]}$$

$$= \frac{[g]}{[J] \cdot \left[\frac{g}{m^3}\right]}$$

$$= \left[\frac{m^3}{J}\right]$$

combined:

$$= \left[\frac{\mathrm{J}}{\mathrm{s}\cdot\mathrm{m}^2}\right]\cdot\left[\frac{\mathrm{m}^3}{\mathrm{J}}\right]$$

$$= \underline{\underline{\left[\frac{\mathrm{m}}{\mathrm{s}}\right]}}$$

The combined effect of thermally induced liquid water and water vapour flow can be calculated by adding eqns 5.10/2 and 5.10/6 together. Therefore the final equation for water flow in soil under non-isothermal conditions would be (see eqns 5.10/2 and 5.10/6):

$$\text{Fw[i]} = \text{AvgKw[i]} \cdot \text{dHPot\_dT[i]} \cdot \text{dT\_dz[i]} + \text{Fw\_vapTemp[i]} \qquad (5.10/7)$$

Application in ModelMaker

The two gradients $\dfrac{\mathrm{dHPot}}{\mathrm{d}T}$ and $\dfrac{\mathrm{d}T}{\mathrm{d}z}$ which are needed for the calculation of thermally induced water movement are calculated separately in the <variables> (global) 'dHPot_dT[i]' and 'dT_dz[i]' respectively. The flow of water out or into the top soil layer is defined by the energy balance calculations, and at the bottom of the soil profile we assumed that no temperature gradients exist (see eqn 3.5/4). Thus, the two variables only have to be defined for soil layers 2–14 (dHPot_dT[2..14], dT_dz[2..14]). For the thermally induced water vapour flow, an additional <variable> 'Fw_Temp[i]', is added. The notation which has to be added to the previous model, *Mod5–1d.mod*, is presented in Fig. 5.11 (bold components). The final model is presented in *Mod5–1e.mod.*

The user can decide via the two user-definable statements in the <independent event> 'Choice' whether the thermally induced liquid water and water vapour flows should be calculated with the model or not. The statements are:

```
Fw_tempChoice=GetChoice("Calculate non-isothermal WATER
        flow?(Y/N)"," Non-Isothermal water
        flow",Fw_tempChoice);
Fw_vapTempChoice=GetChoice("Calculate non-isothermal WATER
        VAPOUR flow?(Y/N)"," Non-Isothermal water vapour
        flow",Fw_vapTempChoice);
```

The <define> 'Fw_TempChoice' holds the decision for the liquid water flow and the <define> 'Fw_vapTempChoice' holds the decision for the water vapour flow. The <variable> 'Fw_vapTemp[2..14]' is made conditional in response to the <define> 'Fw_vapTempChoice' (Yes(1) = calculated according to eqn 5.10/6, No(0) = set to zero).

The <variable> 'Fw[1..15]' is made conditional in response to Fw_TempChoice: if the user decides to calculate thermally induced liquid flow, Fw[i] is calculated according to eqn 5.10/7.

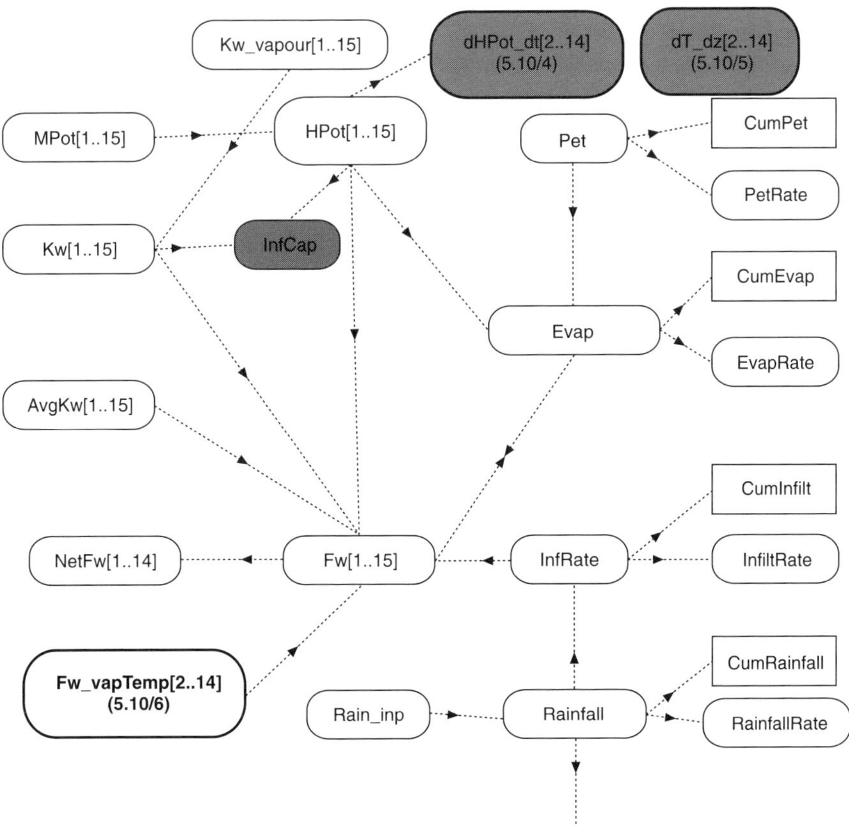

Fig. 5.11. Additional notation (bold) to calculate non-isothermal flow of liquid water and water vapour (*Mod5-1e.mod*).

Running the model

The model can be executed in various ways according to the many user decisions. It is advisable to try out as many user-definable scenarios as possible to see the effect of the various calculation procedures. Non-isothermal calculations have only a negligible effect on temperature and water regimes in the soil. This is most likely to be related to the input variables of the current data file (*Inp5–1b.txt*).

Of the numerous model outputs, the actual and potential evaporation and the related soil volumetric water dynamics are presented for a 10-day simulation period in Fig. 5.12. The user inputs for this simulation run are:

- no rainfall;
- free drainage at bottom boundary;

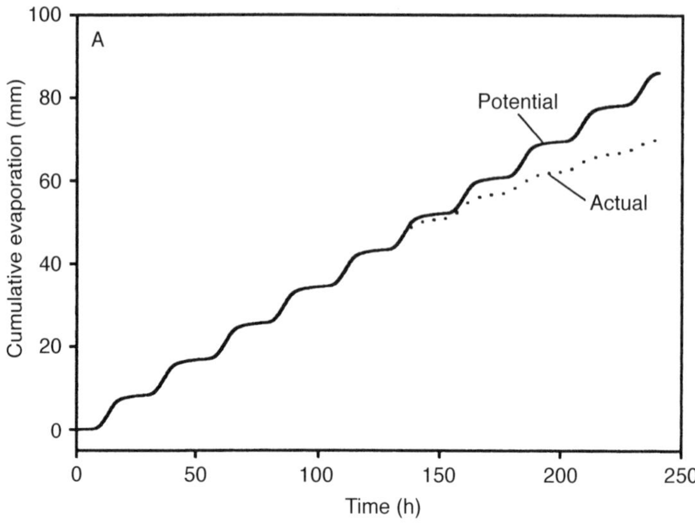

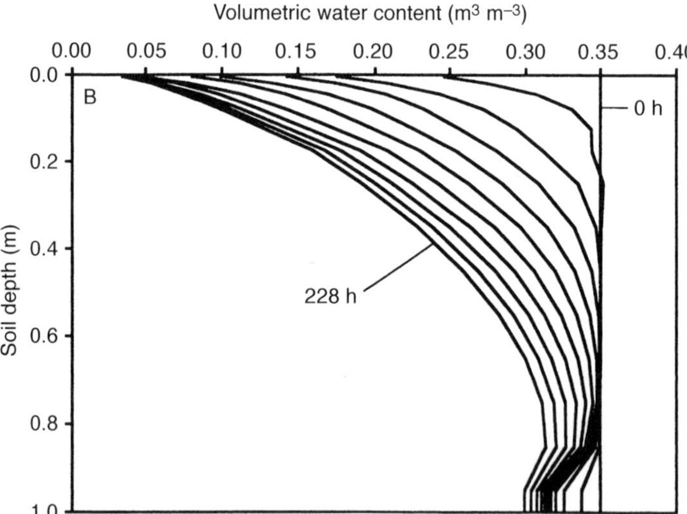

Fig. 5.12. Cumulative evaporation (A) and soil profile volumetric water content (24 h increments) (B) for a 10-day simulation using *Mod5-1e.mod* (for user settings, see text).

- calculate effect of water vapour movement on KT and Kw values;
- calculate non-isothermal movements for liquid water and water vapour.

The soil profile representation for volumetric water content (Fig. 5.12B) was produced according to the procedure described in Section 3.4 (p. 171). The soil volumetric water contents, which are plotted in 24 h increments, represent the data at the middle of each day (1200 h – noon) (Fig. 5.12B). Potential and actual evaporation are plotted cumulatively (Fig. 5.12A).

As already discussed earlier, complex models such as the ones presented in this chapter may be executed very slowly within ModelMaker. To test whether such models are still calculated within an acceptable time, the simulation times for 1 simulated day were determined for all five models developed in this chapter. Three 'Pentium' computers were used for this test: a 100 MHz Laptop with 40 Mb RAM, a 166 MHz MMX with 32 Mb RAM and a 333 MHz Pentium II with 64 Mb RAM. The computing times for 1 simulated day are summarized in Table 5.1. The results show that simulation times can be considerably reduced with fast Pentium computers. If users are interested in running long simulations with one of the models presented in this chapter, it is advisable to use a fast computer. However, the results show that complex mathematics with demanding computations such as those in problems in environmental physics can be tackled with ModelMaker.

5.11 Future Development

Future developments to the model may include refined calculation procedures for the boundary layer dynamics, including the effect of cloudiness on long-wave radiation (Monteith and Unsworth, 1990, pp. 51–53). In addition, an extended energy balance is needed if cropped soils are considered. The necessary additions to the energy balance for cropped soil are outlined in Monteith and Unsworth (1990, pp. 231–263).

However, this last addition requires the understanding of processes such as photosynthesis, plant growth and soil–plant–atmosphere relationships. Particularly important are the crop–water relationships. These processes are described in the next chapter in the context of the development of a crop growth model.

Table 5.1. Times (min:s) for simulating 1 day with the various models developed in Chapter 5.

	Mod5-1a	Mod5-1b	Mod5-1c	Mod5-1d	Mod5-1e
100 Mhz	0:40	1:15	2:10	2:40	2:50
166 Mhz	0:11	0:21	0:35	0:40	0:45
333 Mhz	0:06	0:11	0:19	0:21	0:24

Plant Growth 6

6.1 Introduction

One of the most important processes on this earth is the photosynthetic activity of green plants. Without this process, most life forms on the surface of the earth (including human life) would not exist. Photosynthesis is unique among biological processes in that it converts solar energy into organic compounds. These compounds are subsequently consumed by organisms either directly or indirectly. Furthermore, photosynthesis creates oxygen, which is essential for those life forms with aerobic respiratory metabolism. Photosynthesis can be defined as the light-dependent synthesis of organic metabolites from gaseous CO_2 (Charles-Edwards, 1981, p. 7). However, the acquisition of carbon by the plant is linked to other processes, such as assimilation of nitrogen (N), phosphorus (P), potassium (K) or the numerous micronutrients (Mg, Ca, Cu, Mn, Fe, etc.). In the last century, Justus von Liebig formulated a minimum law which states that plant growth is governed by the concentration of that element which is least available. Plant growth occurs above ground as stem and leaf growth and below ground as root growth. While carbon (in the form of CO_2) is obtained from photosynthesis, the other elements and water are mostly taken up by the roots. However, the various parts of the plants are only able to grow if all elements are present in adequate amounts. Therefore, the process of how plants distribute resources to the above- and below-ground plant parts must be understood. Growth, ageing and senescence of various plant parts lead to plant residues, which can be mineralized and made available to the plants again and which involve an interaction with the soil microbial biomass. Thus, a plant growth model must also consider some of the processes which have been described in the previous chapters.

Temperature and moisture relations within the soil–plant–atmosphere system determine the rates at which processes proceed in the plant. Plants can be considered as an additional 'channel' for water transport between the soil and the atmosphere. The process of water loss by plants, called 'transpiration', occurs in relation to the environmental conditions. Energy relationships at the soil–atmosphere interface (Chapter 5) can only be fully understood once the contribution of plants is included.

6.2 Conceptual Plant Growth Model

The most important processes and relationships which are needed to model vegetative plant growth are presented in Fig. 6.1.

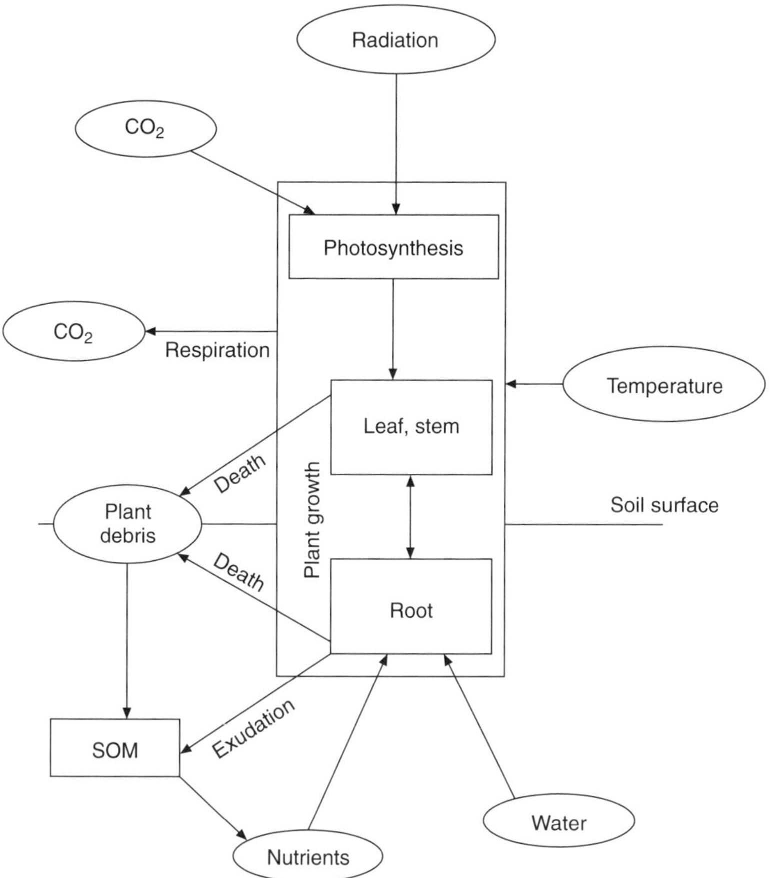

Fig. 6.1. Important processes (squares) and pools (round components) and their relationships to vegetative plant growth.

Even with this simple scheme, it is apparent that there is a need to integrate various processes to model plant growth successfully. Some of these processes have already been considered in detail here. For instance, the mineralization of plant debris, which arises from decaying plant material, and its relationship to soil organic matter (SOM) turnover were the subject of Chapter 2. The soil temperature, soil water and energy relations were considered in Chapters 3–5.

The various processes depicted in the conceptual plant growth model (Fig. 6.1) are considered in the forthcoming sections in detail.

Since photosynthesis is essential for plant growth, we begin with the development of a model for leaf and canopy photosynthesis. The model will then be extended in a stepwise fashion to cater for growth of leaves, stems and roots, as well as decay of plant material and allocation of C and N within the plant. The effect of soil and air temperature and the soil moisture status on the various processes are the subject of the final section.

The model follows closely the plant growth model developed by Thornley and colleagues (see Thornley, 1998). This group, which worked at the Hurley Grassland Research Institute (Hurley, Maidenhead, UK), published numerous papers on the development of a mechanistic pasture model. No other plant growth model has been published in such detail. Moreover, their step-by-step approach in the model development can almost be followed by reading their publications in a chronological order. Over the last 20 years, they added more and more elements to the model and published them bit by bit, which makes it easy for the reader to follow the model development. While working through this chapter, it is possible to go back to the original papers and even incorporate aspects which are not considered in this chapter. The current stage of their model (called the 'Hurley Pasture Model') has recently been summarized in the book *Grassland Dynamics – an Ecosystem Simulation Model* (Thornley, 1998). This book is recommended if the reader wishes to read a detailed description of the entire model.

6.3 Photosynthesis

A key process of any plant growth model is the realistic representation of photosynthesis. Various relationships have been considered over the years, which range from very simple to complex biochemical models (Thornley and Johnson, 1990, Chapter 9). In order to become familiar with the concept of photosynthesis, we first develop a basic model for this process which is used in many plant growth models (e.g. Johnson and Thornley, 1983). This model is also referred to as a 'rectangular hyperbola'. Based on the basic principles, a more realistic photosynthesis model (called a 'non-rectangular hyperbola') is presented, which is used as the 'growth engine' for the plant growth model developed later on. The development of both photosynthesis models is mainly based on the comprehensive and easy-to-follow derivations by Thornley and Johnson (1990, Chapters 9 and 10). For more background information on photosynthesis, the study of this text is highly recommended.

6.3.1 Simple model

Basic photosynthesis model (rectangular hyperbola)
The basic biochemical equation for photosynthesis is:

$$12H_2O + 6CO_2 \xrightarrow{\text{light}} C_6H_{12}O_6 + 6H_2O + 6O_2 \qquad (6.3/1)$$

Some processes occur under light and are therefore referred to as the 'light reactions of photosynthesis'. There are also 'dark reactions of photosynthesis', which are important in our modelling approach for the formation of carbohydrates (right-hand side of eqn 6.3/1) and the respiratory activity associated with the various biochemical reactions. A detailed description of photosynthesis, including the biochemical processes during 'dark' and 'light' reactions, is presented in Thornley and Johnson (1990, Chapter 9).

The development of the basic photosynthetic equation is similar to the derivation of Michaelis–Menten kinetics (Section 1.4.6). A simplified illustration of photosynthesis may be given by considering that light will react with a molecular species (e.g. nicotinamide adenine dinucleotide phosphate ($NADH^+$)) to gain an activated state (NADH) plus energy (ATP), which can be mathematically represented by (see also Thornley and Johnson, 1990, Chapter 9):

$$M + I \xrightarrow{k1} M^* \qquad (6.3/2)$$

where

$M =$ molecular species
$M^* =$ activated form of M
$k1 =$ rate constant for the conversion from M to M^*
$I =$ radiation

The formation rate of the activated molecule is therefore:

$$\frac{dM^*}{dt} = k1 \cdot M \cdot I \qquad (6.3/3)$$

Within the leaf the activated molecules (M^*) react with carbon dioxide CO_2 to form carbohydrates ('CH2O', e.g. $C_6H_{12}O_6$), which can be described by:

$$M^* + Ci \xrightarrow{k2} M + CH2O \qquad (6.3/4)$$

where:

$Ci =$ carbon dioxide concentration (internal) at the reaction site
$CH2O =$ carbohydrate
$k2 =$ rate constant for the conversion from 'M^* + Ci' to 'M + CH2O'

The rate at which the activated molecule is oxidized can be calculated as:

$$\frac{dM^*}{dt} = -k2 \cdot M^* \cdot Ci \tag{6.3/5}$$

The overall change of the activated molecule can be calculated by combining eqns 6.3/3 and 6.3/5:

$$\frac{dM^*}{dt} = k1 \cdot M \cdot I - k2 \cdot M^* \cdot Ci \tag{6.3/6}$$

Assuming that the total amount of the molecule (M) stays constant, i.e.:

$$M_{tot} = M + M^* \tag{6.3/7}$$

we can rewrite eqn 6.3/6 as (eliminating M):

$$\frac{dM^*}{dt} = k1 \cdot \left(M_{tot} - M^*\right) \cdot I - k2 \cdot M^* \cdot Ci \tag{6.3/8}$$

In steady state, when the rate of change of the activated molecule (M^*) is zero (i.e. $\frac{dM}{dt} = 0$) equation 6.3/8 reduces to:

$$k1 \cdot \left(M_{tot} - M^*\right) \cdot I - k2 \cdot M^* \cdot Ci = 0 \tag{6.3/9}$$

Rearranging eqn 6.3/9 for M^* we obtain ($k1 \cdot I \cdot M_{tot} - k1 \cdot I \cdot M^* - k2 \cdot M^* \cdot Ci = 0 \rightarrow k1 \cdot I \cdot M_{tot} - M^* \cdot (k1 \cdot I + k2 \cdot Ci) = 0 \rightarrow k1 \cdot I \cdot M_{tot} = M^* \cdot (k1 \cdot I + k2 \cdot Ci)$):

$$M^* = \frac{k1 \cdot I \cdot M_{tot}}{k1 \cdot I + k2 \cdot Ci} \tag{6.3/10}$$

The maximum rate of photosynthesis (P_{max}) for the formation of carbohydrates can be derived from eqn 6.3/4 as:

$$P_{max} = k2 \cdot M^* \cdot Ci \tag{6.3/11}$$

Not all carbon dioxide might be actually used to build carbohydrates. Therefore we multiply the maximum photosynthesis with a utilization factor (fu) to obtain the actual rate of photosynthesis or formation of carbohydrates (P):

$$P = fu \cdot k2 \cdot M^* \cdot Ci \tag{6.3/12}$$

Substituting eqn 6.3/10 into eqn 6.3/12, we obtain:

$$P = \frac{k1 \cdot I \cdot M_{tot} \cdot fu \cdot k2 \cdot Ci}{k1 \cdot I + k2 \cdot Ci}$$

After multiplication of the nominator and denominator with 'fu • Mtot', we obtain:

$$P = \frac{fu \cdot k1 \cdot M_{tot} \cdot I \cdot Ci \cdot fu \cdot k2 \cdot M_{tot}}{fu \cdot k1 \cdot M_{tot} \cdot I + Ci \cdot fu \cdot k2 \cdot M_{tot}} \tag{6.3/13}$$

To simplify the equation, we introduce the photochemical efficiency (α) which is the initial gradient of the photosynthesis light response:

$$\alpha = fu \cdot k1 \cdot M_{tot} \tag{6.3/14}$$

and the carboxylation resistance (*rx*):

$$rx = \frac{1}{fu \cdot k2 \cdot M_{tot}} \tag{6.3/15}$$

which represents a resistance for CO_2 transport within the cells to the photosynthetic sites.

Assuming that the internal leaf CO_2 concentration (*Ci*) does not affect the rate of photosynthesis, we can define the maximum rate (*Pm*) as (concentration divided by resistance):

$$Pm = \frac{Ci}{rx} = \frac{Ci}{\dfrac{1}{fu \cdot k2 \cdot M_{tot}}} = Ci \cdot fu \cdot k2 \cdot M_{tot} \tag{6.3/16}$$

Equation 6.3/13 can now be simplified using eqns 6.3/14–6.3/16 to:

$$P = \frac{\alpha \cdot I \cdot Pm}{\alpha \cdot I + Pm} \tag{6.3/17}$$

where:

$\alpha =$ photochemical efficiency (in the ModelMaker model it will be called 'alpha') [kg $CO_2 \cdot J^{-1}$] $\{1 \times 10^{-8}\}$
$I =$ radiation (PAR) [$W \cdot m^{-2}$] or [$J \cdot s^{-1} \cdot m^{-2}$]
$Pm =$ maximum rate of photosynthesis [kg $CO_2 \cdot m^{-2}$ leaf $\cdot s^{-1}$] $\{1 \times 10^{-6}\}$

The units for 'Pm' are in kg CO_2 m^{-2} s^{-1}:

$$[P] = \frac{\left[\dfrac{kg\ CO_2}{J}\right] \cdot \left[\dfrac{J}{m^2\ leaf \cdot s}\right] \cdot \left[\dfrac{kg\ CO_2}{m^2\ leaf \cdot s}\right]}{\left[\dfrac{kg\ CO_2}{J}\right] \cdot \left[\dfrac{J}{m^2\ leaf \cdot s}\right] + \left[\dfrac{kg\ CO_2}{m^2\ leaf \cdot s}\right]}$$

$$= \frac{\left[\dfrac{kg\ CO_2}{m^2\ leaf \cdot s}\right] \cdot \left[\dfrac{kg\ CO_2}{m^2\ leaf \cdot s}\right]}{\left[\dfrac{kg\ CO_2}{m^2\ leaf \cdot s}\right] + \left[\dfrac{kg\ CO_2}{m^2\ leaf \cdot s}\right]}$$

$$= \frac{\left[\frac{\text{kg CO}_2}{\text{m}^2\ \text{leaf}\cdot\text{s}}\right]\cdot\left[\frac{\text{kg CO}_2}{\text{m}^2\ \text{leaf}\cdot\text{s}}\right]}{\left[\frac{\text{kg CO}_2}{\text{m}^2\ \text{leaf}\cdot\text{s}}\right]} = \left[\frac{\text{kg CO}_2}{\text{m}^2\ \text{leaf}\cdot\text{s}}\right]$$

Equation 6.3/17 describes a rectangular hyperbola for the relationship between the rate of photosynthesis (*P*) and radiation (*I*). This formula has been used extensively to describe leaf photosynthesis. However, the main disadvantage of this formulation is that it does not take the internal concentration of CO_2 (*Ci*) into account. Furthermore, the internal CO_2 concentration is dependent on the external CO_2 concentration (*Ca*) surrounding the leaf where photosynthesis takes place. However, before we include these additional aspects, we first have a look at how photosynthesis from the leaf level can be extended to the entire plant using the rectangular hyperbola model.

Photosynthesis of the entire canopy

The canopy can be defined as a community of plants with similar features. It is therefore not the single leaf or plant we have to consider but all the plants together. With eqn 6.3/17, we have developed an expression relating light intensity to photosynthetic activity. Everyone who has seen a pasture will immediately comment that the light relations for the various parts of the plants within a canopy are quite different. While on top of the canopy the leaves are exposed to full radiation, the light intensity will gradually decline towards the soil surface. To account for these light relationships within a canopy, as a first approximation we use a Beer's law analogy, which predicts an exponential extinction of light intensity (*I*) through the canopy:

$$I = I_0 \cdot e^{-k \cdot LAI} \qquad (6.3/18)$$

where:

I_0 = light intensity on top of the canopy $[J \cdot s^{-1} \cdot m^{-2}]$
k = canopy extinction coefficient for light [–] {0.5}
LAI = leaf area index $[m^2\ leaf \cdot m^{-2} ground]$

The leaf area index (LAI) is an important feature in this expression, because it will determine the shape of the exponential extinction. The total LAI is determined at the ground and is a measure of the entire leaf area above ground in relation to the ground area. Light incident on a leaf can either be absorbed (*a*), reflected (*r*) or transmitted (*t*), so that:

$$a + r + t = 1$$

The fraction of the light being absorbed by the leaves can be calculated as:

$$a = 1 - (r + t)$$

Adding the fraction of light transmitted (t) and reflected downwards (r) together, we can define a factor 'm' which combines transmittance and downward reflection, so that the expression for absorbed radiation reduces to:

$$a = 1 - m$$
$$\{m = 0.1\}$$

The change in light intensity ΔI over an element of leaf area index ΔLAI can be expressed as:

$$-\Delta I = (1 - m) \cdot I \cdot \Delta LAI$$

Rearranged for I and expressed in differential form, the last equation becomes:

$$I = -\frac{dI}{dLAI} \cdot \frac{1}{(1-m)} \qquad (6.3/19)$$

The differential $\frac{dI}{dLAI}$ is also obtained by differentiating eqn 6.3/18 for I with respect to LAI (see Section 1.2.3 for details on differentiating):

$$\frac{dI}{dLAI} = -k \cdot I_0 \cdot e^{-k \cdot LAI} \qquad (6.3/20)$$

Substituting eqn 6.3/20 into eqn 6.3/19, we obtain:

$$I = \left(\frac{k \cdot I_0}{1-m}\right) \cdot e^{-k \cdot LAI} \qquad (6.3/21)$$

This last expression is also referred to as the Monsi–Saeki relationship (Charles-Edwards, 1981, p. 57).

The photosynthetic activity for all leaves of the entire canopy (Pc) can now be calculated by integrating the photosynthetic activity (eqn 6.3/17) over the canopy depth (LAI ranging from 0 at the top of the canopy to LAI at the bottom of the canopy):

$$Pc = \int_0^{LAI} P \cdot dLAI \qquad (6.3/22)$$

where P is now defined by substitution of eqn 6.3/21 into eqn 6.3/17:

$$P = \frac{\alpha \cdot \left(\frac{k \cdot I_0}{1-m}\right) \cdot e^{-k \cdot LAI} \cdot Pm}{\alpha \cdot \left(\frac{k \cdot I_0}{1-m}\right) \cdot e^{-k \cdot LAI} + Pm}$$

After multiplying the numerator and denominator by $(1 - m)$ we obtain:

$$P = \frac{\alpha \cdot \left(\frac{k \cdot I_0}{1-m}\right) \cdot e^{-k \cdot LAI} \cdot Pm \cdot (1-m)}{\alpha \cdot \left(\frac{k \cdot I_0}{1-m}\right) \cdot e^{-k \cdot LAI} \cdot (1-m) + Pm \cdot (1-m)}$$

or:

$$P = \frac{\alpha \cdot k \cdot I_0 \cdot e^{-k \cdot LAI} \cdot Pm}{\alpha \cdot k \cdot I_0 \cdot e^{-k \cdot LAI} \cdot + Pm \cdot (1-m)} \qquad (6.3/23)$$

Substituting eqn 6.3/23 into eqn 6.3/22, we obtain:

$$Pc = \int_0^{LAI} \frac{\alpha \cdot k \cdot I_0 \cdot e^{-k \cdot LAI} \cdot Pm}{\alpha \cdot k \cdot I_0 \cdot e^{-k \cdot LAI} \cdot + Pm \cdot (1-m)} \cdot d\mathrm{LAI} \qquad (6.3/24)$$

If we assume that all the parameters are constant within the canopy, we can perform the integration of the last equation, which yields (see Thornley and Johnson, 1990, Chapter 10):

$$Pc = -\frac{Pm}{k} \cdot \left\{ \ln \left[k \cdot I_0 \cdot \alpha \cdot e^{-k \cdot LAI} + Pm \cdot (1-m) \right] \right\} \Big|_0^{LAI}$$

or:

$$Pc = -\frac{Pm}{k} \cdot \left\langle \begin{array}{l} \left\{ \ln \left[k \cdot I_0 \cdot \alpha \cdot e^{-k \cdot LAI} + Pm \cdot (1-m) \right] \right\} \\ - \left\{ \ln \left[k \cdot I_0 \cdot \alpha + Pm \cdot (1-m) \right] \right\} \end{array} \right\rangle$$

$$Pc = \frac{Pm}{k} \cdot \left\{ \ln \left[\frac{k \cdot I_0 \cdot \alpha + Pm \cdot (1-m)}{k \cdot I_0 \cdot \alpha \cdot e^{-k \cdot LAI} + Pm \cdot (1-m)} \right] \right\} \qquad (6.3/25)$$

This last expression provides us with an analytical solution to calculate canopy photosynthesis. Observation provided evidence that *Pm* is markedly different for leaves which have grown under various radiation conditions. Thornley and Johnson (1990, pp. 246–250) developed equations to account for these effects on *Pm*. This involves the scaling of the value *Pm* at the top of the canopy (i.e. at LAI = 0) according to the total LAI of the canopy. The following relationship is used (Thornley and Johnson, 1990, eqn 10.7a):

$$Pm = Pm0 \cdot \left[1 - \frac{\lambda}{2} \cdot \left(1 - e^{-k \cdot LAI} \right) \right] \qquad (6.3/26)$$

where:

Pm0 = value of *Pm* for leaves on top of the canopy
[kg $CO_2 \cdot m^{-2}$ leaf $\cdot s^{-1}$] {1×10^{-6}}
λ = constant (in the ModelMaker model: 'tau') [–] {between 0 and 1}
k = extinction coefficient [–] {between 0 and 1}
LAI = canopy leaf area index [m^2 leaf $\cdot m^{-2}$ ground]

Equation 6.3/25 provides us with the instantaneous gross photosynthetic rate (per second) for the entire canopy. If we want to compute the daily gross photosynthetic rate (*Pd*), we have to integrate eqn 6.3/25 over the photoperiod per day:

$$\mathrm{Pd} = \int_0^h \mathrm{Pc} \cdot dt \qquad (6.3/27)$$

where:

h = length of the photoperiod [s $\cdot$ day^{-1}]

Instead of developing an analytical expression for daily photosynthesis (e.g. Thornley and Johnson, 1990, pp. 251–258), we use the daily time course of radiation and integrate eqn 6.3/25 numerically (per second basis) using available radiation data. This is compatible with the soil energy balance, which is also calculated on a per second basis.

Application in ModelMaker

Before we extend the model for leaf and canopy photosynthesis, we solve eqn 6.3/25 in ModelMaker and look at the typical shape of the photosynthetic response curve with leaf area index (LAI). The model is presented in *Mod6–1a.mod* (Fig. 6.2).

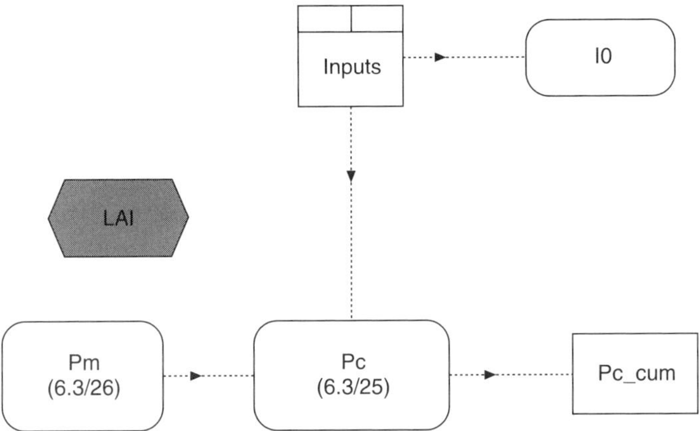

Fig. 6.2. Calculating instantaneous and cumulative canopy photosynthesis with the 'simple' rectangular hyperbola notation (*Mod6-1a.mod*).

In order to show the response to light intensity, a set of light data (increasing from 0 to 1000 W m^{-2}) is supplied to the model via the <input table> 'Inputs'. The various light intensities are read into the <variable> 'I0'. The notations for maximum photosynthetic activity (*Pm*, eqn 6.3/26) and canopy photosynthesis (*Pc*, eqn 6.3/25) are included in the <variables> 'Pm' and 'Pc', respectively. The <compartment> 'Pc_cum' performs the numerical integration according to eqn 6.3/27 (initial value: Pcum_init = 0). In this particular case, not a daily time course of radiation but an arbitrary increase of light intensity is used. This is done to show the effect of light intensity on canopy photosynthesis. In order to see the effect of various values for leaf area index (LAI), an <independent event> 'Choice' is added to the model, which is linked to the global <define> 'LAI'. At the beginning of each run, the user is prompted to supply a value for LAI (between 1–6).

Running the model

The model is set up for 100 time steps (according to the range of data supplied in the 'Input' table). To see the effect of radiation on photosynthesis a graph, 'Pc_I0_graph', is created, which shows the typical photosynthetic response curve (rectangular hyperbola). The output of the model will be presented at the end of the next section (Fig. 6.4A), together with the output of the advanced photosynthesis model. In order to appreciate the effect of various LAI's, a sensitivity analysis (see Section 2.3.3) is included in the graph (thin lines) for the range of LAI values.

6.3.2 Advanced model

The only effect we considered for the basic photosynthesis model (eqn 6.3/17) was the changing light intensity. However, photosynthesis is also affected by the internal CO_2 concentration (*Ci*). This aspect can be included by expressing the value for the maximum photosynthetic rate (*Pm*) with eqn 6.3/16, so that eqn 6.3/17 is written as (see Thornley and Johnson, 1990, eqn 9.7i):

$$P = \frac{\alpha \cdot I \cdot (Ci/rx)}{\alpha \cdot I + (Ci/rx)} \qquad (6.3/28)$$

where:

α = photochemical efficiency [kg $CO_2 \cdot J^{-1}$] $\{1 \times 10^{-8}\}$
I = radiation (PAR) [$W \cdot m^{-2}$] or [$J \cdot s^{-1} \cdot m^{-2}$]
Ci = internal CO_2 concentration [kg $CO_2 \cdot m^{-3}$ leaf]
rx = carboxylation resistance [$s \cdot m^{-1}$]

The internal CO_2 concentration is also influenced by the external CO_2 concentration (*Ca*) and by internal CO_2 sources, such as photorespiration. If we assume that CO_2 is moving from the outside to the inside by diffusion and further assume that the rate of respiration (*R*) is constant, it is possible to define the instantaneous net photosynthetic rate (*Pn*, units: kg CO_2 m^{-2} s^{-1}) either by:

$$Pn = P - R \qquad (6.3/29)$$

or by:

$$Pn = \frac{Ca - Ci}{rd} \qquad (6.3/30)$$

where:

Ca = external CO_2 concentration [kg $CO_2 \cdot m^{-3}$ leaf]
Ci = internal CO_2 concentration [kg $CO_2 \cdot m^{-3}$ leaf]
rd = resistance for diffusion of CO_2 between outside and inside the leaf [$s \cdot m^{-1}$]

This additional effect of the external CO_2 concentration must be taken into account in developing a more realistic model for instantaneous gross leaf photosynthesis. The lengthy derivation is very well documented by Thornley and Johnson (1990, pp. 226–228) and the interested reader is referred to this excellent publication.

The more realistic model for gross leaf photosynthesis which is obtained at the end of the derivation (Thornley and Johnson, 1990, eqn 9.10i) is:

$$P = \frac{1}{2 \cdot \theta} \cdot \left\{ \alpha \cdot I + Pm - \sqrt{(\alpha \cdot I + Pm)^2 - 4 \cdot \theta \cdot \alpha \cdot I \cdot Pm} \right\} \qquad (6.3/31)$$

with:

$$\theta = \frac{rd}{rd + rx}$$

where:

θ = a dimensionless parameter (in ModelMaker: 'theta') [–] {between 0 and 1}
α = photochemical efficiency (in ModelMaker: 'alpha') [kg $CO_2 \cdot J^{-1}$] {1×10^{-8}}
I = radiation (PAR) [$W \cdot m^{-2}$] or [$J \cdot s^{-1} \cdot m^{-2}$]
Pm = maximum rate of photosynthesis (eqt. 6.3/26) [kg $CO_2 \cdot m^{-2}$ leaf $\cdot s^{-1}$]

The parameter θ thus defines the relative importance of the diffusion resistance to the overall resistance of CO_2 to the reaction sites. Under the special condition when the diffusion resistance is zero, the entire parameter reduces to $\theta = 0$ and eqn 6.3/31 reduces to the rectangular hyperbola (eqt. 6.3/17) (see Thornley and Johnson, 1990, p. 228).

In order to obtain an expression for canopy photosynthesis, eqn 6.3/31 has to be integrated over the range of leaf area indices (LAI: from 0 to LAI), according to eqn 6.3/22. It can easily be seen that it is cumbersome to find an analytical solution to eqn 6.3/22 with eqn 6.3/31. However, a solution has already been worked out and can be found in integration tables. The analytical solution of eqn 6.3/22 by using eqn 6.3/31 for *P* is given by Thornley and Johnson (1990, eqn 10.5d). However, to show how an analytical solution can be constructed using integration tables, the derivation is also given below.

Equation 6.3/31 is an expression of leaf photosynthesis with respect to light intensities. However, we want to integrate this equation over the range of LAI (eqn 6.2/22). Therefore, we first have to differentiate this equation with respect to LAI. The relationship of light intensity with LAI is given by eqn 6.3/21. To obtain an expression for the change in LAI, i.e. '*d*LAI', we differentiate eqn 6.3/21 (see rules for differentiation exponential functions in Section 1.2.3):

$$\frac{dI}{dLAI} = -k \cdot \left\{ \left(\frac{k \cdot I_0}{1-m} \right) \cdot e^{-k \cdot LAI} \right\}$$

Rearranged for *d*LAI:

$$dLAI = -\frac{1}{k} \left\{ \frac{dI}{\left(\frac{k \cdot I_0}{1-m} \right) \cdot e^{-k \cdot LAI}} \right\}$$

and combined with eqn 6.3/21, we obtain (see also Thornley and Johnson, 1990, p. 248):

$$dLAI = -\frac{1}{k} \cdot \frac{dI}{I} \tag{6.3/32}$$

(see eqn 6.3/22)

Therefore the integration for canopy photosynthesis can be defined by:

$$Pc = \int_{I(LAI=0)}^{I(LAI=LAI)} P \cdot -\frac{1}{k} \cdot \frac{dI}{I} = -\frac{1}{k} \cdot \int_{I(LAI=0)}^{I(LAI=LAI)} P \cdot \frac{dI}{I} \tag{6.3/33}$$

with *P*, defined by eqn 6.3/31. The constant term $\frac{1}{2 \cdot \theta}$ (of eqn 6.3/31) can be moved in front of the integral, so that the complete equation which we have to solve is given by:

$$Pc = -\frac{1}{k \cdot 2 \cdot \theta} \cdot \int_{I(LAI=0)}^{I(LAI=LAI)} \left\{ \alpha \cdot I + Pm - \sqrt{(\alpha \cdot I + Pm)^2 - 4 \cdot \theta \cdot \alpha \cdot I \cdot Pm} \right\} \cdot \frac{dI}{I} \tag{6.3/34}$$

This equation can be divided into three sections where each one can be integrated separately:

$$Pc = -\frac{1}{k \cdot 2 \cdot \theta} \cdot \left\{ \begin{array}{l} \int_{I(LAI=0)}^{I(LAI=LAI)} \alpha \cdot I \cdot \frac{dI}{I} \\ + \int_{I(LAI=0)}^{I(LAI=LAI)} Pm \cdot \frac{dI}{I} \\ - \int_{I(LAI=0)}^{I(LAI=LAI)} \sqrt{(\alpha \cdot I + Pm)^2 - 4 \cdot \theta \cdot \alpha \cdot Pm \cdot I} \cdot \frac{dI}{I} \end{array} \right\} \quad (6.3/35)$$

While the first two expressions are easy to evaluate, it is considerably more difficult to evaluate the third square-root expression. The following analysis is only concerned with the third expression: $\int_{I(LAI=0)}^{I(LAI=LAI)} \sqrt{(\alpha \cdot I + Pm)^2 - 4 \cdot \theta \cdot \alpha \cdot Pm \cdot I} \cdot \frac{dI}{I}$. Once we have found a solution to this expression, we can compute the overall integral.

The square-root expression can be rewritten as (Johnson and Thornley, 1984):

$$\int_{I(LAI=0)}^{I(LAI=LAI)} \sqrt{\alpha^2 \cdot I^2 + 2 \cdot \alpha \cdot I \cdot Pm + Pm^2 - 4 \cdot \theta \cdot \alpha \cdot Pm \cdot I} \cdot \frac{dI}{I}$$

or:

$$\int_{I(LAI=0)}^{I(LAI=LAI)} \sqrt{Pm^2 + 2 \cdot \alpha \cdot Pm \cdot (1 - 2 \cdot \theta) \cdot I + \alpha^2 \cdot I^2} \cdot \frac{dI}{I}$$

The basic notation of the last expression is in the form:

$$\int \sqrt{a + b \cdot x + c \cdot x^2} \cdot \frac{dx}{x} \quad (6.3/36)$$

with:

$a = Pm^2$
$b = 2 \cdot \alpha \cdot Pm \cdot (1 - 2 \cdot \theta)$
$c = \alpha^2$
$x = I$

Equation 6.3/36 can be integrated via the help of an integration table. For this particular equation, the solution is given by Gradshteyn and Ryzhik (1995, eqn 2.267(1) in combination with eqns 2.261 and 2.266, p. 102) (see also Johnson and Thornley, 1984).

The entire expression is similar to (Gradshteyn and Ryzhik, 1995, eqn 22.67(1)):

$$\int \sqrt{R} \cdot \frac{dx}{x} = \sqrt{R} + a \cdot \int \frac{dx}{x \cdot \sqrt{R}} + \frac{b}{2} \cdot \int \frac{dx}{\sqrt{R}}$$

with:

$$R = a + b \cdot x + c \cdot x^2$$

After applying eqns 2.261 and 2.266 from Gradshteyn and Ryzhik (1995), we obtain:

$$\int \sqrt{R} \cdot \frac{dx}{x} = \sqrt{R} + a \cdot -\frac{1}{\sqrt{a}} \cdot \ln\left[\frac{2 \cdot a + b \cdot x + 2 \cdot \sqrt{a \cdot R}}{x}\right] + \frac{b}{2} \cdot \frac{1}{\sqrt{c}} \cdot \ln\left(2 \cdot \sqrt{c \cdot R} + 2 \cdot c \cdot x + b\right) \quad (6.3/37)$$

Expressing the first two integration terms of eqn 6.3/35 as:

$$\int \alpha \cdot dI = \int \sqrt{c} \cdot dx = \sqrt{c} \cdot x \quad (6.3/38)$$

(note: $c = \alpha^2$; however, only α is needed)

and:

$$\int \frac{Pm}{I} \cdot dI = \int \frac{\sqrt{a}}{x} \cdot dx = \sqrt{a} \cdot \ln(x) \quad (6.3/39)$$

(note: $a = Pm^2$, but only Pm is needed)

The entire analytical solution of eqn 6.3/35 (combining eqns 6.3/37–6.3/39) is:

$$Pc = -\frac{1}{k \cdot 2 \cdot \theta} \left\{ \begin{array}{l} \sqrt{c} \cdot x \\ + \sqrt{a} \cdot \ln(x) \\ - \left[\begin{array}{l} \sqrt{R} + a \cdot -a \cdot \frac{1}{\sqrt{a}} \cdot \ln\left[\frac{2 \cdot a + b \cdot x + 2 \cdot \sqrt{a \cdot R}}{x}\right] \\ + \frac{b}{2} \frac{1}{\sqrt{c}} \cdot \ln\left(2 \cdot \sqrt{c \cdot R} + 2 \cdot c \cdot x + b\right) \end{array} \right] \end{array} \right\}_{I(LAI=0)}^{I(LAI=LAI)}$$

The expression may also be written as:

$$Pc = -\frac{1}{k \cdot 2 \cdot \theta}\left\{\begin{array}{l} \sqrt{c} \cdot x \\ +\sqrt{a} \cdot \ln(x) \\ -\sqrt{R} \\ +a\frac{1}{\sqrt{a}} \cdot \ln\left[\frac{2 \cdot a + b \cdot x + 2 \cdot \sqrt{a \cdot R}}{x}\right] \\ -\frac{b}{2}\frac{1}{\sqrt{c}} \cdot \ln\left(2 \cdot \sqrt{c \cdot R} + 2 \cdot c \cdot x + b\right) \end{array}\right\}_{I(LAI=0)}^{I(LAI=LAI)}$$

This expression is not suitable, because it would not be defined for 'zero' radiation intensities (*x* appears in the denominator and the log is taken from *x*).

Applying some basic transformations (see Section 1.2.3), the last expression can be rewritten (italics) as:

$$Pc = -\frac{1}{k \cdot 2 \cdot \theta}\left\{\begin{array}{l} \sqrt{c} \cdot x \\ +\sqrt{a} \cdot \ln(x) \\ -\sqrt{R} \\ +\sqrt{a} \cdot \mathit{ln}\left(2 \cdot a + b \cdot x + 2 \cdot \sqrt{a \cdot R}\right) \\ -\sqrt{a} \cdot \mathit{ln}(x) \\ -\frac{b}{2}\frac{1}{\sqrt{c}} \cdot \ln\left(2 \cdot \sqrt{c \cdot R} + 2 \cdot c \cdot x + b\right) \end{array}\right\}_{I(LAI=0)}^{I(LAI=LAI)}$$

and subsequently simplified to:

$$Pc = -\frac{1}{k \cdot 2 \cdot \theta}\{X\}\Big|_{I(LAI=0)}^{I(LAI=LAI)} \qquad (6.3/40)$$

with:

$$X = \left\{\begin{array}{l} \sqrt{c} \cdot x \\ -\sqrt{R} \\ +\sqrt{a} \cdot \ln\left(2 \cdot a + b \cdot x + 2 \cdot \sqrt{a \cdot R}\right) \\ -\frac{b}{2}\frac{1}{\sqrt{c}} \cdot \ln\left(2 \cdot \sqrt{c \cdot R} + 2 \cdot c \cdot x + b\right) \end{array}\right\}_{I(LAI=0)}^{I(LAI=LAI)}$$

where:

$$a = Pm^2$$
$$b = 2 \cdot \alpha \cdot Pm \cdot (1 - 2 \cdot \theta)$$
$$c = \alpha^2$$
$$x = I$$
$$R = a + b \cdot x + c \cdot x^2$$

The expression $\{X\}$ has to be evaluated for *I* when LAI = 0 (X_0) and *I* at full LAI (X_L). Therefore the final equation for *Pc* is:

$$Pc = -\frac{1}{k \cdot 2 \cdot \theta} \cdot (X\_L - X\_0) \qquad (6.3/41)$$

The *Pm* values are calculated according to eqn 6.3/26.

Application in ModelMaker

In order to follow the many steps of this rather long integration we define the various components separately and then finally combine them into one expression *Pc* (eqn 6.3/40). Components *a*, *b*, *c*, *R* and *I* will be calculated separately in <variables> either for LAI = 0 (extension of the components with '_0') or for full LAI (extension of the components with '_L'). An additional <variable> 'Pm' is inserted. The calculations for this advanced instantaneous canopy photosynthesis model is presented in *Mod6–1b.mod* (Fig. 6.3).

In order to allow users to change the LAI and θ values (eqn 6.3/31), the <independent event> 'Choice' from *Mod6–1a.mod* has been complemented with a 'GetValue' statement for the θ parameter. The default value for θ is set to the parameter 'theta_default' = 0.85 (recall that the advanced model reduces to the 'rectangular hyperbola' when θ = 0).

Finally, the simple canopy photosynthesis model (Section 6.3.1) and the advanced model from this section are presented together in *Mod6–1c.mod*. This has the advantage that they can be easily compared with each other. For instance, it can be checked whether both models give the same answer if θ is close to zero (see explanation to eqn 6.3/31). Instead of presenting the advanced model in various components (as done in *Mod6–1b.mod*), all the components needed for the calculation are added together in the <variable> 'Pc':

```
Pc = −1/(2 * theta * k_beer) *
     (((sqrt((alpha^2)) * I_L) - sqrt((Pm^2+((2 * Pm * (1-(2 *
     theta)) * alpha) * I_L)+alpha^2 * (I_L^2))) + (sqrt((Pm^2))
     * ln(2 * (Pm^2) + (2 * Pm * (1 - (2 * theta)) * alpha) *
     I_L + 2 * sqrt((Pm^2) * (Pm^2 + ((2 * Pm * (1 - (2 *
     theta)) * alpha) * I_L) + alpha^2 * (I_L^2))))) - ((2 * Pm
     * (1 - (2 * theta)) * alpha)/(2 * sqrt((alpha^2)))) * ln(2
     * sqrt((alpha^2) * (Pm^2 + ((2 * Pm * (1 - (2 * theta)) *
     alpha) * I_L) + alpha^2 * (I_L^2))) + (2 * (alpha^2) * I_L)
     + (2 * Pm * (1 - (2 * theta)) * alpha))) -
```

```
((sqrt((alpha^2)) * I_0) - sqrt((Pm^2 + ((2 * Pm * (1 - (2
* theta)) * alpha) * I_0) + alpha^2 * (I_0^2))) +
(sqrt((Pm^2)) * ln(2 * (Pm^2) + (2 * Pm * (1 - (2 * theta))
* alpha) * I_0 + 2 * sqrt((Pm^2) * (Pm^2 + ((2 * Pm * (1 -
(2 * theta)) * alpha) * I_0) + alpha^2 * (I_0^2))))) - ((2
* Pm * (1 - (2 * theta)) * alpha)/(2 * sqrt((alpha^2)))) *
ln(2 * sqrt((alpha^2) * (Pm^2 + ((2 * Pm * (1 - (2 *
theta)) * alpha) * I_0) + alpha^2 * (I_0^2))) + (2 *
(alpha^2) * I_0) + (2 * Pm * (1 - (2 * theta)) * alpha))))
```

Only the radiation intensity at LAI: 'I_L' and on top of the canopy: 'I_0' are calculated separately.

The model developed so far is useful for evaluating the effect of changing parameters on canopy photosynthesis (*Pc*). Simulations were carried out with the

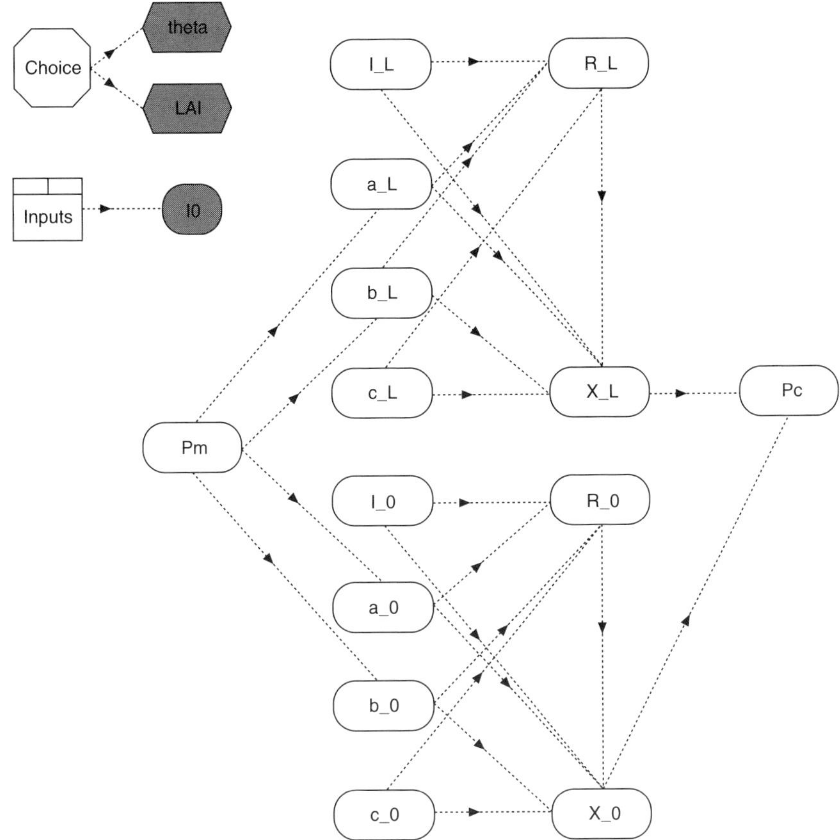

Fig. 6.3. Detailed description of the advanced photosynthesis model showing the steps involved by using a solution from an integration table (*Mod6-1b.mod*).

model (*Mod6–1c.mod,* advanced notation) for radiation ranging from 0 to 1000 W m^{-2} and either changing leaf area indices (LAI) or the θ values. The parameter settings were as follows:

Radiation for both runs: 0 to 1000 W m^{-2}

Run A (Fig 6.4A)
LAI: 1 to 6 (six steps)
θ: fixed to 0.85

Run B (Fig 6.4B)
LAI: fixed to 2
θ: 0 to 1 (six steps)

The results are presented in three-dimensional plots (Fig. 6.4).

The simulations show typical responses of canopy photosynthesis to the parameters (see also Johnson and Thornley, 1984). If the θ value approaches zero, the photosynthetic response curve approaches the solution of the rectangular hyperbola (eqn 6.3/25). Thus the advanced model also includes the possibility of simulating photosynthesis with the simple model by appropriate adjustment of the θ parameter.

To become familiar with the canopy photosynthesis model, it might be a good idea to try out as many user options as possible and see the effect on the output of both the simple and the advanced model. The advanced model for canopy photosynthesis is more complex to implement but is based on a more mechanistic understanding of photosynthesis. In the forthcoming models for plant growth, the advanced canopy photosynthesis model is used as a 'growth engine' to drive a model for vegetative grass growth.

6.4 Plant Growth–Substrate Relationships

6.4.1 Introduction

The description of the plant growth model follows mainly a series of papers by Thornley and colleagues on the development of a vegetative grass growth model. Photosynthesis is described by the advanced model (Section 6.3.2). The papers by Johnson and Thornley (1983, 1985) and Thornley and Verberne (1989) are used extensively and are recommended as background reading.

Section 6.4 is mainly concerned with the substrate relationships (carbon and nitrogen) to plant growth. The effects of the environmental parameters on plant growth are presented in Section 6.5.

6.4.2 Growth above ground

The model in this section is mainly based on the model by Johnson and Thornley (1983), which describes the vegetative growth of grass. The grass crop is divided

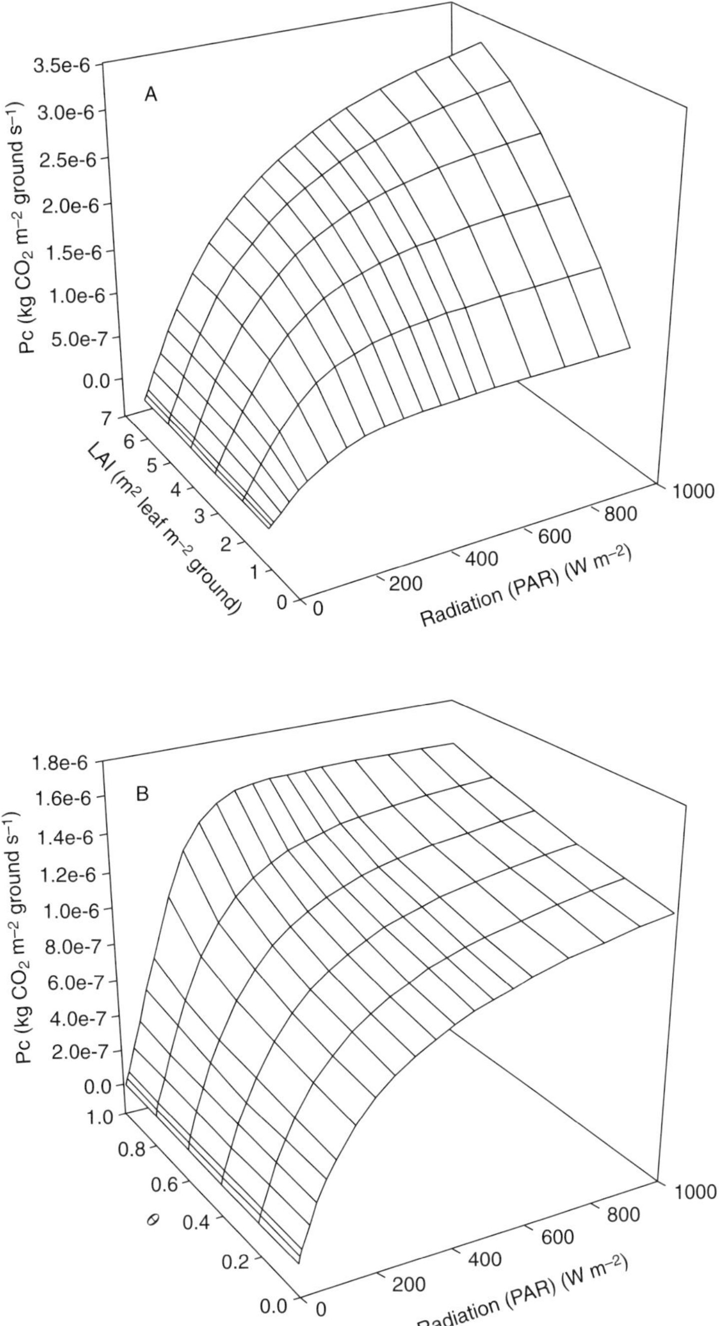

Fig. 6.4. Relationship of radiation–LAI (A) and radiation–θ (B) with canopy photosynthesis (*Mod6-1c.mod*, eqn 6.3/41) (if θ = 0 results correspond to the 'rectangular hyperbola' model, eqn 6.3/25; for parameters, see text).

according to its age into four compartments: 1 = growing leaves, 2 = first fully expanded leaves, 3 = second fully expanded leaves and 4 = senescing leaves. This division is considered to be representative for a growing grass crop. Each compartment is associated with a certain amount of structural material (dry matter) and a certain leaf area index (LAI). The overall dry matter and leaf area index (LAI) are then the sum of the four subcompartments (Fig. 6.5). The carbon produced through photosynthesis will enter a substrate C pool (Wc). This carbon pool supplies growth of new structure (leaves, etc., G) and growth respiration (Rg), which are subtracted from the substrate C pool (Fig. 6.5).

The size of the carbon pool (Wc) can be calculated by integrating the various rates entering and leaving this pool:

$$\frac{d\mathrm{Wc}}{dt} = \mathrm{P} - \mathrm{C\_use} - \mathrm{Rm\_sh} \qquad (6.4/1)$$

where:

P = carbon input into the shoot from photosynthesis [$\mathrm{kg\ C \cdot m^{-2} \cdot s^{-1}}$]
C_use = gross rate of C used for growth plus growth respiration [$\mathrm{kg\ C \cdot m^{-2} \cdot s^{-1}}$]
Rm_sh = shoot maintenance respiration [$\mathrm{kg\ C \cdot m^{-2} \cdot s^{-1}}$]

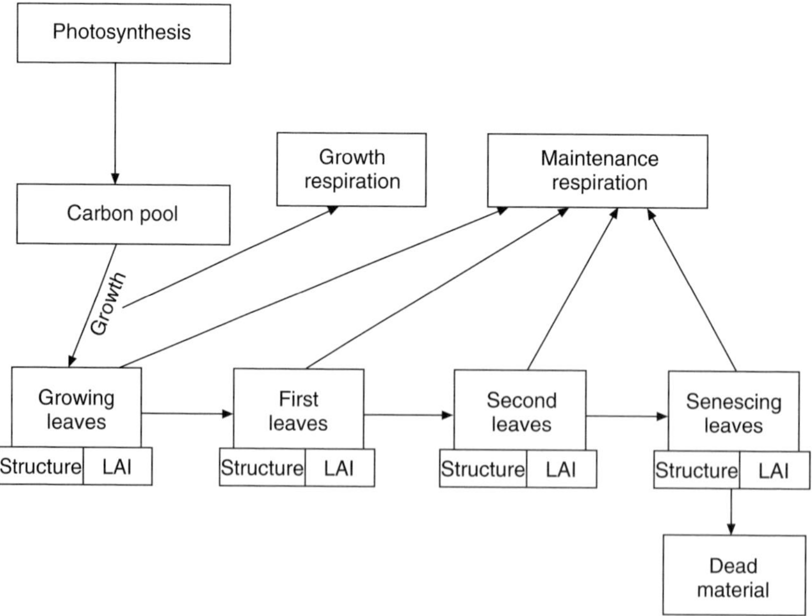

Fig. 6.5. Conceptual model for vegetative grass growth (acccording to Johnson and Thornley, 1983) (see text for details).

The carbon input into the shoot is governed by the rate of canopy photosynthesis (units: kg CO_2 m^{-2} ground s^{-1}). However, only a portion of this overall photosynthetic flow is available for shoot growth; the rest will be used for root growth. Since we are at the moment not interested in root growth, we simply multiply the gross rate of C entering the pool with a parameter (fsh) which determines the fraction of the overall C flux into the shoot. The rate at which carbon (not carbon dioxide!) will enter the C substrate pool for shoot growth (P) is:

$$P = Pc \cdot \frac{Molwt\_C}{Molwt\_CO2} \cdot fsh \qquad (6.4/2)$$

Molwt_C = molecular weight of C [$g \cdot mol^{-1}$] {12}
Molwt_CO2 = molecular weight of CO_2 [$g \cdot mol^{-1}$] {44}
fsh = fraction of C flux available for shoot growth [–] {0.9}
Pc = gross canopy photosynthesis (eqn. 6.3/41) [$kg\ CO_2 \cdot m^{-2} \cdot s^{-1}$]

Note: in contrast to Johnson and Thornley (1983), who calculated daily photosynthetic rates, here only the instantaneous rate of canopy photosynthesis is considered. This means that all rate parameters which are defined on a per day basis by Johnson and Thornley (1983) have to be scaled down to per second values (dividing them by 86,400 [$s \cdot day^{-1}$]).

Shoot growth (Gsh) is related to the overall substrate pool (Wc) and the proportion of the live shoot structure (Wsh) in relation to the overall C content of the system (Wc + Wsh) (see Johnson and Thornley, 1983, eqn 13):

$$Gsh = mu \cdot Wc \cdot \frac{Wsh}{W} \qquad (6.4/3)$$

with

$$W = Wsh + Wc$$

$$mu = \frac{mu20}{86400}$$

where:

mu20 = growth coefficient at 20°C [day^{-1}] {0.5}
Wsh = total structural dry weight of shoot (eqt. 6.4/8) [$kg\ C \cdot m^{-2}$ ground]
Wc = substrate carbon (eqt. 6.4/1) [$kg\ C \cdot m^{-2}$ ground]

The overall rate of carbon extraction (C_use) from the C pool Wc encompasses both the carbon which is used up during shoot growth (Gsh) and the carbon which is respired during growth (Rg). Introducing a growth efficiency or yield factor (Y, between 0 and 1), we can calculate the gross rate of C which has to be extracted from Wc to maintain Gsh and Rg as (see Johnson and Thornley, 1983, eqn 14):

$$C\_use = \frac{Gsh}{Y} \tag{6.4/4}$$

The last expression includes both Gsh and Rg and can therefore be written as:

$$\frac{Gsh}{Y} = Gsh + Rg \tag{6.4/5}$$

The growth respiration Rg can be calculated by rearrangement of the last equation

$$\left(Rg = \frac{Gsh}{Y} - Gsh \rightarrow Rg = Gsh \cdot \left(\frac{1}{Y} - 1\right) \rightarrow Rg = Gsh \cdot \left(\frac{1}{Y} - \frac{Y}{Y}\right)\right)$$

(see also Johnson and Thornley, 1983, eqn 15):

$$Rg = \left(\frac{1 - Y}{Y}\right) \cdot Gsh \tag{6.4/6}$$

where:

Y = yield factor determining the efficiency with which carbon is used for shoot growth [–] {0.75}

Recall that the crop structure is divided into four components: W1 (growing leaves), W2 (first fully expanded leaves), W3 (second fully expanded leaves) and W4 (senescing leaves). The new growth Gsh is primarily used to grow leaves (W1). Rates of leaf appearance (γ, 'gamma_sh') determine the rate at which other leaves appear and finally the rate at which the leaves of compartment W4 die and form debris (Fig. 6.5). The compartment W1 contains growing leaves and therefore the average weight of the leaves in W1 is less than the weight which is passed on to W2. To account for this, the flow from W1 to W2 has to be multiplied with a dimensionless factor (f_W1, > 1). The set of equations describing the four structural compartments are:

$$\begin{aligned} \frac{dW1}{dt} &= Gsh - F1 \\ \frac{dW2}{dt} &= F1 - F2 \\ \frac{dW3}{dt} &= F2 - F3 \\ \frac{dW4}{dt} &= F3 - F4 \end{aligned} \tag{6.4/7}$$

$$\frac{dS}{dt} = F4$$

with flows between compartments:

$$F1 = f\_W1 \cdot gamma\_sh \cdot W1$$
$$F2 = gamma\_sh \cdot W2$$
$$F3 = gamma\_sh \cdot W3$$
$$F4 = gamma\_sh \cdot W4$$

$$gamma\_sh = \frac{gamma\_sh20}{86400}$$

where:

W1 =	structural weight of growing leaves [kg C $\cdot$ m^{-2} ground]
W2 =	structural weight of first fully expanded leaves [kg C $\cdot$ m^{-2} ground]
W3 =	structural weight of second fully expanded leaves [kg C $\cdot$ m^{-2} ground]
W4 =	structural weight of senescing leaves [kg C $\cdot$ m^{-2} ground]
S =	structural weight of plant debris [kg C $\cdot$ m^{-2} ground]
f_W1 =	factor to adjust flow from W1 to W2 [–] {2}
gamma_sh20 =	rate of leaf appearance at 20°C [day^{-1}] {0.15}
W_init =	initial weights of all four compartments [kg C $\cdot$ m^{-2} ground] {0.02}
S_init =	initial weight of plant debris [kg C $\cdot$ m^{-2} ground] {0}

Assuming that the last compartment W4 consists of material which is half alive and half dead (Johnson and Thornley, 1985), the overall weight of live plant structure (Wsh) can then be calculated as:

$$Wsh = W1 + W2 + W3 + \frac{W4}{2} \tag{6.4/8}$$

and the dead shoot material as:

$$Wsh4\_dead = \frac{W4}{2} \tag{6.4/9}$$

Some respiration is also related to maintenance (Rm_sh) of the four compartments and is calculated by Johnson and Thornley (1983, eqn 16) as being proportional to the dry weights of the four compartments (W1 to W4). The related maintenance coefficients (M1 to M4) are decreasing with plant age. The overall respiration from the shoot structure due to maintenance is calculated as:

$$Rm\_sh = Msh1 \cdot W1 + Msh2 \cdot W2 + Msh3 \cdot W3 + Msh4 \cdot W4 \tag{6.4/10}$$

with:

$$\mathrm{Msh1} = \frac{\mathrm{Msh1\_20}}{86400}$$

$$\mathrm{Msh2} = \frac{\mathrm{Msh2\_20}}{86400}$$

$$\mathrm{Msh3} = \frac{\mathrm{Msh3\_20}}{86400}$$

$$\mathrm{Msh4} = \frac{\mathrm{Msh4\_20}}{86400}$$

where:

Msh1_20 = maintenance coefficient for respiration of W1 at 20°C [day^{-1}] {0.02}
Msh2_20 = maintenance coefficient for respiration of W2 at 20°C [day^{-1}] {0.02}
Msh3_20 = maintenance coefficient for respiration of W3 at 20°C [day^{-1}] 0.015}
Msh4_20 = maintenance coefficient for respiration of W4 at 20°C [day^{-1}] {0.01}

Leaf area index (LAI) calculation

A fraction (fract_L) of the new growth Gsh is used for leaf growth. In order to convert this flux of carbon (units: kg C m^{-2} ground s^{-1}) into rates of leaf area index expansion (units: m^2 leaf m^{-2} ground s^{-1}), we have to multiply this rate by a function of specific leaf area index (SLA), which is defined according to Johnson and Thornley (1983, eqn 21). It is assumed that the relative amount of substrate available for growth ($\frac{\mathrm{Wc}}{\mathrm{W}}$) governs the specific leaf area (SLA). The equation has the form:

$$\mathrm{SLA} = \mathrm{SLAm} \cdot \left(1 - \mathrm{ISLA} \cdot \frac{\mathrm{Wc}}{\mathrm{W}}\right) \tag{6.4/11}$$

where:

SLAm = maximum specific leaf area [m^2 leaf • kg^{-1} C] {40}
ISLA = incremental specific leaf area parameter [–] {1}
Wc = available carbon substrate (eqn 6.4/1) [kg C • m^{-2} ground]
W = available carbon plus structural carbon (see eqn 6.4/3) [kg C • m^{-2} ground]

The maximum obtainable specific leaf area is set to SLAm and is reduced according to the ratio: $\frac{Wc}{W}$. This means that if a lot of substrate Wc is available but only a small amount of live shoot structure, the ratio $\frac{Wc}{W}$ approaches 1, with the result that leaf tissue with a high density (low SLA values) is produced. If the amount of live shoot structure dominates the overall carbon content, then specific leaf area close to the maximum value SLAm is produced.

The leaf areas related to the four compartments (W1–W4) are calculated in analogy to eqn 6.4/7:

$$\frac{dL1}{dt} = LG - FL1$$

$$\frac{dL2}{dt} = FL1 - FL2$$

$$\frac{dL3}{dt} = FL2 - FL3 \qquad (6.4/12)$$

$$\frac{dL4}{dt} = FL3 - gamma\_sh \cdot L4$$

with:

$$LG = fract\_L \cdot Gsh \cdot SLA$$
$$FL1 = f\_W1 \cdot gamma\_sh \cdot L1$$
$$FL2 = gamma\_sh \cdot L2$$
$$FL3 = gamma\_sh \cdot L3$$

where:

L1 = leaf area index of component W1 [m^2 leaf · m^{-2} ground]
L2 = leaf area index of component W2 [m^2 leaf · m^{-2} ground]
L3 = leaf area index of component W3 [m^2 leaf · m^{-2} ground]
L4 = leaf area index of component W4 [m^2 leaf · m^{-2} ground]

The entire leaf area index (LAI) can now be calculated as (again assuming that the last compartment consists of material which is half dead and half alive):

$$LAI = L1 + L2 + L3 + \frac{L4}{2} \qquad (6.4/13)$$

In the paper by Johnson and Thornley (1983), the various rate parameters are defined for a temperature of 20°C. Relationships which scale these parameters to other temperatures are introduced in Section 6.5.1.

Application in ModelMaker

Before expanding the model, we bring the first stage of model development together. The notation described in this section is added to the previous photosynthesis model (*Mod6–1c.mod*, only advanced model). The various <compartments> (eqns 6.4/1, 6.4/7 and 6.4/12) and <variables> (all other equations) and the <parameters> are added to the model. The <variables> 'Rm_sh', 'Gsh' and 'LAI' are defined globally, as well as the rate parameters, which have to be scaled down to per second values: 'Msh1', 'Msh2', 'Msh3', 'Msh4', 'gamma_sh' and 'mu'. The model has to be supplied with radiation (PAR) values. Hourly data for 80 days are contained in the file *Inp6–1d.txt* and are supplied to the model via the <lookup file> 'Input'. The model is presented in *Mod6–1d.mod* (Fig. 6.6).

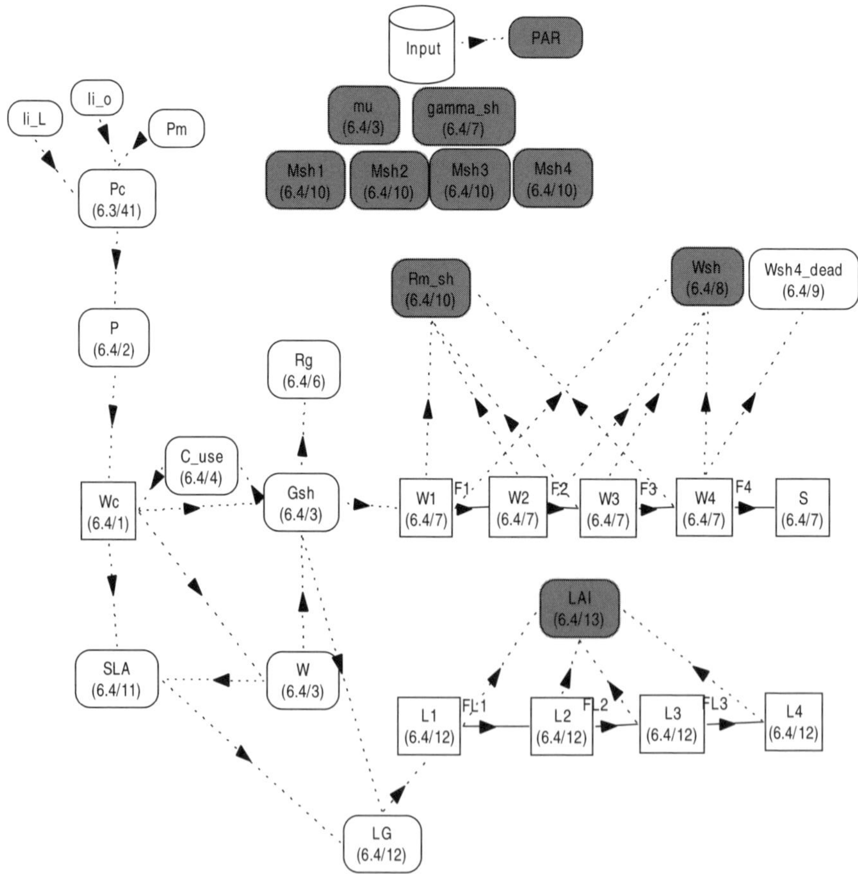

Fig. 6.6. Model for vegetative crop growth based on the paper by Johnson and Thornley (*Mod6-1d.mod*).

The model is set up for a simulation period of 80 days (= 6,912,000 s) with output steps every hour (= 1920 outputs). Currently the step length is variable, but choosing a 'fixed step' length with one step between outputs would approximately halve the simulation time.

Outputs in the form of graphs are included for respiration (Rm and Rg), total dry weight (W), the structural plant components (W1 to W4) and leaf area index LAI. Some of the outputs of the model are illustrated at the end of the next section, once other elements are added to the basic model.

So far, we have considered a plant growth model which describes aboveground growth as a function of C input into the system via photosynthesis. However, plants will also grow in the soil in the form of roots. The roots are needed to pick up nutrients and water from the soil, which are used at the reaction sites to synthesize new material. One of the most important nutrients is nitrogen (N), which is taken up by roots in available forms (ammonium (NH_4^+-N) and nitrate (NO_3^--N)). Finally, the newly synthesized material is allocated to the various plant parts. All of the above mentioned aspects are considered in the next section in an extension to the current model.

6.4.3 Root growth and partitioning

The extension of the model is mainly based on the paper by Johnson and Thornley (1985), which is recommended as background reading. So far, we have considered only the assimilation of carbon by the plant to sustain vegetative growth. However, we made the underlying assumption that nutrients and water which are needed for plant growth are unlimited. While carbon is acquired by the plant through photosynthesis, nitrogen is obtained by root uptake. Therefore a root system has to be added to the previous model. The additional substrate nitrogen which is now available to the plant alters also some of the notations of the previous model regarding the vegetative growth of the plant. Essential for the model are functions which describe the partitioning of the resources (carbon, nitrogen) between shoots and roots.

The growth of the various compartments is expressed in this section in kg structure as opposed to kg C in the previous section.

Root growth and root maintenance respiration

The root system can be derived in analogy to the shoot system (eqn 6.4/7) as a four-compartment notation (Johnson and Thornley, 1985, eqn 1b):

$$\frac{dR1}{dt} = Gr - FR1$$

$$\frac{dR2}{dt} = FR1 - FR2 \qquad (6.4/14)$$

$$\frac{d\text{R3}}{dt} = \text{FR2} - \text{FR3}$$

$$\frac{d\text{R4}}{dt} = \text{FR3} - \text{FR4}$$

with:

FR1 = f_R1 · gamma_r · R1
FR2 = gamma_r · R2
FR3 = gamma_r · R3
FR4 = gamma_r · R4

$$\text{gamma\_r} = \frac{\text{gamma\_r20}}{86400}$$

where:

R1 =	structural weight of growing roots [kg structure · m^{-2} ground]
R2 =	structural weight of first fully expanded roots [kg structure · m^{-2} ground]
R3 =	structural weight of second fully expanded roots [kg structure · m^{-2} ground]
R4 =	structural weight of senescing roots [kg structure · m^{-2} ground]
gamma_r20 =	rate of root appearance at 20°C [day^{-1}]
Gr =	production rate of new root structure (see below, eqn 6.4/40) [kg structure · m^{-2} ground · s^{-1}]
FR1 =	rate of flow from R1 to R2 [kg structure · m^{-2} ground · s^{-1}]
FR2 =	rate of flow from R2 to R3 [kg structure · m^{-2} ground · s^{-1}]
FR3 =	rate of flow from R3 to R4 [kg structure · m^{-2} ground · s^{-1}]
FR4 =	rate of flow from R4 [kg structure · m^{-2} ground · s^{-1}]
f_R1 =	factor to adjust the flow from R1 to R2 [–] {2}

The total live root structure is calculated as:

$$\text{Wr} = \text{R1} + \text{R2} + \text{R3} + \frac{\text{R4}}{2} \tag{6.4/15}$$

and the dead root structure as:

$$\text{Wr4\_dead} = \frac{\text{R4}}{2} \tag{6.4/16}$$

The maintenance respiration from the roots Rm_r is calculated, in analogy to the shoot maintenance respiration, by using maintenance coefficients for root respiration (Mr1 to Mr4). Since the plant dry matter is now expressed in terms of structure and not carbon, we have to multiply the result by the fractional carbon content of the live plant (fractC), which determines the kg carbon per kg structure (note: 'Rm_sh' (eqn 6.4/10) as used in this section also has to be multiplied by 'fract_C'!):

$$Rm\_r = fract\_C \cdot (Mr1 \cdot R1 + Mr2 \cdot R2 + Mr3 \cdot R3 + Mr4 \cdot R4) \qquad (6.4/17)$$

with:

$$Mr1 = \frac{Mr1\_20}{86400}$$

$$Mr2 = \frac{Mr2\_20}{86400}$$

$$Mr3 = \frac{Mr3\_20}{86400}$$

$$Mr4 = \frac{Mr4\_20}{86400}$$

where:

fract_C = fractional carbon content of live plant material [kg C $\cdot$ kg^{-1} structure] {0.45}
Mr1_20 = maintenance coefficient for root respiration of R1 at 20°C [day^{-1}] {0.02}
Mr2_20 = maintenance coefficient for root respiration of R2 at 20°C [day^{-1}] {0.02}
Mr3_20 = maintenance coefficient for root respiration of R3 at 20°C [day^{-1}] {0.015}
Mr4_20 = maintenance coefficient for root respiration of R4 at 20°C [day^{-1}] {0.01}

The total live structural dry matter (Wg) can be defined by adding the shoot dry matter (Wsh, eqn 6.4/8) and the root dry matter (Wr, eqn 6.4/15) together (see also Johnson and Thornley, 1985, eqn 1e):

$$Wg = Wsh + Wr \qquad (6.4/18)$$

Calculation of various plant substrate and structure fractions

In the last section, we already introduced the carbon substrate pool Wc (eqn 6.4/1). Later on in this section, we shall define an additional nitrogen substrate pool (Wn, eqn 6.4/42). We need this additional nitrogen substrate pool to derive variables which are needed in our calculation.

The fractions of shoot and root live structures: Fract_sh and Fract_r (units:

kg shoot structure kg^{-1} total structure) related to the overall dry matter (Wg) are defined by:

$$\text{Fract\_sh} = \frac{\text{Wsh}}{\text{Wg}} \tag{6.4/19}$$

and:

$$\text{Fract\_r} = \frac{\text{Wr}}{\text{Wg}} \tag{6.4/20}$$

The concentrations of carbon substrate C (units: kg C kg^{-1} structure) and nitrogen substrate N in the entire plant (units kg N kg^{-1} structure) (Johnson and Thornley, 1985, eqn 1h) are defined by:

$$\text{C} = \frac{\text{Wc}}{\text{Wg}} \tag{6.4/21}$$

and

$$\text{N} = \frac{\text{Wn}}{\text{Wg}} \tag{6.4/22}$$

Assuming that the carbon and nitrogen substrates are bound into structures of a certain molecular weight, we can also calculate the entire storage weight Ws (units: kg storage m^{-2} ground) by:

$$\text{Ws} = \frac{\text{Molwt\_Csubst}}{\text{Molwt\_C}} \cdot \text{Wc} + \frac{\text{Molwt\_Nsubst}}{\text{Molwt\_N}} \cdot \text{Wn} \tag{6.4/23}$$

where:

Molwt_Csubst = molecular weight of carbon structure (sucrose is assumed to be representative) $[g \cdot mol^{-1}]$ {28.5}
Molwt_Nsubst = molecular weight of nitrogen structure (nitrate is assumed to be representative) $[g \cdot mol^{-1}]$ {62}
Molwt_C = molecular weight of carbon $[g \cdot mol^{-1}]$ {12}
Molwt_N = molecular weight of nitrogen $[g \cdot mol^{-1}]$ {14}
Wc = carbon substrate (eqt. 6.4/1) [kg C $\cdot$ m^{-2} ground]
Wn = nitrogen substrate (see below. eqn 6.4/42) [kg N $\cdot$ m^{-2} ground]

The overall storage dry weight (Ws) can also be divided into shoot (Ws_sh) and root (Ws_r) storage dry weight by multiplication of eqn 6.4/23 with either eqn 6.4/19 or eqn 6.4/20):

$$\text{Ws\_sh} = \text{Fract\_sh} \cdot \text{Ws} \tag{6.4/24}$$

and:

$$Ws\_r = Fract\_r \cdot Ws \tag{6.4/25}$$

Substrate carbon and nitrogen in the shoot (Wc_sh, Wn_sh) and root (Wc_r, Wn_r) expressed on a per area basis (units: kg C m^{-2} ground or kg N m^{-2} ground) are calculated as:

$$Wc\_sh = Wc \cdot Fract\_sh \tag{6.4/26}$$

$$Wc\_r = Wc \cdot Fract\_r \tag{6.4/27}$$

$$Wn\_sh = Wn \cdot Fract\_sh \tag{6.4/28}$$

$$Wn\_r = Wn \cdot Fract\_r \tag{6.4/29}$$

The concentrations of nitrogen and carbon for shoot and root structure (Csh, Cr, Nsh, Nr) expressed per kg structure (units: kg C kg^{-1} structure, or kg N kg^{-1} structure) are calculated as:

$$Csh = \frac{Wc\_sh}{Wsh} \tag{6.4/30}$$

$$Cr = \frac{Wc\_r}{Wr} \tag{6.4/31}$$

$$Nsh = \frac{Wn\_sh}{Wsh} \tag{6.4/32}$$

$$Nr = \frac{Wn\_r}{Wr} \tag{6.4/33}$$

Finally the overall total plant dry mass (W) is calculated by adding the plant structure (Wg, eqn 6.4/18) and the storage (Ws, eqn 6.4/23) together:

$$W = Wg + Ws \tag{6.4/34}$$

Not all plant structural and substrate variables are needed in this section. However, they are needed for output purposes and as input variables for the plant water relationship model (Section 2.5.3).

Partitioning function

During vegetative growth, the shoots assimilate carbon and the roots acquire nutrients (e.g. nitrogen) and water. However, the available substrates are needed in all parts of the plant to sustain growth. Shoots and roots are therefore in constant interaction to distribute the substrates in the most appropriate way. There are various approaches in the literature to define partitioning functions. An excellent overview on partitioning is given in Thornley and Johnson (1990, Chapter 13). Briefly, partitioning functions can be divided into empirical, mechanistic and teleonomic functions.

Empirical functions are based on observations and are generally easy to implement, but they have the disadvantage that they are very inflexible in their response to variable conditions in the plant. They are generally not related to the physiology of the plant. Mechanistic functions are the most complex functions and are based on the mechanism of substrate transport within plants. They are the most flexible functions but are more difficult to implement in a model. Furthermore, some of the parameters of a mechanistic function are difficult to estimate.

Intermediate between the empirical and mechanistic functions are the teleonomic or 'goal-seeking' functions. Teleonomic functions are generally not more complex than empirical functions but take into account the physiological responses of plants to partitioning. The principle of this approach is that plants 'seek' to balance growth by trying to adjust the rate at which carbon is acquired in proportion to the rate of nitrogen uptake. This relationship can be expressed in mathematical terms as (see Thornley and Johnson, 1990, eqn 13.2a):

$$\mathrm{Wsh} \cdot \sigma_C \propto \mathrm{Wr} \cdot \sigma_N \tag{6.4/35}$$

where:

Wsh = structural dry matter of shoots (eqn 6.4/8) [kg structure $\cdot$ m^{-2} ground]
Wr = structural dry matter of roots (eqn 6.4/15) [kg structure $\cdot$ m^{-2} ground]
σ_C = shoot activity [kg C $\cdot$ kg^{-1} structure]
σ_N = root activity [kg N $\cdot$ kg^{-1} structure]

Insufficient C uptake by the shoot leads to an increase in the shoot:root ratio (more shoot growth than root growth). This behaviour can be interpreted as the plant, through shoot growth, seeking to pick up more carbon to balance the N supply. Similarly, insufficient N supply by the soil will decrease the shoot:root ratio (more root growth than shoot growth) to match the C supply by the shoots. The goal of the entire plant can therefore be interpreted as aiming to settle down to a fixed C/N ratio.

The derivation of the teleonomic function which is used here to adjust shoot and root growth is derived in Thornley and Johnson (1990, pp. 372–379). Since they present an excellent and easy-to-follow derivation, we shall not repeat it here.

The final partitioning function (Part) is defined as (see Thornley and Johnson, 1990, eqn 13.5n):

$$\mathrm{Part} = \frac{\mathrm{Fract\_r} \cdot \left(\mathrm{N} / (\mathrm{N} + \mathrm{fract\_N}) \right)}{\mathrm{Fract\_sh} \cdot \left(\mathrm{C} / (\mathrm{C} + \mathrm{fract\_C}) \right)} \tag{6.4/36}$$

where:

Fract_sh = fraction of shoot live material in entire plant material (eqn 6.4/19) [kg shoot structure $\cdot$ kg^{-1} structure]

Fract_r = fraction of root live material in entire plant material (eqn 6.4/20) [kg root structure · kg^{-1} structure]
fract_N = fractional nitrogen content of live plant material [kg N · kg^{-1} structure] {0.025}
fract_C = fractional carbon content of live plant material [kg C · kg^{-1} structure] {0.45}
C = concentration of carbon substrate in entire plant (eqn 6.4/21) [kg C · kg^{-1} structure]
N = concentration of nitrogen substrate in entire plant (eqn 6.4/22) [kg N · kg^{-1} structure]

The partitioning functions for shoot (Part_sh) and root (Part_r) growth are defined by (see Thornley and Johnson, 1990, eqn 13.4d):

$$\text{Part\_sh} = \frac{\text{Part}}{1 + \text{Part}} \tag{6.4/37}$$

and:

$$\text{Part\_r} = \frac{1}{1 + \text{Part}} \tag{6.4/38}$$

This teleonomic partitioning approach has been used in the grassland model by Thornley and Verberne (1989, eqns 7a–7c). The model described in Thornley (1998) uses a mechanistic partitioning function. The partitioning functions Part_sh and Part_r are applied to the rates of shoot (Gsh) and root growth (Gr), respectively.

Shoot (Gsh, units: kg structure m^{-2} ground s^{-1}), and root growth rates (Gr) are defined by (see Thornley and Johnson, 1990, eqns 13.4a, 13.4b):

$$\text{Gsh} = \text{mu} \cdot \text{C} \cdot \text{N} \cdot \text{Part\_sh} \cdot \text{Wsh} \tag{6.4/39}$$

and:

$$\text{Gr} = \text{mu} \cdot \text{C} \cdot \text{N} \cdot \text{Part\_r} \cdot \text{Wr} \tag{6.4/40}$$

with:

$$\text{mu} = \frac{\text{mu20}}{86400}$$

where:

mu20 = growth coefficient at 20°C [$day^{-1} \cdot C^{-1} \cdot N^{-1}$] {100}
C = concentration of carbon substrate in entire plant (eqn. 6.4/21) [kg C · kg^{-1} structure]
N = concentration of nitrogen substrate in entire plant (eqn 6.4/22) [kg N · kg^{-1} structure]

Wsh = shoot structure (eqn 6.4/8) [kg structure · m^{-2} ground]
Wr = root structure (eqn 6.4/15) [kg structure · m^{-2} ground]

Growth respiration (Rg, eqn 6.4/6) should take both the shoot growth rate (Gsh, eqn 6.4/39) and the root growth rate (Gr, eqn 6.4/40) into account and is now defined by (Johnson and Thornley, 1985, eqn 6e):

$$Rg = \left(\frac{1-Y}{Y}\right) \cdot (Gsh + Gr) \cdot fract\_C \qquad (6.4/41)$$

where:

Y = yield factor determining the efficiency with which carbon is used for growth [–] {0.75}
fract_C = fractional carbon content of live plant material [kg C · kg^{-1}structure] {0.45}

Substrate nitrogen and root uptake

So far, we have not defined the pool of nitrogen substrate (Wn, units: kg N m^{-2} ground). The size of the plant-available nitrogen pool depends on uptake of nitrogen (Un), the usage of N during plant growth (fract_N · (Gsh + Gr) and the supply via direct recycling of substrates from dead plant material (Dn, defined by flows: F4, eqn 6.4/7 and FR4, eqn 6.4/14):

$$\frac{dWn}{dt} = Un - fract\_N \cdot (Gsh + Gr) + Dn \qquad (6.4/42)$$

where:

Un = uptake rate for nitrogen by roots (defined below, eqn 6.4/45) [kg N · m^{-2} ground · s^{-1}]
fract_N = fractional nitrogen content of live plant material [kg N · kg^{-1} structure] {0.025}
Gsh = growth rate of shoot structure (eqn 6.4/39) [kg structure · m^{-2} ground · s^{-1}]
Gr = growth rate of root structure (eqn 6.4/40) [kg structure · m^{-2} ground · s^{-1}]
Dn = recycling of nitrogen from degraded shoot and root structures (defined below, eqn 6.4/54) [kg N · m^{-2} ground · s^{-1}]

Nitrogen uptake by the roots (Un) is calculated according to Johnson and Thornley (1985, eqn 3a). Uptake is proportional to the maximum root activity for nitrogen uptake (rootact_m, units: kg dry soil kg^{-1} root structure s^{-1}). This activity parameter is adjusted for available nitrogen in the soil, the variation in activity in the different age categories R1–R4 and the inhibition due to low levels of carbon or high levels of nitrogen substrate. The maximum specific root activity (rootact_m, units: kg dry soil kg^{-1} root structure s^{-1}) is defined by:

$$rootact\_m = \frac{rootact20}{86400} \qquad (6.4/43)$$

where:

rootact20 = root activity at 20°C [kg dry soil · kg^{-1} root structure · day^{-1}] {3000}

The maximum root activity (rootact_m), which can also be interpreted as the activity of the youngest roots (R1), is scaled down in response to various relationships. It is first adjusted for the available soil nitrogen (Ns, units: kg N kg^{-1} dry soil) with the relationship (Johnson and Thornley, 1985, eqn 3b; units: kg N kg^{-1} root structure s^{-1}):

$$root\_act = rootact\_m \cdot Ns \qquad (6.4/44)$$

To express the activity for older roots in relationship to the younger roots, dimensionless activity parameters are introduced (v2–v4), which predict that root activity falls with age. In addition, a competitive weighting function is added, which reduces the nitrogen uptake rate either when carbon is limited or at high nitrogen levels. The final equation for nitrogen uptake by roots (Un) is given by (see Johnson and Thornley, 1985, eqn 3a):

$$Un = root\_act \cdot \frac{R1 + R2 \cdot v2 + R3 \cdot v3 + R4 \cdot v4}{1 + \frac{Kc}{C} \cdot \left(1 + \frac{N}{Kn}\right)} \qquad (6.4/45)$$

where:

R1 to R4 = structural root dry mass (eqn 6.4/14) [kg structure · m^{-2} ground]
v2 to v4 = dimensionless weighting parameters [–] {0.5, 0.25, 0.1}
C carbon concentration of entire plant (eqn 6.4/21) [kg C · kg^{-1} structure]
N nitrogen concentration of entire plant (eqn 6.4/22) [kg N · kg^{-1} structure]
Kn = root activity parameter [kg N · kg^{-1} structure] {0.005}
Kc = root activity parameter [kg C · kg^{-1} structure] {0.05}

The root activity for nitrogen uptake expressed in terms of the entire root structural (Wr, eqn 6.4/15 + Wr4_dead, eqn 6.4/16) and root storage weight (Ws_r, eqn 6.4/25), is calculated as (units: kg N kg^{-1} structure + storage s^{-1}):

$$root\_act\_av = \frac{Un}{Wr + Wr4\_dead + Ws\_r} \qquad (6.4/46)$$

where:

Un = root uptake (eqn 6.4/45) [kg N · m^{-2} s^{-1}]
Wr = live root structure (eqn 6.4/15) [kg structure · m^{-2}]

Wr4_dead = dead root structure (eqn 6.4/16) [kg structure $\cdot$ m^{-2}]
Ws_r = root storage dry weight (eqn 6.4/25) [kg structure $\cdot$ m^{-2}]

A simple representation for the dynamics of soil-available nitrogen (Ns) is added to the model, which takes the combined effect of fertilizer application, leaching losses, mineralization of the biomass and root uptake into account. Fertilizer addition is supplied via user-defined inputs, while mineralization is set to a certain rate and leaching is modelled to be proportional to available nitrogen (Ns). Instead of integrating the sum of all rates (as done by Johnson and Thornley, 1985, eqn 10p), it was decided to integrate the individual rates and then calculate the mass balance. The notation describing Ns is:

$$\text{Ns} = \text{Na\_inc} + \text{BNtot} - \text{Untot} - \text{Nsleach} \tag{6.4/47}$$

where:

Na_inc = accumulated fertilizer input into the soil (from input file) [kg N $\cdot$ kg^{-1} dry soil]
Bntot = cumulative release of nitrogen due to mineralization [kg N $\cdot$ kg^{-1} dry soil]
Untot = cumulative nitrogen uptake by roots [kg N $\cdot$ kg^{-1} dry soil]
Nsleach = cumulative nitrogen lost via leaching [kg N $\cdot$ kg^{-1} dry soil]

with

$$\frac{d\text{BNtot}}{dt} = \text{BN} \tag{6.4/48}$$

$$\text{BN} = \frac{\text{BN20}}{86400}$$

$$\frac{d\text{Untot}}{dt} = \frac{\text{Un}}{\text{dr} \cdot \text{dens\_bulk} \cdot 1000} \tag{6.4/49}$$

$$\frac{d\text{Nsleach}}{dt} = \frac{\text{beta}}{86400} \cdot \text{Ns} \tag{6.4/50}$$

where:

BN20 = mineralization rate of organic matter [kg N $\cdot$ kg^{-1} dry soil $\cdot$ day] {1×10^{-6}}
dr = rooting depth [m] {0.2}
dens_bulk = soil bulk density [g soil $\cdot$ cm^{-3} soil] {1} (has to be multiplied by 1000 to obtain units: kg soil m^{-3} soil)
beta = leaching factor for soil-available nitrogen [day^{-1}] {0.2}

Nitrogen uptake is an active process where energy is required. Therefore some respirational C loss is associated with N uptake by roots (Rn_up, units: kg C m^{-2} ground s^{-1}). The notation for this respiratory loss is directly proportional to the nitrogen uptake rate (Un) (Johnson and Thornley, 1985, eqn 6g):

$$\text{Rn\_up} = \text{alphaN} \cdot \text{Un} \qquad (6.4/51)$$

where:

alphaN = parameter for the respiration cost of nitrogen uptake [kg C · kg^{-1} N] {0.5}

Some of the degraded plant structure from shoot and root (flows: F4 and FR4) will be recycled and made directly available again for plant growth. A recycling function (Fract_recy) is defined with the assumption that the recycled fraction depends on the plant nitrogen concentration (Johnson and Thornley, 1985, eqns 7j and 7k)):

$$\text{Fract\_recyc} = \frac{\text{Kd}}{\text{Kd} + \text{N}} \qquad (6.4/52)$$

where:

Kd = structural degradation parameter [N] {0.002}

The plant debris contains both carbon and nitrogen. Therefore a flow of carbon (Dc) into the carbon substrate pool Wc and a flow of nitrogen (Dn) into the nitrogen substrate Wn can be calculated. The supply of substrate carbon which enters the Wc pool (eqn 6.4/1) is:

$$\text{Dc} = \text{Fract\_recyc} \cdot \frac{\text{fracdegC} \cdot \text{fract\_N}}{\text{fracdegN}} \cdot (\text{F4} + \text{FR4}) \qquad (6.4/53)$$

and the nitrogen flow which enters the Wn pool (eqn 6.4/42):

$$\text{Dn} = \text{Fract\_recyc} \cdot \text{fract\_N} \cdot (\text{F4} + \text{FR4}) \qquad (6.4/54)$$

where:

fracdegC = fractional carbon content of degradable structure [kg C · kg^{-1} degradable structure] {0.4}
fracdegN = fractional nitrogen content of degradable structure [kg N · kg^{-1} degradable structure] {0.15}
fract_N = fractional nitrogen content of live plant material [kg N · kg^{-1} structure] {0.025}
F4 = flow of material from degrading shoots (eqn 6.4/7) [kg structure · m^{-2} ground]
FR4 = flow of material from degrading roots (eqn 6.4/14) [kg structure · m^{-2} ground]

Change of notation from the previous section
The notation of the specific leaf area index (SLA, eqn 6.4/11) is changed to (Johnson and Thornley, 1985, eqn 5d):

$$SLA = SLAm \cdot (1 - ISLA \cdot C) \quad (6.4/55)$$

where:

SLAm = maximum specific leaf area [m^2 leaf · kg^{-1} structure] {25}
ISLA = incremental specific leaf area parameter [C^{-1}] {2.5}
C = concentration of carbon in plant structure (eqn 6.4/21) [kg C · kg^{-1} structure]

Since the live plant material is defined in terms of structure (and not carbon), the new notation for maintenance respiration Rm_sh (eqn 6.4/10) is (see also comments before eqn 6.4/17):

$$Rm\_sh = fract\_C \cdot (Msh1 \cdot W1 + Msh2 \cdot W2 + Msh3 \cdot W3 + Msh4 \cdot W4) \quad (6.4/56)$$

where:

fract_C = fractional carbon content of live plant material [kg C · kg^{-1} structure] {0.45}

After adjusting the substrate carbon (Wc, eqn 6.4/1) for the input flow of carbon from degrading shoots and roots (Dc, eqn 6.4/53) and respiration due to nitrogen uptake (Rn_up, eqn 6.4/51) the notation has to be expanded to (defining components P and C_use (eqn 6.4/4) directly in compartment Wc):

$$\frac{dWc}{dt} = \frac{Pc \cdot Molwt\_C}{Molwt\_CO2} - \frac{fract\_C}{Y} \cdot (Gsh + Gr) - Rm\_sh - Rm\_rt - Rn\_up + Dc \quad (6.4/57)$$

where:

Pc = instantaneous canopy photosynthesis (eqt. 6.3/41)
Rm_sh = shoot maintenance respiration (eqn 6.4/56) [kg C · m^{-2} · s^{-1}]
Rm_r = root maintenance respiration (eqn 6.4/17) [kg C · m^{-2} · s^{-1}]
Rn_up = respiration due to nitrogen uptake (eqn 6.4/51) [kg C · m^{-2} · s^{-1}]
Gsh = growth rate of shoot structure (eqn 6.4/39) [kg structure · m^{-2} ground · s^{-1}]
Gr = growth rate of root structure (eqn 6.4/40) [kg structure · m^{-2} ground · s^{-1}]
Y = yield factor determining the efficiency with which carbon is used for shoot growth [–] {0.75}
fract_C = fractional carbon content of live plant material [kg C · kg^{-1} structure] {0.45}

Dc = flow of carbon due to degrading shoot and root structure (eqn 6.4/53) [kg C · m^{-2} · s^{-1}]
Wc_init = initial amount of carbon substrate [kg C · m^{-2} ground] {0.016}

Some parameters in Section 6.4.2 are redefined according to the settings given in Johnson and Thornley (1985). They are changed to:

alpha = {1×10^{-8}} (used in eqn 6.3/40)
Pm_0 = {6×10^{-7}} (used in eqn 6.3/40)
Wsh_init = {0.05} (used in eqn 6.4/7)
mu20 = {100} (used in eqn 6.4/39 and 6.4/40)

Application in ModelMaker

We have to add rather a lot of notation to the previous model *Mod6–1d.mod*. However, to obtain a working model, all the definitions in the last section have to be added together, because most of them depend on each other. The complete model is presented in *Mod6–1e.mod* (Fig. 6.7). Figure 6.7 indicates which components have to be inserted into the previous model (*Mod6–1d.mod*). Below is a summary of the various components which have to be added (global components are marked by an asterisk '*'):

<compartments>:
'R1', 'R2', 'R3', 'R4','Wc', 'Wn', 'Untot', 'BNtot', 'Nsleach'

<variables> :
'Rm_r', 'Wr', 'Wr4_dead', 'Gr' 'Part', 'Part_sh', Part_r', 'Ws', 'Ws_sh'*, 'Ws_r'*, 'Wg', 'W', 'C'*, 'N'*, 'Fract_sh'*, 'Fract_r'*, 'Wc_sh', Wc_r', 'Csh'*, 'Cr'*, 'Wn_sh', 'Wn_r', 'Nsh'*, 'Nr'*, 'Un', 'Ns', 'Root_act', 'root_act_av', 'Na', 'Rg', 'Rn','Fract_recyc', 'Dc', 'Dn', 'gamma_r'*, 'Bn'*, 'rootact_m'*, 'Mr1'*, 'Mr2'*, 'Mr3'*, 'Mr4'*

<parameters>
'alphaN', 'beta', 'BN20', 'dens_bulk', 'dr', 'f_R1', 'fracdegC', 'fracdegN', 'frac_C', 'frac_N', 'gamma_r20', 'Kc', 'Kd', 'Kn', 'Molwt_Csubst', 'Molwt_N', 'Molwt_Nsubst', 'Mr1_20', 'Mr2_20', 'Mr3_20', 'Mr4_20', 'R1init', 'R2init', 'R3init', 'R4init', 'rootact20', 'v2', 'v3', 'v4'

<define>
'Na_inc'

<component event>
'Na_cum'

The only ModelMaker component which has not been used so far is a <component event> 'Na_cum'. This component is needed to accumulate the fertilizer input to the system (supplied via an input file). The statement in the <component event> is:

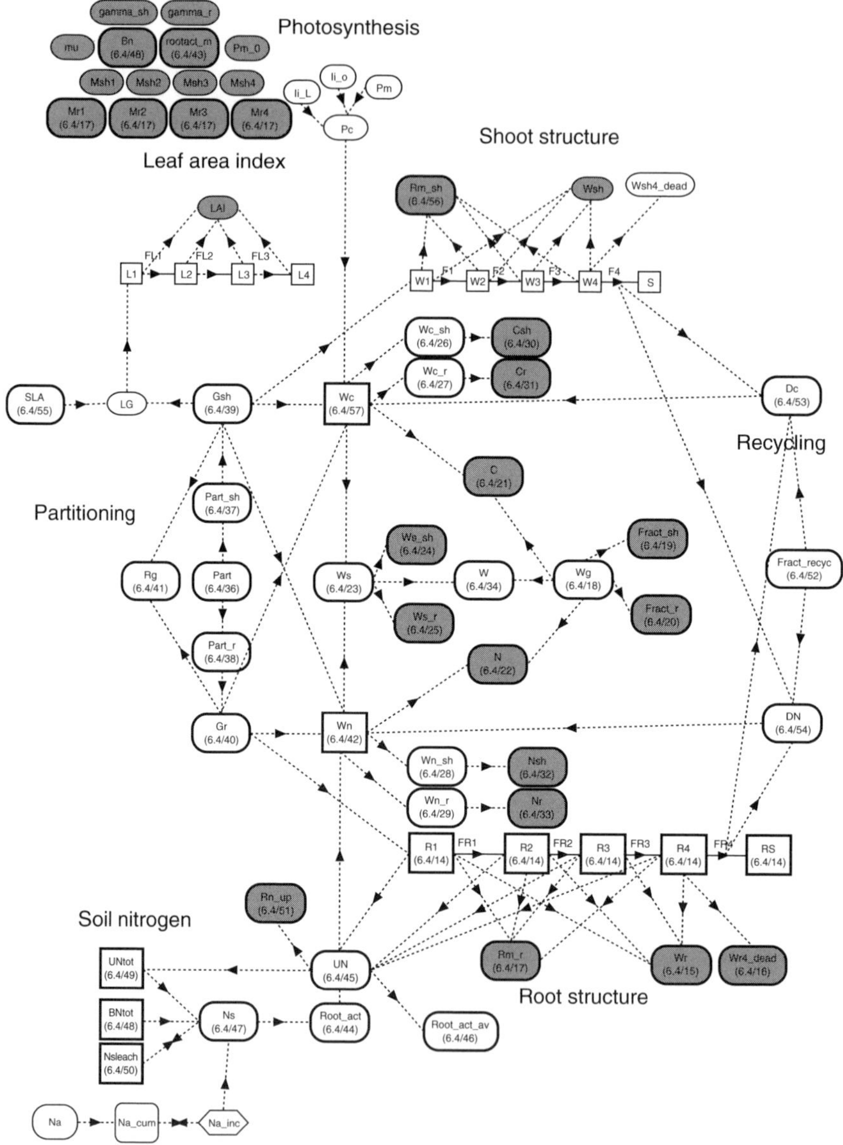

Fig. 6.7. Plant growth model including shoot and root growth, and substrate relationships (according to Johnson and Thornley, 1985) (*Mod6-1e.mod*) (see also Fig. 6.6).

Na_inc = Na_inc + Na;

The <define> 'Na_inc' holds the total amount of fertilizer applied.

Running the model

The model is set up, in the same way as the previous model, for 80 days and hourly outputs. There are numerous output options for the model, which cannot be demonstrated here. It is up to the user to discover the dynamics of the model. However, as an example, the output for leaf area index is presented. Recall that some parameters were changed in this current model (*Mod6–1e.mod*), compared with the previous model (*Mod6–1d.mod*). The changes included the Pm_0 value from 1.5×10^{-6} (*Mod6–1d.mod*) to 6×10^{-7} (*Mod6–1e.mod*). To see the effect of this crucial parameter for plant growth, both models (*Mod6–1d.mod* and *Mod6–1e.mod*) are executed with both Pm_0 values. The effect of the Pm_0 parameter on both models is illustrated, with the output of the leaf area index (LAI), in Fig. 6.8.

Model *Mod6–1d.mod* (Fig. 6.8A) predicts for both Pm_0 values higher LAI values than with *Mod6–1e.mod*, which is clearly an effect of the non-existing root system, where some of the carbohydrates would otherwise have gone. Generally, a higher maximum photosynthetic rate leads to higher LAI values. This tendency also exists for the second model (*Mod6–1e.mod*, Fig. 6.8B). However, the additional root system and the teleonomic approach, especially during rapid plant growth (high Pm_0 value), have an effect. In the second model, fertilizer was applied on day 1 and day 40 (Fig. 6.8B indicated by the arrows), each at rates of 100 kg N ha^{-1}. To interpret the LAI results (Fig. 6.8B), it is helpful to have a look at the structural dry weight for roots and shoots under the two settings for Pm_0 (Fig. 6.9) which reflects the dynamics of the teleonomic partitioning approach.

The root system under the high Pm_0 value does not seem to be large enough to match the carbon input through photosynthesis (Fig. 6.9B). The consequence is that the root system has to grow to meet the requirements of the teleonomic partitioning approach. However, the situation is different for the low Pm_0 value, because the root system is even slightly declining (Fig. 6.9A), indicating that the root structural live weight is more than adequate to match the carbon input through photosynthesis. At times, when the soil nitrogen is depleting, the growth of the root system under the high Pm_0 regime (Fig. 6.9B) is even more rapid. However, there is a point where the roots simply cannot grow fast enough to meet this ever-increasing shoot live structure. This is the point where the shoot live structure actually starts to decline (Fig. 6.9B). If nitrogen is applied to the soil again, the shoots may grow a bit further, but the same situation occurs again, once the soil nitrogen concentration has obtained a low value. We can interpret the results as an indication that carbon is going into roots to 'seek' nitrogen to match the high C input through the upper plant parts. The high Pm_0 value (Fig. 6.9B) probably leads to some oscillation in the shoot structure. This feature is also reflected in the LAI dynamics (Fig. 6.8B).

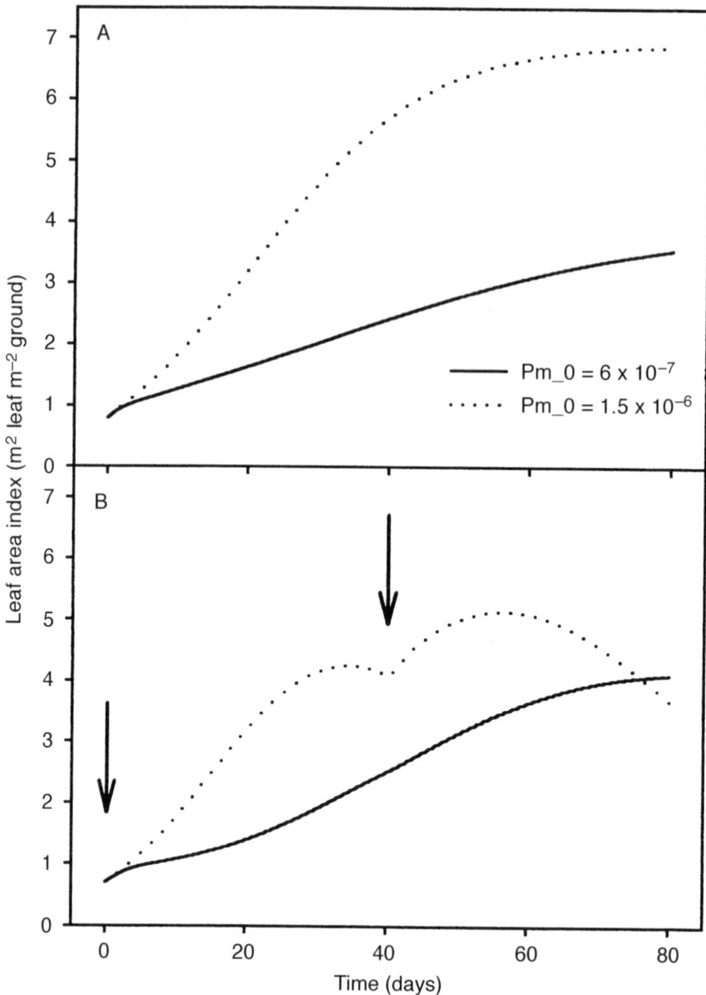

Fig. 6.8. Leaf area index (LAI) at two settings for Pm_0 calculated with the model *Mod6-1d.mod* (A) and *Mod6-1e.mod* (B) (arrows indicate the times of fertilizer applications).

This little simulation experiment already shows that the outcome of the model leads us to think about the simulation results, which may lead to a better understanding of the system. The result of this exercise is that the lower Pm_0 is probably more suitable than the high value, because it avoids oscillation of the output (possibly this was the reason why Johnson and Thornley (1985) changed Pm_0 to a lower value).

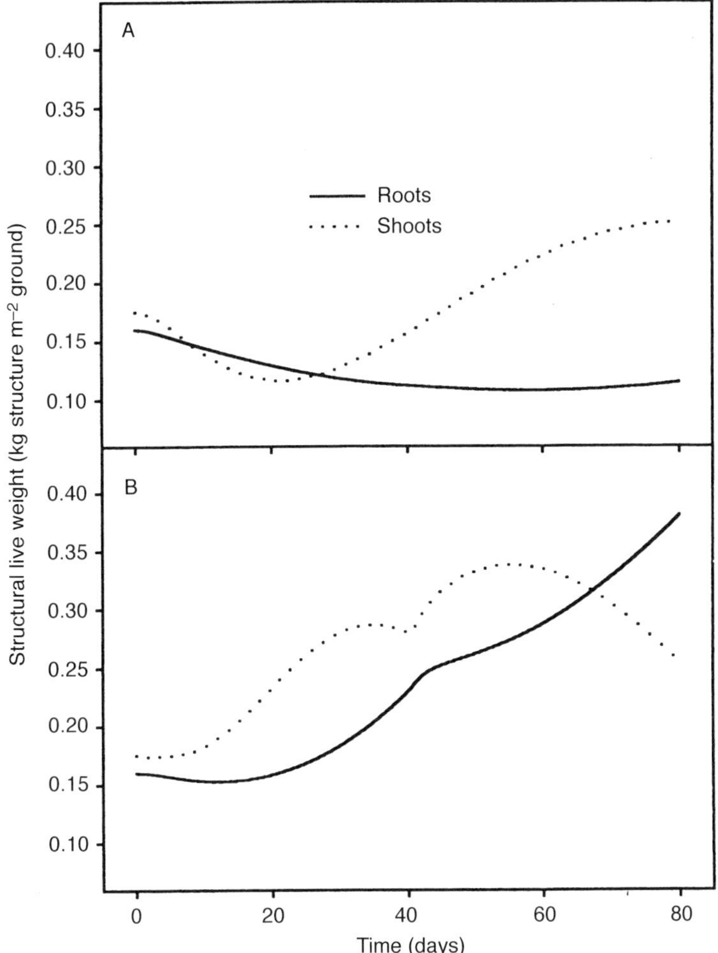

Fig. 6.9. Structural shoot and root live weight at two different settings for Pm_0 (A: Pm_0 = 6×10^{-7}; B: Pm_0 = 1.5×10^{-6}) using *Mod6-1e.mod*.

It is advisable to carry out more of these simulation exercises in order to become familiar with the dynamic behaviour of the model and to observe the effect on various parts of the plant growth simulator when certain parameters are changed.

So far, we have neglected any environmental influence, such as temperature or moisture, on plant growth. However, a plant growth model would not be complete without such considerations. Therefore, in the next section, we shall complement the model by the addition of a temperature function and a plant–water relationship model.

6.5 Plant Growth – Relationships with Environmental Factors

6.5.1 Introduction

In this section, we consider the influence of environmental factors on plant growth. The main foci are the role of temperature on the various processes and the development of an additional submodel to simulate plant water status. The model for plant–water relations was developed as one of the last additions to the grassland ecosystem model by Thornley and colleagues (Thornley, 1996). The plant water model is also described by Thornley (1998). The model is designed to link the notations for internal plant substrates and variable root:shoot partitioning and root density to the physical chemistry of plant water. The model also simulates the effect of environmental conditions on the energy relationships and the related loss of water via transpiration.

6.5.2 Temperature

In Section 6.4, we defined some rate parameters for plant growth (e.g. mu20, gamma_sh20), with values being representative for a temperature of 20°C (extension: '_20'). These parameters change if the temperatures are different from the reference temperature. The plant parameters are divided into parameters which operate above and below ground. Therefore temperature adjustments are calculated with either soil temperature (*Tso*) for below-ground parameters, or air temperature (*Ta*) for above-ground parameters.

To calculate the effect of temperature, the dimensionless function described by Thornley (1998, p. 54, eqn 3.11a) is used:

for T_0min < T < T_0max:

$$\mathrm{fT} = \mathrm{mfT} \cdot \frac{(\mathrm{T} - \mathrm{T\_0\,min})^{\mathrm{qfT}} \cdot (\mathrm{T\_0\,max} - \mathrm{T})}{(\mathrm{Tref} - \mathrm{T\_0\,min}) \cdot (\mathrm{T\_0\,max} - \mathrm{Tref})} \qquad (6.5/1)$$

else fT = 0

where:

T_0min =	lower temperature limit for function [°C] {0}
T_0max =	upper temperature limit for function (fT = 0) [°C] {45}
Tref =	reference temperature at which the rate parameters are defined [°C] {20}
qfT =	parameter defining the skewness of the temperature function [–] {2}
mfT =	parameter for the temperature function [-] {1}
T =	temperature: either air temperature (Ta) or soil temperature (Tso) is used [°C]

At the reference temperature (*Tref*), the temperature function has a value of 1. Below and above this temperature, the function takes on values either below or above unity. For above-ground parameters, the function (fT_sh) is calculated with the air temperature (*Ta*) and, for below-ground parameters, the function (fT_r) is calculated with the soil temperature (*Tso*). The function ft_sh is applied by multiplication to the following <variables>:

'gamma_sh', 'Pm_0', 'Msh1', 'Msh2', 'Msh3', 'Msh4', 'Gsh'

the function 'fT_r' is applied to the following <variables>:

'gamma_r', 'Bn', 'rootact_m', 'Mr1', 'Mr2','Mr3','Mr4', 'Gr'

Application in ModelMaker

The air and soil temperatures are supplied to the model via the input file 'Input'. These additional variables are contained in the input file *Inp6–1f.txt*. The two <variables> 'fT_sh' and 'fT_r' are added to the previous model (*Mod6–1e.mod*), with the common parameters specified by eqn 6.5/1. In order to allow the user to perform the simulation either with or without the temperature adjustment, an <independent event> 'Choice' has been added to the model, which contains the statement:

```
fT_choice=GetChoice("Calculate temperature adjustment?
    (Y/N)","Temperature adjustment",fT_choice);
```

The two variables fT_sh and fT_r are calculated according to the value of the <define> 'fT_choice'. If the user decides to calculate temperature adjustments, the calculation is carried out according to eqn 6.5/1, otherwise the two temperature functions are set to unity. The final model is contained in the file *Mod6–1f.mod.*

The effect of the temperature adjustment on the simulation output is presented in Fig. 6.10 for the leaf area index (LAI). The model is executed with and without this temperature adjustment (Fig. 6.10). The results show an increased LAI without the calculation of temperature adjustments. This result shows the effect of the air and soil temperature, which are, on average, above the reference temperature (*Tref* = 20°C) (see input file *Inp6–1f.txt*). This means that the temperature function obtains values above unity and that therefore the growth rates, the maintenance respiration rates and the transformation rates between the leaf compartments are enhanced. This combination leads to overall reduced LAI values. It is also interesting to observe the change in the respiration (especially maintenance respiration for both shoots Rm_sh and roots Rm_r), which shows a diurnal pattern if the temperature adjustments are calculated. This reflects the temperature influence on the maintenance respiration parameters (Msh1–4 and Mr1–4).

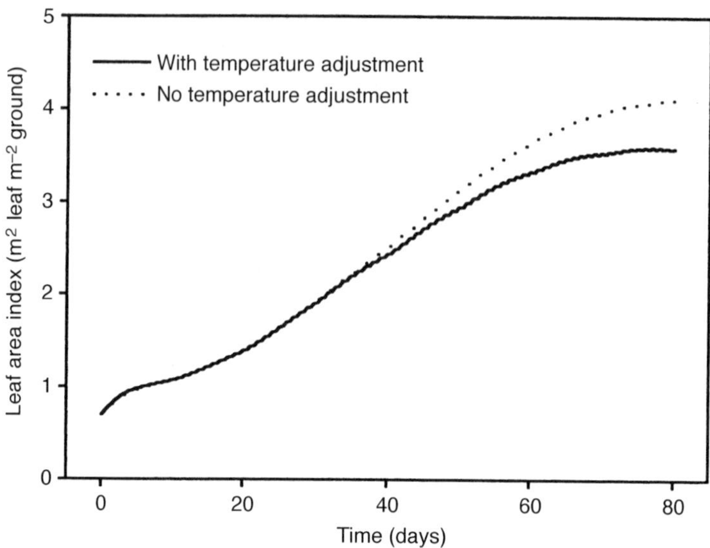

Fig. 6.10. Leaf area index calculated with and without temperature adjustment (*Mod6-1f.mod*).

6.5.3 *Plant water status*

The soil–plant–atmosphere continuum

An essential element for the various biochemical reactions in green plants is water. In the plant growth models developed so far, it was assumed that water was not a limiting factor for plant growth. However, plants will lose water to the atmosphere as a result of the evaporative demand of the atmosphere. This loss of water, from plants in the form of water vapour, is referred to as transpiration. The combined loss of water from soil (evaporation, see Chapter 5) and transpiration is referred to as 'evapotranspiration'. Plants can be regarded as an additional interface for water transfer between the soil and the atmosphere. In order to grow uninhibitedly, the plant has to balance the rate of water loss by uptake of water through the root system. Any imbalance in the rates of water movement within this soil–plant–atmosphere system will result in reduced rates of biochemical processes (reduced growth) and subsequent closure of the main pathways for water loss to the atmosphere (closure of stomata). However, plants try to keep the stomata open as long as possible, because these are also the elements where CO_2 is entering the system, CO_2 being needed for photosynthesis. One of the positive effects while plants transpire water may be the prevention of excessive heating of the plant tissue through radiation (Hillel, 1998, p. 552). The interaction between soil, plant and atmosphere is often referred to as the soil–plant–atmosphere continuum (SPAC). An excellent overview of the SPAC system is presented by Hillel

(1998). In this system, a continuous flow of water occurs from the soil to the roots and through the xylem to the leaves and is finally transpired through the stomata in the leaves to the atmosphere.

The flow of water in the SPAC system happens, in analogy to the soil water movement (Chapter 4), according to water potential gradients and the hydraulic properties within the various parts of the system (e.g. hydraulic conductivity in soil). The basic notation for flow has already been introduced in Chapter 1 (eqn 1.3/1), where the proportionality factor 'k' defines the hydraulic conductivity of soil. However, the flows in the SPAC system are usually expressed by water potential gradients ($d\Psi$) divided by appropriate resistances (R) to water flow (Hillel, 1977, p. 156; Campbell, 1985, p. 122). Such a notation was already used for the description of the evaporation rate from bare soil to the atmosphere (Chapter 5, eqn 5.6/4). The general notation for flow with the resistance term is:

$$F = -\frac{d\Psi}{R} \qquad (6.5/2)$$

However, either the resistance or the conductance term may be used because they are related via eqn 5.5/1. If we consider flows of water from soil to root (Fso_rt), root to shoot (Frt_sh) and shoot to atmosphere, the SPAC system can be described by the water potential and resistances, as illustrated in Fig. 6.11.

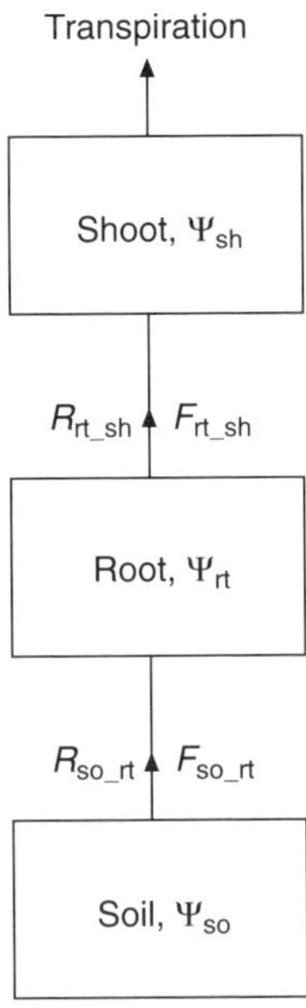

Fig. 6.11. Schematic representation of the flows of water (*F*) between the various parts of the SPAC system (see text for details).

Plants always try to keep the various flows of water in balance in order to maximize plant growth. During a period of high transpiration, the flow of water from the root to the shoot and soil to the root will have to be adjusted to replace the transpired water. The main mechanism in plants to achieve this is the lowering of the tissue water potential in the various parts of the plants.

Plant water potentials

The two main components of the water potential in plants are the osmotic potential (see Section 4.2) and the pressure potential (Thornley, 1996). Usually the matric potential in plant tissue is considered as either controversial or negligible (Thornley and Johnson, 1990, p. 442; Hillel, 1998, p. 553). Therefore the overall tissue water potential (Ψ_{pl}) is calculated as:

$$\Psi_{pl} = \Psi_{os} + \Psi_{p} \tag{6.5/3}$$

where:

Ψ_{os} = osmotic potential $[J \cdot kg^{-1}]$
Ψ_{p} = pressure potential $[J \cdot kg^{-1}]$

The dimensions for potential in this section are the energy units: J kg^{-1} water (for conversions to other units, see Table 4.1). This is mainly done to keep the notation as close as possible to the original by Thornley (1996).

The osmotic potential (Ψ_{os}) is mainly related to the action of osmotically active plant storage material (mainly ions, such as K^+, Cl^-, anions of organic acids) and the plant water content. The osmotic potential is defined by the Van't Hoff equation (see Thornley, 1996, eqn 1b) for shoot (POTos_sh) and root (POTos_r) separately:

$$\mathrm{POTos\_sh} = -\frac{R \cdot (Ta + 273.16) \cdot \mathrm{Fos\_sh} \cdot \mathrm{Ws\_sh}}{{}^{\mathrm{molstorage}}\!/_{1000} \cdot \mathrm{Mw\_sh}} \tag{6.5/4}$$

and:

$$\mathrm{POTos\_r} = -\frac{R \cdot (Tso + 273.16) \cdot \mathrm{Fos\_r} \cdot \mathrm{Ws\_r}}{{}^{\mathrm{molstorage}}\!/_{1000} \cdot \mathrm{Mw\_r}} \tag{6.5/5}$$

where

R = gas constant $[J \cdot mol^{-1} \cdot K^{-1}]$ {8.314}
Ta = air temperature (Section 6.5.2) [°C] {input}
Tso = soil temperature (Section 6.5.2) [°C] {input}
Fos_sh = fraction of osmotically active shoot storage mass [–] {2}
Fos_r = fraction of osmotically active root storage mass [–] {2}
Ws_sh = shoot storage mass (eqn 6.4/24) $[kg \cdot m^{-2}$ ground]
Ws_r = root storage mass (eqn 6.4/25) $[kg \cdot m^{-2}$ ground]

molstorage = molecular weight of storage compounds (division by 1000 is needed to convert to kg mol^{-1}) [g · mol^{-1}] {20}
Mw_sh = water content of shoots (see below, eqn 6.5/40) [kg water · m^{-2} ground]
Mw_r = water content of roots (see below eqn 6.5/39) [kg water · m^{-2} ground]

The pressure potential (Ψ_p) refers to the positive pressure, called turgor, within the cells, which is essential for cell division and cell expansion. The pressure (positive potential) in cells is a function of the water content and the structural dry mass of the plant. The elasticity of the cells, the cell wall thickness and the cell size are important aspects which determine the pressure potential within the cells. A detailed derivation of the various components defining the pressure potential in plants mathematically is presented by Thornley and Johnson (1990, pp. 440–447). The pressure potential of shoot (POTpr_sh) and root (POTpr_r) are calculated by (Thornley, 1996, eqn 1c):

$$\mathrm{POTpr\_sh} = \frac{e \cdot \left(\frac{\mathrm{Cpr} \cdot \mathrm{Mw\_sh}}{\mathrm{Wsh}} - 1\right)}{\mathrm{dens\_H2O} \cdot 1000} \tag{6.5/6}$$

and:

$$\mathrm{POTpr\_r} = \frac{e \cdot \left(\frac{\mathrm{Cpr} \cdot \mathrm{Mw\_r}}{\mathrm{Wr}} - 1\right)}{\mathrm{dens\_H2O} \cdot 1000} \tag{6.5/7}$$

where:

e = cell wall rigidity parameter [Pa] {2000000}
Cpr = parameter affecting pressure component of plant water potential [kg structure · kg^{-1} water] {0.2}
Mw_sh = water content of shoots (see below, eqn 6.5/40) [kg water · m^{-2} ground]
Mw_r = water content of roots (see below, eqn 6.5/39) [kg water · m^{-2} ground]
Wsh = structural dry live weight of shoots (eqn 6.4/8) [kg structure · m^{-2} ground]
Wr = structural dry live weight of roots (eqn 6.4/15) [kg structure · m^{-2} ground]
dens_H2O = density of water [g · cm^{-3}] {1} (has to be multiplied by 1000 to obtain units: kg m^{-3})

(Note: parameter values *e* and *Cpr* are taken from Thornley (1998, p. 128). In the original publication (Thornley, 1996), *e* had a value of 100,000 and *Cpr* a value of 0.32).

The total water potential in the shoot (POTsh) and root (POTr) is calculated, according to eqn 6.5/3 as:

$$\text{POTsh} = \text{POTos\_sh} + \text{POTpr\_sh} \tag{6.5/8}$$

and:

$$\text{POTr} = \text{POTos\_r} + \text{POTpr\_r} \tag{6.5/9}$$

If the water-supply of the plant is adequate, the pressure potential is high enough to keep the cells under full turgor. At the same time, the high tissue water content will cause a dilution to the osmotically active substances, which results in an overall less negative plant potential. Under conditions when the water-supply to the cells is not fast enough to counterbalance the transpiration rate, the tissue water content and the turgor will decrease. At the same time, the concentration of osmotically active substances will increase, which leads to the overall effect that the plant water potential obtains a more negative value. Water stress is therefore indicated by wilting, a sign that the plant is trying to lower its overall potential to maintain water flow through the SPAC system.

The water stress of the plant is reflected by the actual water content relative to the water content when the tissue would be fully turgid (i.e. plant water potential $\Psi_{pl} = 0$). The relative water content of the shoot (MwRel_sh) and root tissue (MwRel_r) is calculated by (Thornley, 1996, eqns 1e and 1f):

$$\text{MwRel\_sh} = \frac{\text{Mw\_sh}}{\text{Mw\_sh\_0}} \tag{6.5/10}$$

and:

$$\text{MwRel\_r} = \frac{\text{Mw\_r}}{\text{Mw\_r\_0}} \tag{6.5/11}$$

The shoot and root water contents at zero plant water potential (Mw_sh_0 and Mw_r_0) have to be calculated. First, the total tissue water potentials are calculated via eqns 6.5/8 and 6.5/9 (including eqn 6.5/4–6.5/7). After multiplication of the resulting expressions with the water contents Mw_sh (eqn 6.5/40) and Mw_r (eqn 6.5/39) and rearranging, we obtain the following quadratic equations (Thornley, 1996, eqn 1f):

$$0 = \left\{ \begin{array}{l} \dfrac{\text{e} \cdot \text{Cpr}}{\text{dens\_H2O} \cdot 1000 \cdot \text{Wsh}} \cdot \text{Mw\_sh}^2 \\ -\left(\text{POTsh} + \dfrac{\text{e}}{\text{dens\_H2O} \cdot 1000} \right) \cdot \text{Mw\_sh} \\ -\dfrac{\text{R} \cdot (\text{Ta} + 273.16) \cdot \text{Fos\_sh} \cdot \text{Ws\_sh}}{\text{molstorage}/1000} \end{array} \right\}$$

and:

$$0 = \left\{ \begin{array}{l} \dfrac{e \cdot Cpr}{dens\_H2O \cdot 1000 \cdot Wr} \cdot Mw\_r^2 - \left(POTr + \dfrac{e}{dens\_H2O \cdot 1000} \right) \cdot Mw\_r \\ -\dfrac{R \cdot (Tso + 273.16) \cdot Fos\_r \cdot Ws\_r}{molstorage/1000} \end{array} \right\}$$

After setting the shoot (POTsh) and root (POTr) water potential to zero the above equations can be solved for Mw_sh and Mw_r with the procedure for solving quadratic equations described in Section 1.2.3, eqns 1.2/2 and 1.2/3. For instance the shoot water content (Mw_sh) can be obtained by setting:

$$A = \frac{e \cdot Cpr}{dens\_H2O \cdot 1000 \cdot Wsh};$$

$$B = POTsh + \frac{e}{dens\_H2O \cdot 1000};$$

$$C = -\frac{R \cdot (Ta + 273.16) \cdot Fos\_sh \cdot Ws\_sh}{molstorage/1000}; \; POTsh = 0.$$

The results for the shoot and root water contents at zero water potential are:

$$Mw\_sh\_0 = \left\{ \begin{array}{l} \dfrac{e}{dens\_H2O * 1000} + \\ \sqrt{\left(\dfrac{e}{dens\_H2O \cdot 1000} \right)^2 - \dfrac{4 \cdot e \cdot Cpr}{dens\_H2O \cdot 1000 \cdot Wsh} \cdot \dfrac{R \cdot (Ta + 273.16) \cdot Fos\_sh \cdot Ws\_sh}{molstorage/1000}} \end{array} \right\} \quad (6.5/12)$$

and

$$Mw\_r\_0 = \left\{ \begin{array}{l} \dfrac{e}{dens\_H2O * 1000} + \\ \sqrt{\left(\dfrac{e}{dens\_H2O \cdot 1000} \right)^2 - \dfrac{4 \cdot e \cdot Cpr}{dens\_H2O \cdot 1000 \cdot Wr} \cdot \dfrac{R \cdot (Tso + 273.16) \cdot Fos\_r \cdot Ws\_r}{molstorage/1000}} \end{array} \right\} \quad (6.5/13)$$

e =	cell wall rigidity parameter [Pa] {2000000}
Cpr =	parameter affecting pressure component of plant water potential [kg structure · kg^{-1} water] {0.2}
Mw_sh =	water content of shoots (see below, eqn 6.5/40) [kg water · m^{-2} ground]
Mw_r =	water content of roots (see below eqn 6.5/39) [kg water · m^{-2} ground]
Wsh =	structural dry live weight of shoots (eqn 6.4/8) [kg structure · m^{-2} ground]
Wr =	structural dry live weight of roots (eqn 6.4/15) [kg structure · m^{-2} ground]
Ws_sh =	dry weight of shoot storage material (eqt. 6.4/24) [kg storage · m^{-2} ground]
Ws_r =	dry weight of root storage material (eqt. 6.4/25) [kg storage · m^{-2} ground]
dens_H2O =	density of water [g · cm^{-3}] {1} (has to be multiplied by 1000 to obtain units: kg m^{-3})
molstorage =	molecular weight of storage compounds (division by 1000 is needed to convert to kg mol^{-1}) [g · mol^{-1}] {20}
Fos_sh =	fraction of osmotically active shoot storage mass [–] {2}
Fos_r =	fraction of osmotically active root storage mass [–] {2}

Stomatal control

In the dark, stomata are generally closed. During the daytime, when photosynthetic active radiation (PAR) is above zero, the key factor for opening and closing the stomata is the turgor pressure of the guard cells. Under low turgor, the stomata close and at high turgor the stomata are open (Hillel, 1998, p. 554). Therefore it is reasonable to relate the stomatal control to the relative shoot water potential (MwRel_sh). The conductance of water vapour through the stomata will range between a minimum conductance value (gs_min), when the relative shoot water is low, to a maximum conductance value (gs_max), when the stomata are fully open. Thornley (1996) uses a linear relationship between the minimum and maximum stomatal conductance, which can also be defined by the corresponding range in relative shoot water content: Mw_gs_min at gs_min and Mw_gs_max at gs_max. Stomatal conductances below Mw_gs_min are set to gs_min and above Mw_gs_max to gs_max. In addition to the turgor, the concentration of CO_2 affects the stomatal conductance (Hillel, 1998). The notation for the stomatal conductance (gs), related to radiation, relative shoot water content and carbon dioxide, is defined as (see also Thornley, 1998, eqn 3.2u):

if MwRel_sh < Mw_gs_min or PAR = 0:

gs = gs_min

if Mw_gs_min < MwRel_sh < Mw_gs_max and PAR > 0:

$$gs = gs\_min + \frac{(MwRel\_sh - Mw\_gs\_min) \cdot (gs\_max - gs\_min)}{Mw\_gs\_max - Mw\_gs\_min}$$

if MwRel_sh > Mw_gs_max and PAR > 0:

$$gs = gs\_max \tag{6.5/14}$$

with:

$$gs\_min = gs\_min\_350 \cdot fgs\_CO2$$
$$gs\_max = gs\_max\_350 \cdot fgs\_CO2$$

$$fgs\_CO2 = \frac{1 + CgsCO2}{1 + CgsCO2 \cdot \dfrac{CO2conc}{350}}$$

where:

gs_min_350 = minimum stomatal conductance at 350 ppm CO_2 concentration $[m \cdot s^{-1}]$ {0.0005}
gs_max_350 = maximum stomatal conductance at 350 ppm CO_2 concentration $[m \cdot s^{-1}]$ {0.005}
fgs_CO2 = function to adjust stomatal conductance for CO_2 concentration [–]
CgsCO2 = parameter describing the dependence of stomatal conductance on CO_2 [–] {2}
CO2conc = ambient CO_2 concentration [ppm] {350}
Rn = net radiation (input, see below) $[W \cdot m^{-2}]$

For the entire canopy, we can derive a canopy conductance (gc), which Thornley (1996) calculates by:

$$gc = gs \cdot LAI \tag{6.5/15}$$

where

gs = stomatal conductance (eqn 6.5/14) $[m \cdot s^{-1}]$
LAI = leaf area index (eqn 6.4/13) $[m^2\ leaf \cdot m^{-2}\ ground]$

Transpiration

Transpiration from the canopy can be calculated in a similar way to evaporation (Chapter 5) by introducing energy balance considerations. In order to calculate the combined flow of evaporation and transpiration, the well-known Penmann–Monteith equation can be used, which is described in detail by Monteith and Unsworth (1990, p. 247) and Hillel (1998, pp. 589ff.). This equation provides a notation for the latent energy flux (water vapour flow) from the surface to the atmosphere. Thornley (1996) used this equation to calculate transpiration from the canopy. He made the assumption that the ground heat flux (G) (and therefore evaporation from the soil) is zero, which is close to reality if we are dealing with a dense canopy, as in grassland (see Thornley, 1998, p. 131). The Penman–Monteith equation is defined by (see Monteith and Unsworth, 1990, eqn 15.4; Thornley, 1998, eqn 6.6d the psychrometer constant is missing in the numerator!):

$$\mathrm{LE} = \frac{\mathrm{Del\_vp} \cdot \mathrm{Rn\_abs} + \mathrm{Psy\_const} \cdot \mathrm{L} \cdot \mathrm{Kh} \cdot \mathrm{VPD}}{\mathrm{Del\_vp} + \mathrm{Psy\_const} \cdot \left(1 + \frac{\mathrm{Kh}}{\mathrm{gc}}\right)} \tag{6.5/16}$$

where:

Del_vp = change of saturation vapour pressure (eqn 5.8/7) $[\mathrm{Pa} \cdot \mathrm{K}^{-1}]$
Rn_abs = net radiation absorbed by the canopy (the ground heat flux, G, is set to zero) $[\mathrm{W} \cdot \mathrm{m}^{-2}]$
Psy_const = psychrometer constant (see below eqn 6.5/19) $[\mathrm{Pa} \cdot \mathrm{K}^{-1}]$
L = latent heat of vaporization (eqn 5.8/8) $[\mathrm{J} \cdot \mathrm{g}^{-1}]$
Kh = boundary layer conductance (see below, eqn 6.5/23) $[\mathrm{m} \cdot \mathrm{s}^{-1}]$
VPD = vapour pressure difference (see below, eqn 6.5/18) $[\mathrm{g} \cdot \mathrm{m}^{-3}]$
gc = canopy conductance (eqn 6.5/15) $[\mathrm{m} \cdot \mathrm{s}^{-1}]$

The Penmann–Monteith equation (eqn 6.5/16) is expressed in the units given by Monteith and Unsworth (1990) as J $\mathrm{m}^{-2}\,\mathrm{s}^{-1}$:

$$[\mathrm{LE}] = \frac{\left[\frac{\mathrm{Pa}}{\mathrm{K}}\right] \cdot \left[\frac{\mathrm{W}}{\mathrm{m}^2}\right] + \left[\frac{\mathrm{Pa}}{\mathrm{K}}\right] \cdot \left[\frac{\mathrm{J}}{\mathrm{g}}\right] \cdot \left[\frac{\mathrm{m}}{\mathrm{s}}\right] \cdot \left[\frac{\mathrm{g}}{\mathrm{m}^3}\right]}{\left[\frac{\mathrm{Pa}}{\mathrm{K}}\right] + \left[\frac{\mathrm{Pa}}{\mathrm{K}}\right]}$$

simplified to:

$$\left[\frac{\mathrm{W}}{\mathrm{m}^2}\right] + \left[\frac{\mathrm{J}}{\mathrm{g}}\right] \cdot \left[\frac{\mathrm{m}}{\mathrm{s}}\right] \cdot \left[\frac{\mathrm{g}}{\mathrm{m}^3}\right]$$

$$\left[\frac{\mathrm{W}}{\mathrm{m}^2}\right] + \left[\frac{\mathrm{J}}{\mathrm{s} \cdot \mathrm{m}^2}\right] = \left[\frac{\mathrm{W}}{\mathrm{m}^2}\right] \quad \left(\text{since} \left[\frac{\mathrm{J}}{\mathrm{s}}\right] = [\mathrm{W}]\right)$$

The Penmann–Monteith equation provides a measure of the latent heat flow from the canopy to the atmosphere. In order to convert this latent heat flow (LE) to a transpiration rate (units: kg water $\mathrm{m}^{-2}\,\mathrm{s}^{-1}$; this unit is compatible with the rest of the notation for Mw_sh, Mw_r and Mw_so (see below)), we have to divide LE by the latent heat of vaporization L and further divide it by 1000 to obtain the desired units. The flow from the plant to the atmosphere Fwsh_atm is therefore defined by:

$$\mathrm{Fwsh\_atm} = \frac{\mathrm{LE}}{\mathrm{L} \cdot 1000} \tag{6.5/17}$$

where:

LE = latent heat flow (Penmann–Monteith equation, eqn 6.5/16) $[J \cdot m^{-2} \cdot s^{-1}]$ or $[W \cdot m^{-2}]$

L = latent heat of vaporization (eqn 5.8/8) $[J \cdot g^{-1}]$

The resulting units are:

$$[\text{Fwsh\_atm}] = \frac{\left[\frac{J}{s \cdot m^2}\right]}{\left[\frac{J}{g}\right]} = \left[\frac{g}{m^2 \cdot s}\right]$$

Division by 1000 leads to units:

$\left[\frac{kg}{m^2 \cdot s}\right]$ (this unit is compatible with the rest of the notation)

The various expressions needed for the Penmann–Monteith equation (eqn 6.5/16) are derived below.

Vapour pressure deficit (VPD) which is a measure of the 'drying power' (units: $g\ m^{-3}$) of the atmosphere is calculated by (Monteith and Unsworth, 1990, p. 10):

$$\text{VPD} = \text{Hs\_sat} - \text{Hs} \qquad (6.5/18)$$

where:

Hs_sat = saturated water vapour concentration of the atmosphere at air temperature (TA) (using eqn 5.6/7 with TA) $[g \cdot m^{-3}]$

Hs = actual water vapour concentration of the atmosphere (at dew-point temperature, DP) (using eqn 5.6/7, with DP) $[g \cdot m^{-3}]$

The change in saturation vapour pressure (Del_vp, eqn 5.8/7) and the latent heat of vaporization (L, eqn 5.8/8) were already defined in Chapter 5. The psychrometer constant (Psy_const, units: Pa K^{-1}) is usually tabulated (e.g. Monteith and Unsworth, 1990, Table A.3) but can also be approximated by the relationship given in Monteith and Unsworth (1990, p. 181):

$$\text{Psy\_const} = \frac{\text{C\_air} \cdot \text{Pressure}}{\text{L} \cdot \text{eps}} \qquad (6.5/19)$$

where:

C_air = specific heat of air $[J \cdot g^{-1} \cdot K^{-1}]$ {1.01}

Pressure = atmospheric pressure [Pa] {99000} (comparison of the calculated with the tabulated psychrometer values showed that best results were obtained with this pressure setting)

eps = ratio of molecular weight of water vapour and air [–] {0.622}

L = latent heat of vaporization (eqn 5.8/8) $[J \cdot g^{-1}]$

The net radiation absorbed by the canopy (Rn_abs), i.e. the energy which is driving the flux LE, is calculated via an empirical relationship (recall that the assumption was made that the ground heat flux G = 0). The fraction of net radiation which is absorbed by the canopy is calculated according to the light relations in the canopy (see eqn 6.3/21):

$$\text{Rn\_abs} = \text{Rn} \cdot \text{fnr\_ac} \tag{6.5/20}$$

with:

$$\text{Rn} = \text{PAR} \cdot \text{Rnet\_par} \tag{6.5/21}$$

$$\text{fnr\_ac} = \frac{1 - \text{rcan} - \text{m}}{1 - \text{m}} \cdot \left(1 - e^{-\text{k\_beer} \cdot \text{LAI}}\right) \tag{6.5/22}$$

where:

Rnet_par =	ratio of net radiation to photosynthetic radiation (PAR, input) [–] {1.4}
fnr_ac =	fraction of instant net radiation absorbed by canopy (similar to eqn 6.3/21) [–]
rcan =	canopy reflection coefficient [–] {0.15}
m =	light transmission coefficient [–] {0.1}
k_beer =	light extinction coefficient through the canopy [–] {0.5}
LAI =	leaf area index (eqn 6.4/13) [m^2 leaf · m^{-2} ground]

An expression for the boundary layer conductance (Kh) has already been derived in Chapter 5 (eqns 5.5/2–5.5/6). The derivation there involved the calculation of a stability parameter, which again was dependent on the sensible heat flow (C). However, the entire expression could only be evaluated if the calculations were performed in an iteration procedure to find the soil surface temperature. This was needed to find the values for the stability corrections for heat (Fh) and momentum transfer (Fm). The expression for the boundary layer conductance used by Thornley (1996) is very similar to the one derived in Chapter 5. However, the main difference is that Thornley is not considering any stability corrections for momentum or heat. This avoids the complexity of running an entire energy balance, which is also not the objective of this section. However, it must be remembered that, if we want to calculate the boundary layer conductance (Kh) based on sound physical principles, we cannot avoid this additional complexity. A detailed derivation of the boundary layer conductance which is used in this section is given by Thornley and Johnson (1990, pp. 414–416) and will therefore not be repeated here.

The final expression for Kh is (Thornley and Cannell, 1996, eqn 2d):

$$\text{Kh} = \frac{\text{Kaman}^2 \cdot \text{U}}{\ln\left(\frac{\text{Zmeas} + \text{zh} - \text{d}}{\text{zh}}\right) \cdot \ln\left(\frac{\text{Zmeas} + \text{zm} - \text{d}}{\text{zm}}\right)} \tag{6.5/23}$$

where:

kaman = von Karman's constant [–] {0.41}
Zmeas = height where weather data are measured [m] {2}
zm = roughness length for momentum transfer (see eqn 5.5/3) [m]
zh = roughness length for heat and vapour transfer (see eqn 5.5/5) [m]
d = zero plane displacement (see eqn 5.5/3) [m]
U = wind speed $[m \cdot s^{-1}]$

The canopy height (*h*) is derived from an empirical relationship with leaf area index (LAI):

$$h = 0.026 \cdot LAI \qquad (6.5/24)$$

So far, we have focused on the plant–water relations. However, they are also linked to the soil–water relationships.

Soil water status

A detailed description of soil–water relations has already been presented in Chapters 4 and 5. The approach in the previous chapters involved a layered approach and the calculation of soil–water dynamics in response to potential gradients in the soil. The model by Thornley (1996) considers only a single, uniform, soil compartment, which makes the code for the plant–water relationships more transparent. In addition, the root system, which is essential for water uptake, has only been developed for a single soil compartment model (Section 6.4.3). As long as the soil is at a more or less uniform soil moisture, such an approach is likely to be sufficiently accurate. However, under conditions when the root system is exposed to a wide range of soil moisture conditions (e.g. topsoil is dry, and deeper in the soil there is a water-table), the water uptake by the roots in the various parts of the system may be quite different and could not accurately be approximated by a one-compartment soil model. In this section, we use the one-soil compartment approach, which makes it easier to understand the basic relationships between soil and the plant. However, a multilayered soil system may represent field conditions better.

A one-compartment soil water model was presented by Thornley (1996, 1998). This approach is slightly modified by using some of the elements from Chapter 4.

The volumetric water content (VWC) in the Thornley model is calculated from the soil water content (Mw_so (see below, eqn 6.5/38), units: kg m^{-2}) by:

$$VWC = \frac{Mw\_so}{dsoil \cdot (dens\_H2O \cdot 1000)} \qquad (6.5/25)$$

where:

Mw_so = soil water content (see below, eqn 6.5/38) [$kg \cdot m^{-2}$]
dsoil = depth of soil profile [m] {1}
dens_H2O = density of water (multiply by 1000 to obtain units: kg m^{-3}) [$g \cdot cm^{-3}$] {1}

From the volumetric water content (VWC) the hydraulic conductivity (Kw, eqn 4.4/2) and the matric potential (MPOT, eqn 4.3/1) are calculated, according to the procedure described in Chapter 4. The units of Kw are m s^{-1} and for MPOT m. These variables are expressed by Thornley with different dimensions: Kwso (units: kg s m^{-3}) and MPOTso (units: J kg^{-1}); therefore the following transformations are needed (see also Table 4.1):

$$Kwso = \frac{Kw \cdot (dens\_H2O \cdot 1000)}{gravity} \quad (6.5/26)$$

and:

$$MPOTso = MPOT \cdot gravity \quad (6.5/27)$$

Note: the saturated volumetric water content (VWC_sat), which is needed for the calculation of Kw and MPOT, is calculated according to eqn 2.3/40 (dens_bulk = 1; dens_solid = 2.6). In the notation by Thornley, the volumetric water content at field capacity (VWC_FC) is also needed. Here we divert slightly from his notation and actually calculate this value by solving eqn 4.3/1 for volumetric water content. To perform this operation, a value for the matric potential at field capacity (POT_FC = −0.6 m) has to be provided:

$$VWC\_FC = e^{\left\{\frac{\ln(-POT\_FC)-A}{B}\right\}} \cdot VWC\_sat \quad (6.5/28)$$

where:

POT_FC = matric potential at field capacity [m] {−0.6}
A = parameter of the soil moisture characteristic curve [–] {−2.5}
B = parameter of the soil moisture characteristic curve [–] {−3}
VWC_sat = saturated volumetric water content (eqn 2.3/40) [$m^3 \cdot m^{-3}$]

The flow of water into the soil is dependent on the precipitation (rain) and the fraction of precipitation water which is not reaching the soil because it is intercepted by the canopy and directly evaporated again. The amount intercepted is assumed to be directly related to the canopy structure. The rate of precipitation (rain) is often measured in hourly intervals and is supplied to the calculations (units: mm h^{-1}). The rate at which intercepted water is lost via evaporation is calculated by (Thornley, 1996, eqn 2h):

$$Fwrain\_atm = rain \cdot frain\_i \cdot frain\_evap \quad (6.5/29)$$

with:

$$\text{frain\_i} = 1 - e^{-\text{k\_beer} \cdot \text{LAI}} \qquad (6.5/30)$$

where:

frain_evap = fraction of intercepted water which is directly evaporated again [–] {0.45}

The various flows of water (all in units of kg m^{-2} s^{-1}) entering and leaving the soil determine how the soil water content (Mw_so, units: kg m^{-2}) will change. The overall flow (Fw_soil) is calculated in this simple water model by:

$$\text{Fw\_soil} = \text{rain} - \text{Fwrain\_atm} - \text{Fwso\_r} \qquad (6.5/31)$$

where:

rain = rainfall [kg · m^{-2} · s^{-1}]
The conversion from mm h^{-1} involves the division by 1000 and 3600 (to units: m s^{-1}) and multiplication by water density (kg m^{-3}) to obtain units: kg m^{-2} s^{-1}.
Fwrain_atm = water lost from direct evaporation of intercepted water [kg · m^{-2} · s^{-1}]
Fwso_r = rate of root uptake [kg · m^{-2} · s^{-1}]

The root uptake Fwso_r is calculated as:

$$\text{Fwso\_r} = \frac{\text{MPOTso} - \text{POTr}}{\text{Rso\_r}} \qquad (6.5/32)$$

with the resistance for water flow between soil and root (Rso_r, units: m^4 kg^{-1} s^{-1}) calculated by (Thornley, 1996, eqn 2i):

$$\text{Rso\_r} = \frac{\text{Cso\_r} \cdot \text{dens\_rt}}{\text{Kwso} \cdot \text{Wr}} + \frac{\text{Crs\_r} \cdot}{\text{dens\_rt}} \cdot \left(\frac{\text{Wr} + \text{Kwrs\_r}}{\text{Wr}}\right) \qquad (6.5/33)$$

where:

Cso_r = parameter affecting the resistance between soil and root [m^2] {80}
Crs_r = parameter affecting the resistance between soil and root [m · s^{-1}] {50000}
Kwrs_r = soil root resistance parameter [kg structure · m^{-2}] {1}
Kwso = soil hydraulic conductivity (eqn 6.5/26) [kg · m^{-2} · s^{-1}]
Wr = live root structure (eqn 6.4/15) [kg · m^{-2}]
dens_rt = root density (see below, eqn 6.5/34) [kg structure · m^{-3}]

The resistance term (Rso_r) represents both the resistance between the soil and the root surface (the first term in eqn 6.5/33) and the resistance between the root surface and the entire root (second term in eqn 6.5/33). A detailed description of the resistance term is given in Thornley (1998, pp. 124–125).

New structural root growth (dens_rt_new) is assumed to have a density which lies between a maximum and a minimum value as a function of the soil moisture. Highest root density is obtained if the soil is at field capacity (VWC_FC). The overall rooting density is considered to be a function of the root growth rate (Gr), the structural root live weight (Wr) and the density of the newly synthesized material (see also Thornley, 1998, p. 52). The root density (dens_rt) is calculated by the differential equation (Thornley, 1996, eqn 4):

$$\frac{d\text{dens\_rt}}{dt} = \text{Gr} \cdot \left(\frac{\text{dens\_rt\_new} - \text{dens\_rt}}{\text{Wr}} \right) \qquad (6.5/34)$$

with:

$$\text{dens\_rt\_new} = \text{dens\_rt\_max} - \left\{ \begin{bmatrix} (\text{dens\_rt\_max} - \text{dens\_rt\_min}) \\ \cdot \left(\frac{\text{VWC\_FC} - \text{VWC}}{\text{VWC}} \right) \end{bmatrix} \right\} \qquad (6.5/35)$$

where:

Gr = production rate of new root structure (eqn 6.4/40) [kg structure · m^{-2}]
Wr = live root structure (eqn 6.4/15) [kg · m^{-2}]
VWC_FC = volumetric water content at field capacity (eqn 6.5/28) [$m^3 \cdot m^{-3}$]
VWC = volumetric water content (eqn 6.5/25) [$m^3 \cdot m^{-3}$]
dens_rt_max = maximum value for root density [kg structure · m^{-3}] {1}
dens_rt_min = minimum value for root density [kg structure · m^{-3}] {0.1}
dens_rt_init = initial root density [kg structure · m^{-3}] {1}

We have defined the water flows from the shoot to the atmosphere (Fwsh_atm, eqn 6.5/17) and from the soil to the root (Fwso_r, eqn 6.5/32). Therefore the only missing flow of water is the one from the root to the shoot (Fwr_sh).

Water flow from root to shoot (Fwr_sh)

The flow of water from the root to the shoot (Fwr_sh) is calculated in analogy to the flow from the soil to the root by the potential difference multiplied by the conductance. The conductance between the root and shoot (gwr_sh, units: kg s m^{-4}) is a function of the shoot and root tissue water content and a plant water conductivity parameter. However, the value of the conductance is dominated by the tissue water content, which is currently at a minimum. The flow from the root to the shoot and the tissue conductance is calculated as (Thornley, 1996, eqns 2k and 2l):

$$\text{Fwr\_sh} = \text{gwr\_sh} \cdot (\text{POTr} - \text{POTsh}) \qquad (6.5/36)$$

with:

$$gwr\_sh = Cw \cdot \left(\frac{Mw\_r \cdot Mw\_sh}{Mw\_r + Mw\_sh} \right) \quad (6.5/37)$$

where:

Cw = plant water conductivity constant $[s \cdot m^{-2}]$ $\{5.787 \times 10^{-8}\}$
POTsh = water potential in the shoot (eqn 6.5/8) $[J \cdot kg^{-1}]$
POTr = water potential in the root system (eqn 6.5/9) $[J \cdot kg^{-1}]$

The water contents of the soil: Mw_so, root: Mw_r and shoot: Mw_sh are defined by the differential equations:

soil:

if Fw_soil > 0 and VWC > VWC_FC:

$$\frac{dMw\_so}{dt} = 0$$

else: (6.5/38)

$$\frac{dMw\_so}{dt} = Fw\_soil$$

root:

$$\frac{dMw\_r}{dt} = Fwso\_r - Fwr\_sh \quad (6.5/39)$$

shoot:

$$\frac{dMw\_sh}{dt} = Fwr\_sh - Fwsh\_atm \quad (6.5/40)$$

where:

Fw_soil = flow of water into the soil (eqn 6.5/31) $[kg \cdot m^{-2} \cdot s^{-1}]$
Fwso_r = flow of water from the soil to the root (eqn 6.5/32) $[kg \cdot m^{-2} \cdot s^{-1}]$
Fwr_sh = flow of water from the root to the shoot (eqn 6.5/36) $[kg \cdot m^{-2} \cdot s^{-1}]$
Fwsh_atm = flow of water from the shoot to the atmosphere (eqn 6.5/17) $[kg \cdot m^{-2} \cdot s^{-1}]$

Water status functions for plant and soil processes

The water status in the various parts of the SPAC system has an influence on the processes for plant growth. The functions developed here take on values between

0 and 1 and are therefore similar in their nature to the environmental functions derived in Chapter 2 for the C/N transformation processes. Four areas are considered where the water status will have an influence:

1. Photosynthesis: on the parameters 'Pm' and 'alpha' (with: 'fw_ph').
2. Shoot: on the shoot growth rate 'Gsh' (eqn 6.4/39) (with: 'fw_sh'); on the specific leaf area index 'SLA' (eqn. 6.4/55) (with: 'fw_SLA); on the turnover constants 'gamma_sh' (eqn 6.4/7) (with: '1/fw_sh').
3. Root: on the root growth rate 'Gr' (eqn 6.4/40) (with: 'fw_rt'); on the turnover constants 'gamma_r' (eqn 6.4/14) (with: '1/fw_rt').
4. N uptake: on the nitrogen uptake rate 'Un' (eqn 6.4/45) (with: 'fw_un'); organic matter turnover: on the rate constant 'BN' (eqn 6.4/48) (with: 'fw_so').

The function for the adjustment of the specific leaf area index (fw_SLA) is the relative water content of the shoot ('MwRel_sh', eqn 6.5/10):

$$fw\_SLA = MwRel\_sh \tag{6.5/41}$$

The other adjustment functions take the chemical activities of the water in the shoot, root and soil into account (see Thornley, 1996, eqns 4a and 4b):

$$\begin{aligned} fw\_ph &= aw\_sh^{qw\_ph} \\ fw\_sh &= aw\_sh^{qw\_pl} \\ fw\_rt &= aw\_rt^{qw\_pl} \\ fw\_un &= aw\_so^{qw\_un} \\ fw\_so &= aw\_so^{qw\_so} \end{aligned} \tag{6.5/42}$$

with:

$$aw\_sh = e^{\frac{\left(molwt\_H2O/1000\right)\cdot POTsh}{R\cdot(TA+273.16)}}$$

$$aw\_rt = e^{\frac{\left(molwt\_H2O/1000\right)\cdot POTr}{R\cdot(Tso+273.16)}}$$

$$aw\_so = e^{\frac{\left(molwt\_H2O/1000\right)\cdot MPOTso}{R\cdot(Tso+273.16)}}$$

where:

qw_ph = parameter for the effect of water stress on photosynthesis [–] {5}
qw_pl = parameter for the effect of water stress on plant biochemistry [–] {10}
qw_so = parameter for the effect of water stress on soil biochemistry [–] {10}
qw_un = parameter for the effect of water stress on nitrogen uptake [–] {3}

Application in ModelMaker

The components which have to be added to the previous model defining the plant–water relations are presented in Fig. 6.12 and the complete model is presented in *Mod6–1g.mod.*

The functions describing the effect of water stress on the various parts of the SPAC system (eqns 6.5/41 and 6.5/42) are added as 'global' and conditional <variables> to the model. While adding the notations, great care should be taken to avoid mistakes.

In addition to the notation, two statements are included in the <independent event> 'Choice', so that the user can choose to carry out simulations with or without the additional plant–water relation model. The new statements are:

```
crop_water_choice=GetChoice("Include crop water relations
    (Y/N)?","Crop water relation",crop_water_choice);
rain_choice=GetChoice("Include rain (Y/N)?","Rainfall
    decision",rain_choice);
```

All the 'fw_' variables are made conditional in response to the value held in the <define> 'crop_water_choice'. If the user decides not to include the plant–water relations, all 'fw_' variables will take on a value of 1.

In addition, a choice 'rain_choice' is added to allow rainfall or not.

The input values to the model, which are supplied to the model via the <Lookup file > 'Input', contain additional columns of data for dew point temperature, DP, wind speed, U, precipitation, rain, and solar radiation, St.

Running the model

Since the model runs on a per second basis, it is possible to investigate the diurnal water dynamics. Figure 6.13 shows the output of the soil, root and shoot potentials over 5 days. This shows the increasing potential from soil $\rightarrow$ root $\rightarrow$ shoot which is needed to drive the water flow within the SPAC system.

The effect of the plant–water dynamics on the leaf area index (LAI) is shown in Fig. 6.14, where an 80-day simulation is executed, either including or not including plant–water relations in the output. This example shows that the plant water status has a tremendous effect on the overall plant growth and cannot be neglected in plant growth simulations. We should also remember that the simulation is closely related to the input data, which, in this example, are only included for illustration purpose. It will be interesting to see the effects when the model is run with 'real' data.

6.6 Future Development

The possible link of this entire plant growth model to an energy balance model (e.g. Chapter 5) has already been mentioned. This requires modelling the additional flow of energy between the soil and the plant together with the atmosphere.

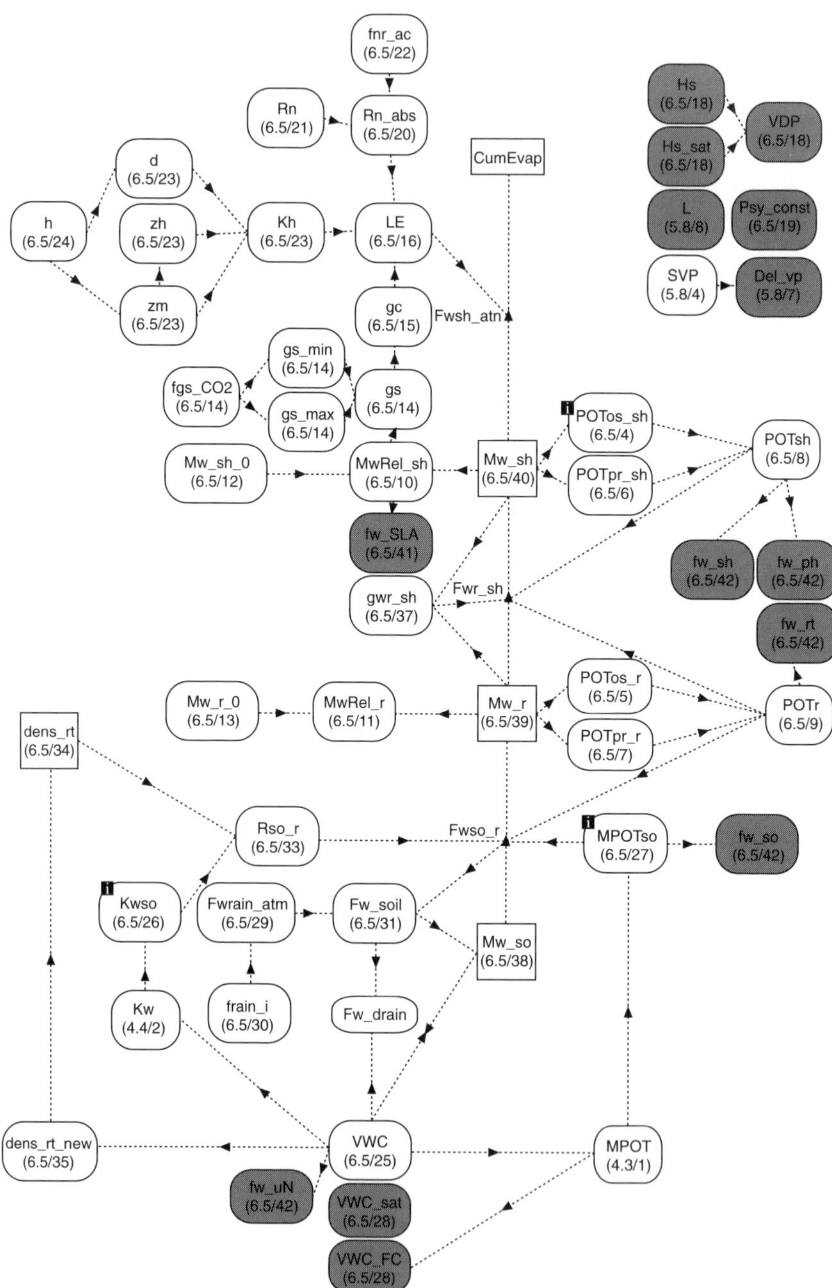

Fig. 6.12. Plant–water relationship notation as a part of the plant growth model (*Mod6-1g.mod*) (see also Figs 6.6 and 6.7).

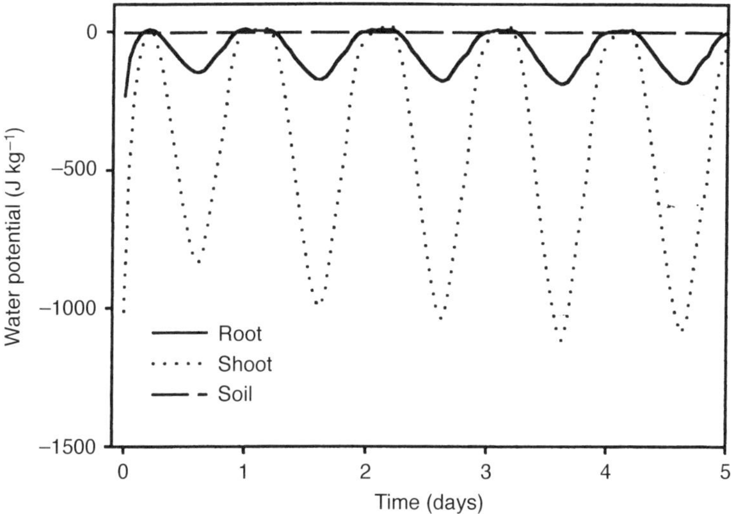

Fig. 6.13. Shoot, root and soil water potential in the first 5 days of an 80-day simulation (*Mod6-1g.mod*).

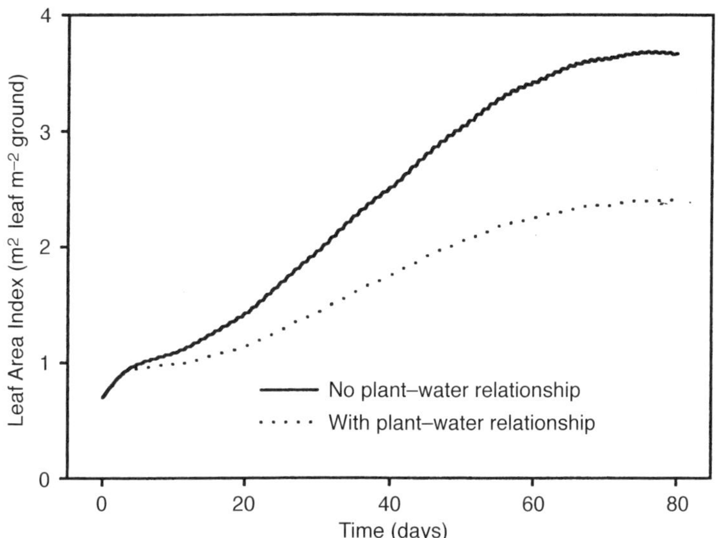

Fig. 6.14. Simulated leaf area index with and without plant–water relationship (*Mod6-1g.mod*).

The items which have to be included are outlined in Monteith and Unsworth (1990, Chapter 15). However, before this is done, it is necessary to include a multilayered soil notation in the plant growth model. To do this, a more complex root growth model, which is able to grow from one layer to the other in response to nutrients, water and temperature, would be needed. A model such as that presented by Brugge (1985) and Brugge and Thornley (1985) could be a starting point for such a multilayered soil–root model. Once a multilayered root model is in place, the soil–water–temperature model developed in Chapters 3–5 could be linked to the plant growth model. If one replaces the simple treatment of organic matter turnover with the models developed in Chapter 2, one could create an ecosystem model by assembling all the submodels together.

The book by Thornley (1998) contains such an ecosystem model. Numerous aspects which he describes in this book have not been considered here. If the reader wishes to go into more depth with plant growth modelling in the context of a wider ecosystem approach, this excellent publication is strongly recommended.

Leaching 7

7.1 Introduction

With the increase in environmental awareness in the last 20 years, it has become apparent that, in many locations, our drinking-water resources are gradually being contaminated with harmful substances. In many areas of the world, drinking-water is extracted from the groundwater, with water pumped up from considerable depth generally being considered to be pure and therefore an ideal source for human consumption. However, controls in the last years showed that groundwater is becoming more and more enriched with substances which have been applied to the soil. These include fertilizers, such as nitrate ($NO_3^-{-}N$), as well as many kinds of pesticides. Moreover, substances such as nitrate, which are mainly present in soil solution, were found in higher amounts than others which are retained by soil particles. Positively charged ions, such as ammonium ($NH_4^+{-}N$), can be retained by negatively charged soil particles.

The translocation of substances is generally unwanted in many respects, because it not only creates problems in terms of groundwater pollution but also depletes valuable resources for plant growth. It is, therefore, also an economic problem for the farmer, should expensive fertilizers aimed at stimulating crop growth be lost without having the desired effect. The simple soil nitrogen-leaching model linked to the plant growth model (Section 6.4.3, eqns 6.4/47–6.4/50) includes a consideration of nitrogen leaching. There it was assumed that a fixed proportion of the available nitrogen leached out of the soil and was lost for root uptake. However, this description of the transport process is clearly a gross oversimplification and does not take account of the various processes which occur in soil.

A detailed overview of the various transport processes is given by Hillel (1998, Chapter 9). The main transport vehicle appears to be the moving water, and translocation into deeper regions of the soil is therefore only possible if the main direction of water movement is downwards. This translocation applies in particular to substances which are dissolved in the soil water (solutes). This transport process is referred to as convective transport. While the solute is transported with the convective stream, a simultaneous interaction may occur with the solid matrix which is mainly related to the concentration gradients of the solutes. This second transport process is therefore referred to as diffusion. A third mode of transport is induced by the water flow characteristics within soil pores. The flow velocities in a pore are highest in the middle of the pore and gradually decline towards the pore edge in response to frictional forces (similar to the frictional forces on horizontal wind velocities over the soil surface (Section 5.5)). Some concentration differences of solutes within the soil pores are associated with these non-uniform flow velocities and are highest under large flow velocities (e.g. high rainfall rates). The process is referred to as hydrodynamic dispersion and its effect (not its mechanism) is similar to that of diffusion. Therefore, these two transport processes are generally combined in one dispersion–diffusion notation.

The characteristic pore flow velocities (in the middle of the pore high and towards the edge low) in connection with interactions with the solid matrix are also responsible for the different translocation characteristics of cations (e.g. NH_4^+-N) and anions (e.g. NO_3^--N). Cations are attracted by soil particles (e.g. clay minerals) and are located, therefore, in the vicinity of the pore fringe, where the flow velocities are very low. Anions are repelled by the generally negatively charged solid matrix and are situated more in the middle of the pores. Due to the large differences in pore flow velocities, anions are generally translocated much more quickly than cations.

Since the movement of substances is linked to the water flow dynamics of the soil, the understanding of these processes (described in Chapters 4 and 5) is essential. However, before we extend the models presented in Chapter 4 and 5, we need to develop the mathematical notations for the three transport processes in soil. The description of the various processes is based on the publications by Kirkham and Powers (1972), Hillel (1977, Chapter 5) and Hillel (1998, Chapter 9).

7.2 Mathematical Notation of Transport Processes

7.2.1 Convection

The mass flow or convective flow of solutes (Fc, units: kg m^{-2} s^{-1}) is proportional to the soil water flow (Fw) which is defined by:

$$\mathrm{Fc} = \mathrm{Conc} \cdot \mathrm{Fw} \tag{7.2/1}$$

where:

Conc = concentration of solutes in the solution [$g \cdot l^{-1}$] or [$kg \cdot m^{-3}$]
Fw = flow of water (see eqn 4.5/4) [$m \cdot s^{-1}$]

The apparent average water flow velocity (Fw_avg) is:

$$Fw\_avg = \frac{Fw}{VWC} \tag{7.2/2}$$

where:

Fw = flow of water (see eqn 4.5/4) [$m \cdot s^{-1}$]
VWC = volumetric water content (see eqn 4.5/3) [m^3 water $\cdot$ m^{-3} soil]

7.2.2 Diffusion

Flow of solutes by diffusion (*Fd*, units: kg m^{-2} s^{-1}) occurs in response to concentration gradients ($\frac{d\text{Conc}}{dz}$, from high to low concentration) and can be formulated by a notation analogous to eqn 1.3/1:

$$Fd = -Ds \cdot \frac{d\text{Conc}}{dz} \tag{7.2/3}$$

where:

Ds = diffusion coefficient for a particular solute in the soil solution [$m^2 \cdot s^{-1}$]
Conc = concentration of solutes in the solution [$g \cdot l^{-1}$] or [$kg \cdot m^{-3}$]
z = depth (d*z* = diffusion distance) [m]

The diffusion coefficient (*Ds*) is usually calculated from the diffusion coefficient in bulk water (Ds_0), adjusted to the volumetric water content (*VWC*), and the tortuosity of the soil, which is itself a function of soil volumetric water content. The equation to calculate the diffusion coefficient is (see also Hillel, 1977, chapter 5):

$$Ds = Ds\_0 \cdot VWC \cdot ftort \tag{7.2/4}$$

with the tortuosity factor (ftort) linearly related to volumetric water content:

$$ftort = Ds\_tort\_A + Ds\_tort\_B \cdot VWC$$

where:

Ds_tort_A = intercept of the relationship 'VWC_ftort' [–] {0.25}
Ds_tort_B = slope of the relationship 'VWC_ftort' [–] {0.93}

The relationship between volumetric water content and the tortuosity factor is soil-specific. The parameters presented above were obtained by regressing data from Hillel (1977, pp. 162–165).

7.2.3 Hydrodynamic dispersion

The flow of solutes in response to hydrodynamic dispersion (Fh, units: kg m^{-2} s^{-1}) is formulated, in analogy to the diffusion equation (eqn 7.2/3), as:

$$\text{Fh} = -\text{Dh} \cdot \frac{d\text{Conc}}{dz} \tag{7.2/5}$$

where:

Dh = dispersion coefficient [$m^2 \cdot s^{-1}$]

The dispersion coefficient is mainly related to the average flow velocity (Fw_avg, eqn 7.2/2) and can therefore be calculated as (see also Hillel, 1977, Chapter 5):

$$\text{Dh} = \text{Fw\_avg} \cdot \text{Dh\_0} \tag{7.2/6}$$

where:

Dh_0 = empirical parameter for hydrodynamic dispersion [m] {0.02}

7.2.4 Combined notation for solute transport

The combined flow of solutes (Fs) can be calculated, by combining eqns 7.2/1, 7.2/3 and 7.2/5, as:

$$\text{Fs} = \text{Fc} + \text{Fd} + \text{Fh}$$

or:

$$\text{Fs} = \text{Fw} \cdot \text{Conc} - \text{Ds} \cdot \frac{d\text{Conc}}{dz} - \text{Dh} \cdot \frac{d\text{Conc}}{dz} \tag{7.2/7}$$

After combining the two coefficients for diffusion (Ds) and dispersion (Dh) in one diffusion–dispersion coefficient (Dsh):

$$\text{Dsh} = \text{Ds} + \text{Dh} \tag{7.2/8}$$

eqn 7.2/7 simplifies to:

$$\text{Fs} = \text{Fw} \cdot \text{Conc} - \text{Dsh} \cdot \frac{d\text{Conc}}{dz} \tag{7.2/9}$$

This last expression defines the instantaneous flow rate for solutes with depth. However, we are interested in the solute flow characteristics in space and in time. For this, the possible change of the volume flux in each soil layer (i.e. $\frac{\mathrm{d}VWC}{\mathrm{d}t}$, eqn 4.5/15) and concentration (i.e. $\frac{\mathrm{d}Conc}{\mathrm{d}t}$) must be taken into consideration. The resulting continuity equation for the convective–diffusive–dispersive flow of solutes is derived in analogy to the continuity equation for heat flow (see Section 3.2).

Consider a volume of soil defined by thickness Δz and area A and a flow of solutes in (Fs_in) and out (Fs_out) of this volume of soil (Fig. 7.1). The change of the solute concentration in the soil volume due to convective–diffusive–dispersive flow ($\Delta z \cdot A \cdot \frac{\partial VWC \cdot Conc}{\partial t}$) is given by the flow rate across the area A into the volume of soil:

$$\text{Fs\_in} = \text{A} \cdot \text{Fs} \tag{7.2/10}$$

minus the flow rate out of the volume of soil across the area A. The flow out of the soil (Fs_out) also contains a possible contribution related to the change in the solute flux across the element of thickness Δz:

$$\text{Fs\_out} = \text{A} \cdot \text{Fs} + \frac{\partial \text{Fs}}{\partial \text{z}} \cdot \text{A} \cdot \Delta \text{z} \tag{7.2/11}$$

The solute transferred into the volume of soil minus the solute transferred out of the volume of soil is given by:

$$\text{Fs\_in} - \text{Fs\_out} = \text{A} \cdot \text{Fs} - \left(\text{A} \cdot \text{Fs} + \frac{\partial \text{Fs}}{\partial \text{z}} \cdot \text{A} \cdot \Delta \text{z} \right)$$

or simplified:

$$\text{Fs\_in} - \text{Fs\_out} = -\frac{\partial \text{Fs}}{\partial \text{z}} \cdot \text{A} \cdot \Delta \text{z} \tag{7.2/12}$$

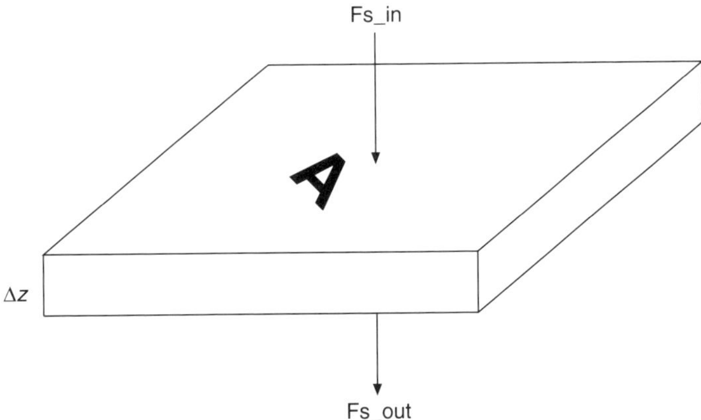

Fig. 7.1. Schematic description of the solute flows into and out of a volume of soil, leading to the continuity equation.

The change of quantity of solute in a volume element of soil is equal to the difference between the flux entering and leaving this soil section (Hillel, 1998, p. 252), i.e.:

$$\Delta z \cdot A \cdot \frac{\partial VWC \cdot Conc}{\partial t} = -\frac{\partial Fs}{\partial z} \cdot A \cdot \Delta z$$

or:

$$\frac{\partial VWC \cdot Conc}{\partial t} = -\frac{\partial Fs}{\partial z} \qquad (7.2/13)$$

Combining eqn 7.2/13 with eqn 7.2/9 yields:

$$\frac{\partial VWC \cdot Conc}{\partial t} = -\frac{\partial \left[Fw \cdot Conc - Dsh \cdot \frac{\partial Conc}{\partial z} \right]}{\partial z}$$

or:

$$\frac{\partial VWC \cdot Conc}{\partial t} = \frac{\partial}{\partial z} \cdot \left[Dsh \cdot \frac{\partial Conc}{\partial z} \right] - \frac{\partial Fw \cdot Conc}{\partial z} \qquad (7.2/14)$$

Under the conditions that *VWC*, *Dsh* and *Fw* are constant, eqn 7.2/14 can be simplified to (Hillel, 1998, p. 252):

$$\frac{\partial Conc}{\partial t} = \frac{Dsh}{VWC} \cdot \frac{\partial^2 Conc}{\partial z^2} - Fw\_avg \cdot \frac{\partial Conc}{\partial z} \qquad (7.2/15)$$

where:

Dsh = diffusive–dispersive coefficient $[m^2 \cdot s^{-1}]$
Conc = solute concentration $[kg \cdot m^{-3}]$
VWC = volumetric water content of a certain soil volume $[m^3 \text{ water} \cdot m^{-3} \text{ soil}]$
Fw_avg = average flow velocity (eqn 7.2/2) $[m \cdot s^{-1}]$
z = depth [m]

The solute transport equation, both with constant parameter settings (eqn 7.2/15) and with variable settings (eqn 7.2/14), is solved in the next two sections.

7.3 Leaching Model with Constant Soil Parameters

The first model involves the solution to eqn 7.2/15. A graphical display of the layered soil system which is to be solved is presented in Fig. 7.2.

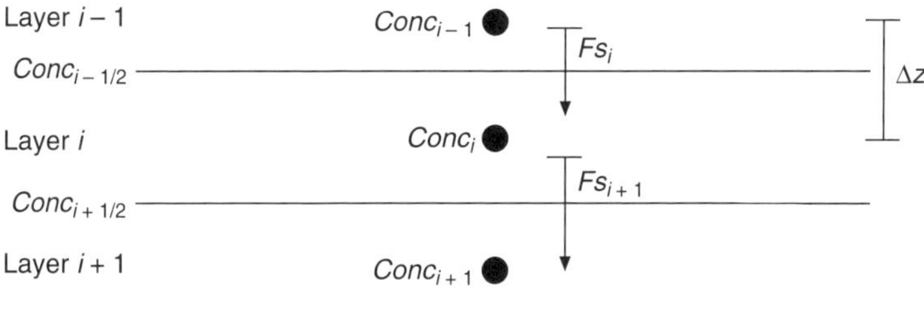

Fig. 7.2. Layered soil system for solute transport.

The solute concentrations midway between layer $i - 1$ and i (*Conc*[i–1/2]) and between layer i and $i + 1$ (*Conc*[i + 1/2]) are calculated by:

$$\mathrm{Conc}[i-1/2] = \frac{\mathrm{Conc}[i] + \mathrm{Conc}[i-1]}{2}$$
$$\mathrm{Conc}[i+1/2] = \frac{\mathrm{Conc}[i] + \mathrm{Conc}[i+1]}{2} \qquad (7.3/1)$$

The diffusion–dispersion term of eqn 7.2/15 (first term on the right-hand side) is calculated by:

$$\left.\frac{\partial \mathrm{Conc}[i]}{\partial t}\right|_{\mathrm{diff\_disp}} = \frac{\mathrm{Dsh}}{\mathrm{VWC}} \cdot \frac{\left(\dfrac{\mathrm{Conc}[i+1] - \mathrm{Conc}[i]}{z}\right) - \left(\dfrac{\mathrm{Conc}[i] - \mathrm{Conc}[i-1]}{z}\right)}{z}$$

or:

$$\left.\frac{\partial \mathrm{Conc}[i]}{\partial t}\right|_{\mathrm{diff\_disp}} = \frac{\mathrm{Dsh}}{\mathrm{VWC}} \cdot \left(\frac{\mathrm{Conc}[i+1] - 2 \cdot \mathrm{Conc}[i] + \mathrm{Conc}[i-1]}{z^2}\right) \qquad (7.3/2)$$

and the convection term (second term on the right-hand side of eqn 7.2/15) is defined by:

$$\left.\frac{\partial \mathrm{Conc}[i]}{\partial t}\right|_{\mathrm{convection}} = \mathrm{Fw\_avg} \cdot \left(\frac{\mathrm{Conc}[i+1/2] - \mathrm{Conc}[i-1/2]}{z}\right)$$

or combined with eqn 7.3/1:

$$\left.\frac{\partial \mathrm{Conc}[i]}{\partial t}\right|_{\mathrm{convection}} = \mathrm{Fw\_avg}\cdot\left(\frac{\left(\mathrm{Conc}[i+1]+\mathrm{Conc}[i]\right)/2 - \left(\mathrm{Conc}[i]+\mathrm{Conc}[i-1]\right)/2}{z}\right)$$

After rearrangement, we obtain:

$$\left.\frac{\partial \mathrm{Conc}[i]}{\partial t}\right|_{\mathrm{convection}} = \mathrm{Fw\_avg}\cdot\frac{\mathrm{Conc}[i+1]-\mathrm{Conc}[i-1]}{2\cdot z} \tag{7.3/3}$$

Combining eqns 7.3/2 and 7.3/3 according to eqn 7.2/15 gives the final equation for convective–diffusive–dispersive flow of solutes with constant parameters:

$$\frac{\partial \mathrm{Conc}[i]}{\partial t} = \frac{\mathrm{Dsh}}{\mathrm{VWC}}\cdot\frac{\mathrm{Conc}[i+1]-2\cdot\mathrm{Conc}[i]+\mathrm{Conc}[i-1]}{z^2} - \mathrm{Fw\_avg}\cdot\frac{\mathrm{Conc}[i+1]-\mathrm{Conc}[i-1]}{2\cdot z} \tag{7.3/4}$$

with:

$$z = \frac{L}{\mathrm{Layers}} \tag{7.3/5}$$

where:

Dsh =	diffusion–dispersion coefficient $[\mathrm{m}^2 \cdot \mathrm{s}^{-1}]$ $\{1.3 \times 10^{-8}\}$
VWC =	volumetric water content $[\mathrm{m}^3 \cdot \mathrm{m}^{-3}]$ $\{0.4\}$
Fw_avg =	average flow velocity (see eqn 7.2/2) $[\mathrm{m} \cdot \mathrm{s}^{-1}]$
Fw =	flow velocity (see eqn 7.2/2) $[\mathrm{m} \cdot \mathrm{s}^{-1}]$ $\{5 \times 10^{-10}\}$
Conc =	solute concentration $[\mathrm{kg} \cdot \mathrm{m}^{-3}]$
L =	soil profile depth [m] {1}
Layers =	soil layers {10}
Conc_init =	initial solute concentration in soil profile $[\mathrm{kg} \cdot \mathrm{m}^{-3}]$

Boundary conditions

For the flow at the soil surface, the solute concentration $\mathrm{Conc}[i-1] = \mathrm{C0}$ is unknown and has to be supplied as an upper boundary. The first soil layer ($i = 1$) is defined by (according to eqn 7.3/4):

$$\frac{\partial \mathrm{Conc}[1]}{\partial t} = \frac{\mathrm{Dsh}}{\mathrm{VWC}}\cdot\frac{\mathrm{Conc}[2]-2\cdot\mathrm{Conc}[1]+\mathrm{C0}}{z^2} - \mathrm{Fw\_avg}\cdot\frac{\mathrm{Conc}[2]-\mathrm{C0}}{2\cdot z} \tag{7.3/6}$$

For the unknown solute concentration C0, we consider two situations:

1. Input of solution with a constant concentration (C0).
2. Input of solution with high concentration initially and then change to solute-free solution. This approach is often referred to as a pulse application.

A supply of solution with a constant concentration (above zero) will gradually increase the soil solution concentration from above until the entire soil profile contains just the concentration of the inflowing solution. Supplying a pulse of solution with a high solute concentration will increase the soil concentration while the solution is applied, but it decreases again due to movement into deeper soil regions and by the 'washing out' effect of solute-free solution applied to the soil surface. It might be of interest to notice the similarity of the system to the 'first model' (Section 1.6).

On the lower boundary, it is assumed that the flow occurs only via convection. The last soil layer ($i = 10$) is therefore defined by (right-hand side of eqn 7.3/3):

$$\frac{\partial \mathrm{Conc}[10]}{\partial t} = -\mathrm{Fw\_avg} \cdot \frac{(\mathrm{Conc}[10] - \mathrm{Conc}[9])}{z} \qquad (7.3/7)$$

Application in ModelMaker

The transport model with constant parameters is defined by the <compartment> 'Conc[1..10]', which also contains the notation for the boundary conditions ($i = 1$ and $i = 10$). The <define> 'z' (eqn 7.3/5) and 'Fw_avg' (eqn 7.2/2) and a <variable> 'C0', defining the concentration of the solution entering the soil profile, are added. The entire model is presented in *Mod7–1a.mod* (Fig. 7.3).

Two simulation scenarios for the input of solute solution across the top boundary can be chosen:

- Constant supply of solute solution with concentration C0.
- Supply of a pulse of solute solution at C0 for a certain amount of time and then changed to a concentration of C0 = 0.

An <independent event> 'Choice' is added which contains statements allowing the user to choose between the two options. If the option with constant supply of solute solution is chosen, the user is also asked to input the concentration of the solution which enters the top soil layer.

The pulse option is regulated via C0 values supplied via a <lookup file> 'Input'.

The statements in the <independent event> 'Choice' are:

```
GetValue("Use constant high solute concentration =0; pulse of
    high solute concentration (input) = 1:","Solute
    input",C0_choice, C0_choice = 0 or C0_choice = 1);
if(C0_choice=0){
 GetValue("Input value for solute concentration
    (0-100):","Solute concentration",C0_constant, C0_constant >0
    and C0_constant < 100);
}
```

The <variable> 'C0' is made conditional according to the <define> 'C0_choice'.

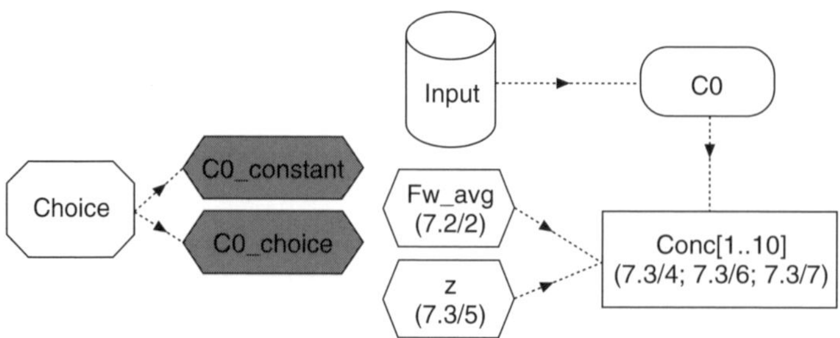

Fig. 7.3. Model for solute transport under constant parameter settings (*Mod7-1a.mod*).

Running the model

The model performs the calculations on a per second basis and is currently set up for 10 days (864,000 s) with outputs every 2 h (120 output steps). The input file *Inp7–1a.txt* contains data of the concentration C0 for the pulse application (currently, for the first hour = 3600 s, a solution containing 100 kg m^{-3} is supplied).

The working of the model is illustrated by the output with the constant supply of solute solution (C0 = 10 kg m^{-3}) compared with the output with a pulse application (Fig. 7.4). With the constant C0 supply (Fig. 7.4A), the soil solution is gradually replaced by the inflowing solution. Under the pulse application option, the concentration of the soil surface solution increases initially but declines once the inflowing solution is solute-free. It can be seen that the solute peak is slowly moving down the profile. Even the initial solute concentration, set to a value slightly higher than zero, is replaced (Fig. 7.4B).

The model derived in this section with constant parameters can provide an idea of the underlying principles for transport process but is rather unrealistic, since the parameters Fw and VWC and, with them, the additional parameters Dsh and Fw_avg are seldom constant. We have already derived detailed models for the soil–water dynamics and will reuse them in the next section, where the solute transport processes under variable parameter settings is presented.

7.4 Leaching Model with Dynamic Soil Parameters

The model with dynamic parameter settings (i.e. Dsh, VWC, Fw_avg, Fw) involves the solution of eqn 7.2/14. For the solution of this equation, we make use of eqn 7.2/9 and first compute the flux of solute at the upper boundary of layer *i* (see Fig. 7.2) by:

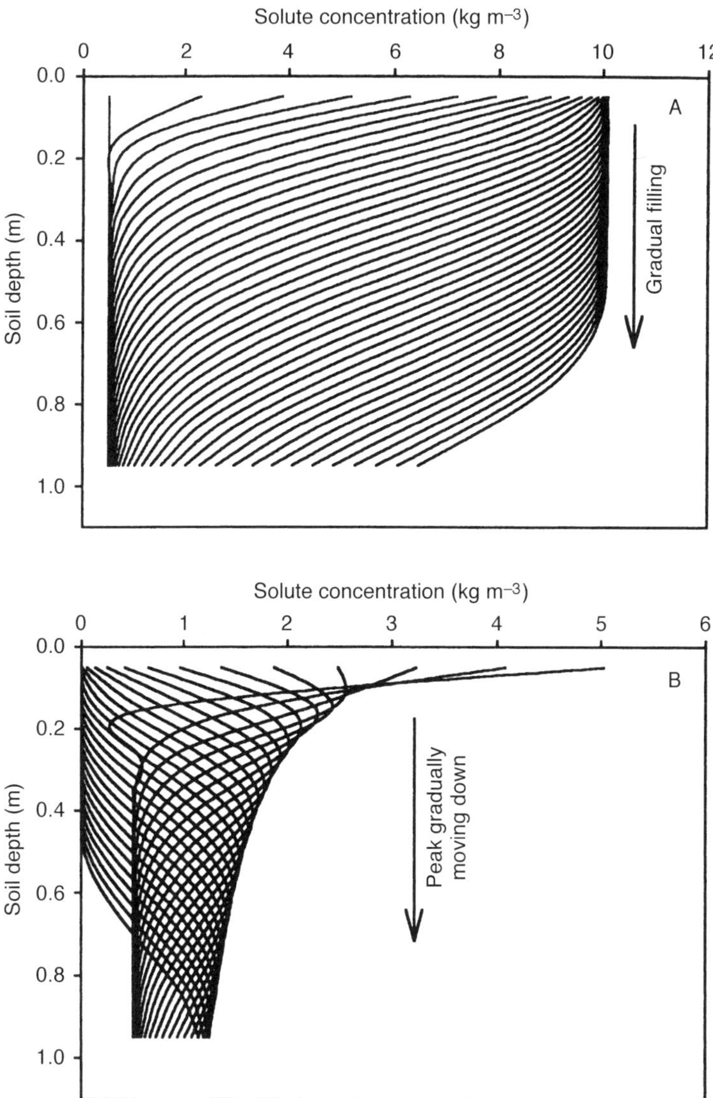

Fig. 7.4. Simulation of solute transport with constant *C0* supply (A) or with pulse *C0* supply (B) (*Mod7-1a.mod*).

$$\mathrm{Fs[i]} = \mathrm{Fw[i]} \cdot \frac{(\mathrm{Conc[i-1]} + \mathrm{Conc[i]})}{2} - \mathrm{Dsh[i]} \cdot \frac{\mathrm{Conc[i]} - \mathrm{Conc[i-1]}}{\mathrm{Dist[i]}} \quad (7.4/1)$$

with:

$$\mathrm{Dsh[i]} = \mathrm{Ds[i]} + \mathrm{Dh[i]}$$

$$\mathrm{Ds[i]} = \mathrm{Ds\_0} \cdot \left(\frac{\mathrm{VWC[i-1]} + \mathrm{VWC[i]}}{2}\right) \cdot \left[\mathrm{Ds\_tort\_A} + \mathrm{Ds\_tort\_B} \cdot \left(\frac{\mathrm{VWC[i-1]} + \mathrm{VWC[i]}}{2}\right)\right]$$

$$\mathrm{Dh[i]} = \mathrm{Fw\_avg[i]} \cdot \mathrm{Dh\_0}$$

$$\mathrm{Fw\_avg[i]} = \frac{\mathrm{Fw[i]}}{\left(\frac{\mathrm{VWC[i-1]} + \mathrm{VWC[i]}}{2}\right)}$$

where:

Fw[i] = flow of water into layer i (eqn 4.5/4) [$\mathrm{m \cdot s^{-1}}$]
Dsh[i] = diffusion–dispersion coefficient (eqn 7.2/8) [$\mathrm{m^2 \cdot s^{-1}}$]
Ds[i] = diffusion coefficient (eqn 7.2/4) [$\mathrm{m^2 \cdot s^{-1}}$]
Dh[i] = dispersion coefficient (eqn 7.2/6) [$\mathrm{m^2 \cdot s^{-1}}$]
VWC[i] = volumetric water content (eqn 4.5/15) [$\mathrm{m^3 \cdot m^{-3}}$]
(the volumetric water content at the transition from layer $i-1$ to i is computed by: '$\left(\frac{\mathrm{VWC[i-1]} + \mathrm{VWC[i]}}{2}\right)$')
Conc[i] = solute concentration (see eqn 7.4/5) [$\mathrm{m^3 \cdot m^{-3}}$]
(the average concentration between layers $i-1$ and i is computed by: '$\left(\frac{\mathrm{Conc[i-1]} + \mathrm{Conc[i]}}{2}\right)$')

Dist[i] = flow distance between adjacent layers (eqn 3.5/1) [m]

The flow of solute out of layer i between layer i and $i+1$ (Fig. 7.2) is computed in exactly the same manner (only replacing index $i-1$ with i and i with $i+1$).

The amount of solutes in layer i (Ams[i]) is calculated by integrating the difference between the flow of solute entering layer i (Fs[i]) minus the flow of solute leaving layer i (Fs[$i+1$]), i.e.:

$$\frac{d\mathrm{Ams[i]}}{dt} = \mathrm{NetFs[i]} \quad (7.4/2)$$

with:

$$\mathrm{NetFs[i]} = \mathrm{Fs[i]} - \mathrm{Fs[i+1]} \quad (7.4/3)$$

The initial amount of solutes in layer i (Ams_init[i]) is given by:

$$\text{Ams\_init[i]} = \text{VolVWC[i]} \cdot \text{Conc\_init[i]} \tag{7.4/4}$$

where:

VolVWC[i] = volume of water in layer i (eqn 4.5/13) [m^3 water · m^{-2} soil]
Conc_init[i] = initial concentration of solutes in layer i [kg · m^{-3}]{0}

Finally, the concentration of solute in layer i (Conc[i], units: kg m^{-3}) is calculated by dividing the amount of solute (Ams[i]) with the thickness of the layer i (z[i]) and further dividing it by the volumetric water content (VWC[i]):

$$\text{Conc[i]} = \frac{\text{Ams[i]}}{\text{z[i]} \cdot \text{VWC[i]}} \tag{7.4/5}$$

Boundary conditions

With eqns 7.4/1–7.4/5, we have defined the solute leaching model with dynamic parameters. However, the entire system of equations cannot be solved yet, because the flows at the boundaries (Fs[1] and Fs[n]) are unknown. For the upper boundary (soil surface), it is assumed that the flow of solutes is given by the flow of water into the soil (Fw[1], eqn. 4.6/7) multiplied by the concentration of the inflowing solution (C0). The actual flow into the soil takes the evaporation rate into account. If no solution enters the soil (Fw[1] = 0) or if evaporation extracts water from the soil (Fw[1] < 0), the flow of solutes across the soil surface is set to zero. The notation for the solute flow into the soil (Fs[1]) is, therefore, defined by:

$$\begin{aligned} &\text{if Fw[1]} \leq 0\text{:} \\ &\text{Fs[1]} = 0 \\ &\text{if Fw[1]} > 0\text{:} \\ &\text{Fs[1]} = \text{Fw[1]} \cdot \text{C0} \end{aligned} \tag{7.4/6}$$

where:

C0 = solute concentration in the solution entering the soil surface [kg · m^{-3}] (according to the considerations for C0 described in connection with eqn 7.3/6)

The solute flow at the bottom boundary (Fs[n + 1]) varies according to the bottom boundary condition of the soil water model. Earlier we considered three bottom boundary conditions: impermeable layer (eqn 4.5/10), free drainage (eqn 4.6/8) and water-table (eqn 4.6/9). For the options ‘impermeable layer’ and ‘water-table’ (upward movement of solute-free water) the flux across the bottom boundary can be considered zero, and under ‘free drainage’ the flow of solute is defined by the convection of solutes with the drainage flow (flow of water out of the last soil layer multiplied by the solute concentration of the last soil layer). Therefore, the solute flows at the bottom boundary are:

Free drainage:
Fs[n + 1] = Fw[n+1] · Conc[n]

Water-table, impermeable layer: (7.4/7)
Fs[n + 1] = 0

Equations for solute leaching (eqns 7.4/1–7.4/7) must be combined with a water model where flows Fw[i] as well as the volumetric water content VWC[i] are calculated. Therefore, the equations for solute leaching with dynamic soil parameters are added to one of the previously developed water models (Chapters 4 and 5).

Application in ModelMaker

It is possible to add the leaching module to any of the dynamic soil water models developed earlier. However, the most comprehensive one is *Mod5-1a.mod*, which combines water and heat flow in soil. The diagram for solute leaching combining eqns 7.4/1 to 7.4/7 with *Mod5-1a.mod* is defined in *Mod7-1b.mod* (Fig. 7.5).

The <variable> 'VWC[1..14]' is used in various parts of the leaching model and is therefore defined as <global>. The additional solute leaching notation links to the dynamic water model and also uses the same soil layer definitions. The solute flow into the soil surface (Fs[1]) is calculated separately so that the <variables> Fw_avg, Dh, Ds and Dsh are only needed for soil layers 2–14.

The concentration of the solution entering the soil surface is either defined by a fixed concentration (user input) or a pulse application (from input file). The statements required in the <independent event> 'Choice' have already been described for the previous model (*Mod7-1a.mod*). The file *Inp7-1b.txt* contains concentrations for the pulse input option. In this file, the concentration is set to 100 kg m^{-3} for the first 10 h and thereafter to zero. The additional user statements are added to the existing statements of *Mod5–1a.mod*, which allows the definition of various scenarios for evaporation, rainfall or soil surface temperature. To allow the rainfall rate to be defined more easily, the parameter Rain (*Mod5-1a.mod*), is replaced by a <define> 'Rain', which takes on a value according to a 'GetValue' statement in the <independent event> 'Choice', i.e.:

```
if(RainfallChoice=1 or RainfallChoice=2){
 GetValue("Input value for Rainfall in mm/day:","Daily rainfall
    rate",Rain, Rain >0);
}
```

For leaching of solutes to occur in the model, the flow rate of water entering the soil (Fw[1]) must be above zero. This can be achieved by simulating rainfall and either setting the evaporation rate to zero or at least setting the evaporation rate (<parameter> 'Evaporation'; e.g. eqn 4.6/1) lower than the rainfall rate (<define> 'Rain'; e.g. eqn 4.6/2).

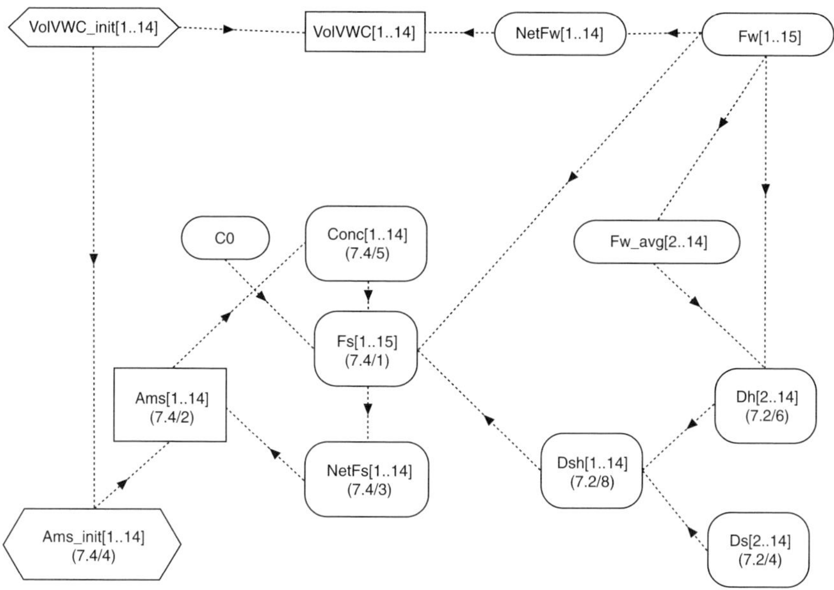

Fig. 7.5. Notation for solute flow with dynamic soil parameters (bold notation) (*Mod7-1b.mod*) (see also Fig. 5.1).

Running the model

The model is set up for 5 days (432,000 s), with outputs every 10 min and five fixed steps between outputs. The output of *Mod7–1b.mod* is similar to that of *Mod7-1a.mod*. However, the link with the soil water model makes the leaching section much more dynamic. To illustrate the difference between the constant versus the pulse C0 supply, simulations were carried out with the two C0 user options. All other user definable parameters were kept the same.

The parameters for the two simulation runs were:

Evaporation = zero (option 3)
Rainfall = diurnally changing (option 1)
Rainfall (mm day^{-1}) = 15
Bottom layer = free drainage (option 3)
Soil surface temp. = diurnally changing (option 2)

The only parameter which is changed between runs is the concentration of C0 on top of the soil:

Run 1 = constant C0 set to 100 kg m^{-3}
Run 2 = pulse C0 input set to 100 kg m^{-3} for the first 10 h, thereafter set to zero.

The concentration of the top soil layer (Conc[1]) for both simulation runs is presented in Fig. 7.6.

The diurnal behaviour of the solute flow entering the soil is well illustrated by the output for the constant C0 supply (Fig. 7.6A). The dynamics of solute concentration are clearly related to the flow rate of water entering the soil surface (Fig. 7.6B). During times when no solutes are entering the soil, the concentration will decline, due to the internal drainage within the soil profile translocating some of the material into deeper parts of the profile.

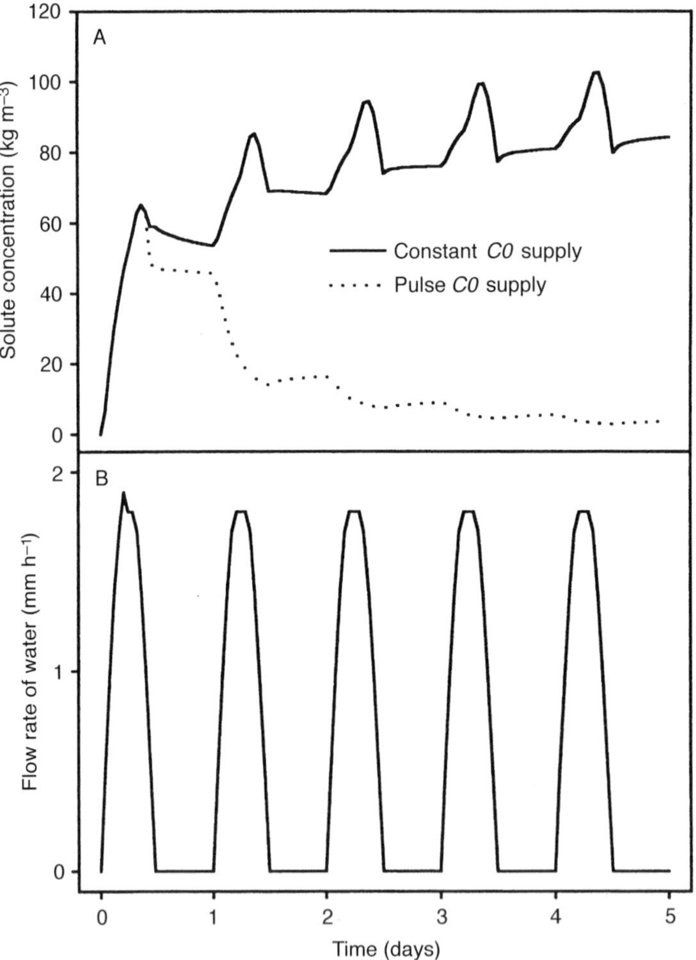

Fig. 7.6. Solute concentration of the top soil layer (Conc[1]) with a constant or a pulse *C0* supply (A) and flow rate of water entering the soil surface (B) simulated with *Mod7-1b.mod* (for parameter settings, see text).

This translocation into deeper soil regions is even quicker if solute-free solution is applied after the initial pulse application (Fig. 7.6A). The translocation into deeper soil regions is due to internal drainage and the additional 'washing out' effect of solute-free solution which enters the soil surface.

Many more such investigations can be performed with this model. It is therefore a good idea to test other user-definable options and to get a 'feeling' for how the solute-leaching section interacts with the rest of the dynamic water model.

7.5 Future Development

Not only does the soil–water dynamics determine the behaviour of solute leaching in soil but there is also a reverse effect of the solute concentration on soil water movement. The solute concentration gives rise to an additional osmotic potential in soil, which lowers the overall soil water potential (see Section 4.2) and leads to an additional water flow from soil regions of low concentration to soil regions of high solute concentration. The calculation of soil osmotic potentials is similar to the calculation of osmotic potentials in the plant tissues (described in Section 6.5.3). An excellent overview of this topic and the calculation procedure is given in Hillel (1998, Chapter 9).

Once the soil water leaching model is connected to a plant growth model, the uptake of certain plant-available solutes by roots needs to be taken into account. However, as already outlined at the end of Chapter 6, this requires the development of a layered root system model. If such a root system model is available, an approach such as that presented by Hillel (1977, Chapter 5) may be used to simulate the additional depletion of the solute concentrations due to root uptake.

Final Comments 8

In this book, processes in the area of soil–biosphere interactions have been considered. These include applications in the area of soil physics, environmental physics, soil biology, soil ecology and plant physiology. This wide range already provides an indication that processes in the soil–biosphere are very complex and can only be understood in a multidisciplinary approach. This also makes it obvious that it is insufficient to just concentrate on one particular field of research to understand the entire picture. Moreover, in order to understand even a particular process, it is necessary to understand its dynamics in relationship to a range of other interrelated processes. However, this task is too complex to be analysed by classical analytical approaches. For specialists in a particular field of research, it has become increasingly frustrating, due to the large body of knowledge, not to be able to consider all the aspects needed to understand the process in question. In such a situation, the benefit of mathematical modelling shows its full potential. The advances in computer technology have made it possible to combine several submodels and therefore gain a much more complete picture of processes in the soil–biosphere.

To understand, for instance, the dynamics of a grassland ecosystem, it is necessary to combine all the models developed in this book so that the influence that one part of the system (e.g. the soil water model) has on another part of the system (e.g. C/N transformations, plant growth) can be fully understood. So-called ecosystem models exist which are essentially performing the task of interrelating the various components (e.g. Thornley, 1998). However, the creation of such ecosystem models is usually a full-time task and generally left to mathematicians and modellers.

I personally think that the task of combining all the various areas of research should not be left only to mathematicians and modellers. Moreover, a multidisciplinary mind will come more or less naturally to the conclusion that the combination of the various parts is needed to fully understand soil–biosphere interactions as a whole. Modelling may be seen as a natural step in encompassing a particular area of research with the wider body of knowledge. This task is therefore not only of interest for researchers but also for students who want to understand the interactions in the soil–biosphere.

Through the study of this book, it is my hope that the reader has not only gained a basic understanding of how conceptual ideas of certain processes can be translated into working mathematical models, but also that a 'feeling' has been created that this task of bringing together all the ideas and concepts is not impossible.

With the help of the advanced modelling package 'ModelMaker', the final step of performing computer-based simulations has become a task which is achievable for many who previously thought that they did not have time or the confidence to create models in the classical way (e.g. specialized computer code). It has been demonstrated that even sophisticated mathematical notations in soil physics and environmental physics, usually requiring advanced computer code, can be solved without too much effort and mathematical background.

This publication should, therefore, be seen as a starting-point for the creation of models in the reader's own area of interest.

Appendix

Taylor polynomials

In order to approximate mathematical functions, it is often possible to fit them to a polynomial function in the form:

$$p(x) = a_0 + a_1 \cdot x + a_2 \cdot x^2 + . \dots + a_n \cdot x^n$$

The nth Taylor polynomial of a function f(x) at $x = a$ is the polynomial $p_n(x)$, defined by:

$$p_n(x) = f(a) + \frac{f'(a)}{1!} \cdot (x - a) + \frac{f''(a)}{2!} \cdot (x - a)^2 + ... + \frac{f^{(n)}(a)}{n!} \cdot (x - a)^n$$

This polynomial coincides with f(x) up to the nth derivative at $x = a$ in the sense that:

$$p_n(a) = f(a),\ p_n'(a) = f'(a),\ \dots,\ p_n^{(n)}(a) = f^{(n)}(a)$$

Note: Taylor polynomials are usually calculated at $x = 0$. Therefore, if $x = a \neq 0$, we can bring it back to the 'normal' notation by substituting for x the value of $(x - a)$.

EXAMPLE

Find all Taylor polynomials of $f(x) = 3 \cdot x^2 - 17$ at $x = 3$.

$f(x) = 3 \cdot x^2 - 17$, $f(3) = 10$
$f'(x) = 6 \cdot x$, $f'(3) = 18$
$f''(x) = 6$, $f''(3) = 6$
$f^{(3)}(x) = 0$, $f^{(3)}(3) = 0$

Derivatives higher than $f^{(3)}(x)$ for the above funcation are all zero; therefore, we obtain:

$$p_1(x) = 10 + 18 \cdot (x-3)$$
$$p_2(x) = 10 + 18 \cdot (x-3) + \frac{6}{2!} \cdot (x-3)^2$$
$$p_3(x) = 10 + 18 \cdot (x-3) + \frac{6}{2!} \cdot (x-3)^2 + \frac{0}{3!}(x-a)^3$$

For all $n \geq 3$we have:

$$p_n(x) = 10 + 18 \cdot (x-3) + \frac{6}{2!} \cdot (x-3)^2$$

Newton–Raphson algorithm

One application of Taylor polynomials is the calculation of zeros of functions. Some of the more basic methods for finding zeros of functions were presented in Chapter 1, Section 1.2.3. However, in many realistic applications there is no simple way of finding zeros.

One method which approximates zeros of functions to any desired degree is the Newton–Raphson algorithm. The idea behind this method is as follows. Starting from a given guess value for the zero, gradually improved zeros are calculated by replacing the function f(x) by its first Taylor polynomial at x_0, i.e. replacing it by:

$$p(x) = f(x_0) + \frac{f'(x)}{1} \cdot (x - x_0)$$

Since p(x) is an approximation of f(x) near $x = x_0$, the zero of f(x) should be close to the zero of p(x). Solving p(x) = 0 for x, we obtain:

$$f(x_0) + f'(x_0) \cdot (x - x_0) = 0$$
$$x = x_0 - \frac{f(x_0)}{f'(x_0)}$$

where the value x provides an improved approximation of the 'real' zero. This solution can be approximated to any desired degree by replacing x_0 with x and repeating the calculations again. The basic principle is similar to the iteration procedure presented in Chapter 5 for finding the soil surface temperature *TS*.

Application in ModelMaker

The method is illustrated with the model *ModA-1.mod* (Fig. A.1), which approximates square roots of given values via the Newton–Raphson algorithm. These approximations are then compared with the output of the predefined square root function.

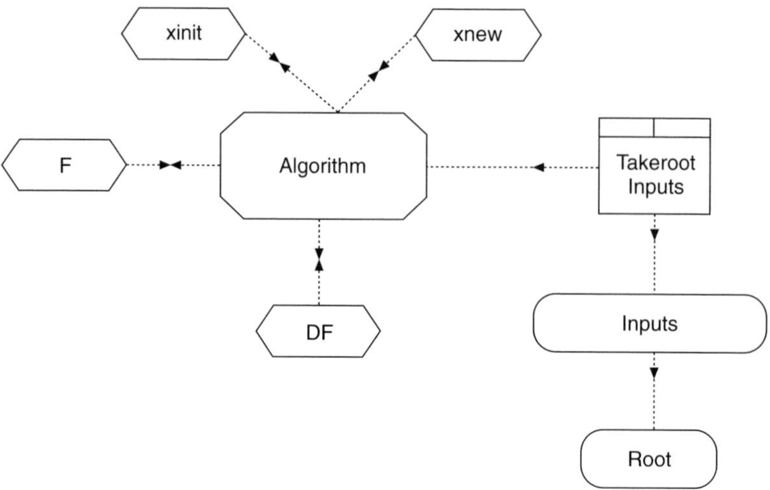

Fig. A.1. Model to calculate square roots via the Newton–Raphson algorithm compared with the inbuilt function for square root calculation of ModelMaker (*ModA-1.mod*).

The calculation procedure in *ModA-1.mod* is as follows. The aim is to find square roots of any value: $\sqrt{A}$. If we rewrite the value as $A = x \cdot x$ then the desired solution is the value 'x' $\left(\sqrt{x \cdot x} = x\right)$. The function which has to be minimized (from where we want to find the zero) is therefore:

$$\mathrm{f}(x) = x \cdot x - A$$

The first derivative of the function $\mathrm{f}(x) = x^2$ which is needed in the Newton–Raphson algorithm is given by (applying rule eqn 1.2/5):

$$\mathrm{f}'(x) = 2 \cdot x$$

In the calculation, we have to set the conditions when the iteration should stop. In this example, this value is set to:

$\mathrm{f}(x) = 0.0001$ (the calculated value of the square root is accurate to the fourth decimal place)

The calculations are performed with a 'do..while' loop. The loop calculations stop as soon as the argument is false (for further background and use of loops in ModelMaker, see Walker, 1997).

The actual calculation is performed in the <independent event> 'Algorithm' with the statement:

```
DO{
  xinit = xnew;
  F = Xinit*Xinit-Inp;
  DF = 2*xinit;
  xnew = xinit-F/DF;
}
while ( abs(F)>accuracy);
```

To start the calculation we have to provide an initial guess of the square root (xinit). (By the way, this can be any number – the program always finds the correct solution!) After calculating the function value *F* and the first derivative *DF*, the improved approximation of the square root is calculated in 'xnew'. The calculation is repeated till the F value drops below a predefined value (<parameter> 'accuracy').

One application of the Newton–Raphson algorithm in biosphere research is the use within the context of soil water flow problems (e.g. Campbell, 1985).

References

Anlauf, R., Kersenbaum, K.C., Ya Ping, L., Nuske-Schüler, A., Richter, J., Springob, G., Syring, K.M. and Utermann, J. (1988) *Modelle für Prozesse im Boden.* Ferdinand Enke Verlag, Stuttgart.

Betlach, M.R. and Tiedje, J.M. (1981) Kinetic explanation for accumulation of nitrite, nitric oxide, and nitrous oxide during bacterial denitrification. *Applied and Environmental Microbiology* 42, 1074–1084.

Blackmer, A.M. and Bremner, J.M. (1978) Inhibitory effect of nitrate on reduction of N_2O to N_2 by soil microorganisms. *Soil Biology and Biochemistry* 10, 187–191.

Blagodatsky, S.A. and Richter, O. (1998) Microbial growth in soil and nitrogen turnover: a theoretical model considering the activity state of microorganisms. *Soil Biology and Biochemistry* 30, 1743–1755.

Brugge, R. (1985) A mechanistic model of grass root growth and development dependent upon photosynthesis and nitrogen uptake. *Journal of Theoretical Biology* 116, 443–467.

Brugge, R. and Thornley, J.H.M. (1985) A growth model of root mass and vertical distribution, dependent on carbon substrate from photosynthesis and with non-limiting soil conditions. *Annals of Botany* 55, 563–577.

Campbell, G.S. (1985) *Soil Physics with BASIC Transport Models for Soil – Plant Systems*, Elsevier, Amsterdam.

Charles-Edwards, D.A. (1981) *The Mathematics of Photosynthesis and Productivity*, Academic Press, London.

Cho, C.M. and Mills, J.G. (1979) Kinetic formulation of the denitrification process in soil. *Canadian Journal of Soil Science* 59, 249–257.

Coleman, K. and Jenkinson, D.S. (1996) RothC- 26.3 – a model for the turnover of carbon in soil. In: Powlson, D.S., Smith, P. and Smith, J.U. (eds) *Evaluation of Soil Organic Matter Models Using Existing Long-term Datasets*, Vol. 38. Springer, Berlin, Heidelberg, pp. 237–246.

Conrad, R. (1995) Soil microbial processes involved in production and consumption of atmospheric trace gases. *Advances in Microbial Ecology* 14, 207–250.

Conrad, R. (1996a) Metabolism of nitric oxide in soil and soil microorganisms and regulation of flux into the atmosphere. In: Murrell, J.C. and Kelly, D.P. (eds) *Microbiology of Atmospheric Trace Gases*, Vol. 1. Springer Verlag, Berlin, pp. 167–203.

Conrad, R. (1996b) Soil microorganisms as controllers of atmospheric trace gases (H_2, CO, CH_4, N_2O, and NO). *Microbiological Reviews* 60, 609–640.

Davidson, E.A. (1991) Fluxes of nitrous oxide and nitric oxide from terrestrial ecosystems. In: Rogers, J.E. and Whitman, W.B. (eds) *Microbial Production and Consumption of Greenhouse Gases: Methane, Nitrogen Oxides, and Halomethanes*. American Society for Microbiology, Washington, DC, pp. 219–235.

Delwiche, C.C. (1981) The nitrogen cycle and nitrous oxide. In: Delwiche, C.C. (ed.) *Denitrification, Nitrification, and Atmospheric Nitrous Oxide*. John Wiley & Sons, New York, pp. 1–15.

Dendooven, L. and Anderson, J.M. (1994) Dynamics of reduction enzymes involved in the denitrification process in pasture soil. *Soil Biology and Biochemistry* 26, 1501–1506.

Dendooven, L. and Anderson, J.M. (1995a) Maintenance of denitrification potential in pasture soil following anaerobic events. *Soil Biology and Biochemistry* 27, 1251–1260.

Dendooven, L. and Anderson, J.M. (1995b) Use of a 'least square' optimization procedure to estimate enzyme characteristics and substrate affinities in the denitrification reactions in soil. *Soil Biology and Biochemistry* 27, 1261–1270.

Dendooven, L., Splatt, P., Anderson, J.M. and Scholefield, D. (1994) Kinetics of the denitrification process in a soil under permanent pasture. *Soil Biology and Biochemistry* 26, 361–370.

Fillery, I.R.P. (1983) Biological denitrification. In: Freney, J.R. and Simpson, J.R. (eds) *Gaseous Loss of Nitrogen from Plant–Soil Systems*. Martinus Nijhoff/Dr W. Junk Publishers, The Hague, pp. 33–64.

Firestone, M.K. (1982) Biological denitrification. In: Stevenson, F.J. (ed.) *Nitrogen in Agricultural Soils*. Agronomy Monograph 22, ASA–CSSA–SSSA, Madison, Wisconsin, pp. 289–326.

Focht, D.D. (1974) The effect of temperature, pH, and aeration on the production of nitrous oxide and gaseous nitrogen – a zero order kinetic model. *Soil Science* 118, 173–179.

Frissel, M.J. and van Veen, J.A. (1981) Simulation model for nitrogen immobilization and mineralization. In: Iskander, I.K. (ed.) *Modeling Wastewater Renovation Land Treatment*. Wiley-Interscience Publications, New York, pp. 359–381.

Goldstein, L.J., Lay, D.C. and Schneider, D.I. (1990) *Calculus and its Applications*. Prentice Hall, Englewood Cliffs, New Jersey.

Gradshteyn, I.S., and Ryzhik, I.M. (1995) *Tables of Integrals, Series, and Products*, 5th edn. Academic Press, New York.

Granli, T. and Bøckman, O.C. (1994) Nitrous oxide from agriculture. *Norwegian Journal of Agricultural Sciences*, Suppl. No. 12, 1–128.

Grant, R.F., Juma, N.G., and McGill, W.B. (1993) Simulation of carbon and nitrogen transformations in soil: mineralization. *Soil Biology and Biochemistry* 25, 1317–1329.

Gunnewiek, H.K. (1996) Organic matter dynamics simulated with the 'Verberne' model. In: Powlson, D.S., Smith, P. and Smith, J.U. (eds) *Evaluation of Soil Organic Matter Models Using Existing Long-term Datasets*. Vol. 38. Springer, Berlin, pp. 255–261.

Heinke, M. and Kaupenjohann, M. (1997) Effects of soil solution on dynamics of N_2O emissions. In: Becker, K.H. and Wiesen, P. (eds) *7th International Workshop on*

Nitrous Oxide Emissions, Vol. 41. Bergische Universität Gesamthochschule Wuppertal – Physikalische Chemie, Cologne, pp. 337–344.

Hillel, D. (1977) *Computer Simulation of Soil–Water Dynamics: a Compendium of Recent Work.* International Development Research Centre, Ottawa.

Hillel, D. (1982a) Soil temperature and heat flow. In: Hillel, D. (ed.) *Introduction to Soil Physics.* Academic Press, London, pp. 155–175.

Hillel, D. (1982b) Evaporation from bare-surface soils. In: Hillel, D. (ed.) *Introduction to Soil Physics.* Academic Press, London, pp. 268–287.

Hillel, D. (1998) *Environmental Soil Physics.* Academic Press, San Diego.

Hunt, H.W. (1977) A simulation model for decomposition in grasslands. *Ecology* 58, 469–484.

Jansson, S.L., Hallam, M.J. and Bartholomew, W.V. (1955) Preferential utilization of ammonium over nitrate by micro-organisms in the decomposition of oat straw. *Plant and Soil* 6, 382–390.

Jenkinson, D.S. and Parry, L.C. (1989) The nitrogen cycle in the Broadbalk wheat experiment: a model for the turnover of nitrogen through the microbial biomass. *Soil Biology and Biochemistry* 21, 535–541.

Jenkinson, D.S. and Rayner, J.H. (1977) The turnover of soil organic matter in some of the Rothamsted classical experiments. *Soil Science* 123, 298–305.

Johnson, I.R., and Thornley, J.H.M. (1983) Vegetative crop growth model incorporating leaf area expansion and senescence, and applied to grass. *Plant, Cell and Environment* 6, 721–729.

Johnson, I.R. and Thornley, J.H.M. (1984) A model of instantaneous and daily canopy photosynthesis. *Journal of Theoretical Biology* 107, 531–545.

Johnson, I.R. and Thornley, J.H.M. (1985) Dynamic model of the response of a vegetative grass crop to light, temperature and nitrogen. *Plant, Cell and Environment* 8, 485–499.

Kirkham, D. and Powers, W. (1972) Miscible displacement. In Kirkham, D. and Powers, W. (eds) *Advanced Soil Physics.* Wiley Interscience, New York, pp. 379–427.

Knowles, G., Downing, A.L. and Barrett, M.J. (1965) Determination of kinetic constants for nitrifying bacteria in mixed culture, with the aid of an electronic computer. *Journal of General Microbiology* 38, 263–278.

Kohl, D.H., Vithayathil, F., Whitlow, P., Shearer, G. and Chien, S.H. (1976) Denitrificaiton kinetics in soil systems: the significance of good fits of data to mathematical forms. *Soil Science Society of America Journal* 40, 249–253.

Ladd, J.N., Oades, J.M. and Amato, M. (1981) Microbial biomass formed from ^{14}C, ^{15}N-labelled plant material decomposing in soils in the field. *Soil Biology and Biochemistry* 13, 119–126.

Laudelout, H., Lambert, R., Firiat, J.L. and Pham, M.L. (1974). Effet de la température sur la vitesse d'oxydation de l'ammonium en nitrate par des cultures mixtes de nitrifiants. *Annals of Microbiology (Institut Pasteur)* 125B, 75–84.

Laudelout, H., Lambert, R. and Pham, M.L. (1976). Influence du pH et de la pression partielle d'oxygène sur la nitrification. *Annals of Microbiology (Institut Pasteur)* 127A, 367–382.

Leggett, D.C. and Iskandar, I.K. (1981) Evaluation of a nitrification model. In: Iskander, I.K. (ed.) *Modeling Wastewater Renovation Land Treatment.* Wiley-Interscience Publications, New York, pp. 313–358.

Li, C. (1996) The DNDC model. In: Poulson, D.S., Smith, P. and Smith, J.U. (eds)

Evaluation of Soil Organic Matter Models Using Existing Long-term Datasets, Vol. 38, Springer, Berlin, Heidelberg, pp. 263–267.

Marshall, T.J. and Holmes, J.W. (1988) *Soil Physics*, 2nd edn. Cambridge University Press, Cambridge.

Monteith, J.L. and Unsworth, M.H. (1990) *Principles of Environmental Physics*, 2nd edn. Edward Arnold, London.

Mueller, T., Jensen, L.S., Hansen, S. and Nielsen, N.E. (1996) Simulating soil carbon and nitrogen dynamics with the soil–plant–atmosphere system model DAISY. In: Powlson, D.S., Smith, P. and Smith, J.U. (eds) *Evaluation of Soil Organic Matter Models Using Existing Long-term Datasets*, Vol. 38. Springer, Berlin, pp. 275–281.

Müller, C. (1996) *Nitrous Oxide Emission from Intensive Grassland in Canterbury, New Zealand*. Tectum Verlag, Marburg.

Müller, C., Kammann, C., Burger, S., Ottow, J.C.G., Grünhage, L., and Jäger, H.-J. (1997) Nitrous oxide emission from frozen grassland soil and during thawing. In: Becker, K.H. and Wiesen, P. (eds) *7th International Workshop on Nitrous Oxide Emissions*. Bergische Universität Gesamthochschule Wappertal – Physikalische Chemie, Cologne, pp. 327–335.

Nömmik, H. (1956) Investigations on denitrification in soil. *Acta Agriculturæ Scandinavica* 6, 195–228.

Panikov, N.S. (1995) *Microbial Growth Kinetics*. Chapman & Hall, London.

Panikov, N.S. (1996) Mechanistic mathematical models of microbial growth in bioreactors and in natural soils: explanation of complex phenomena. *Mathematics and Computers in Simulation* 42, 179–186.

Parton, W.J. (1996). The CENTURY model. In: Powlson, D.S., Smith, P. and Smith, J.U. (eds) *Evaluation of Soil Organic Matter Models Using Existing Long-term Datasets*, Vol. 38. Springer Verlag, Berlin, pp. 283–293.

Parton, W. J., Schimel, D.S., Cole, C.V., and Ojima, D.S. (1987) Analysis of factors controlling soil organic matter levels in great plains grasslands. *Soil Science Society of America Journal* 51, 1173–1179.

Parton, W.J., Mosier, A.R., Ojima, D.S., Valentine, D.W., Schimel, D.S., Weier, K. and Kulmala, A.E. (1996) Generalized model for N_2 and N_2O production from nitrification and denitrification. *Global Biogeochemical Cycles* 10, 401–412.

Paul, E.A. and Clark, F.E. (1996) Ammonification and nitrification. In: Paul, E.A. and Clark, F.E. (eds) *Soil Microbiology and Biochemistry*. Academic Press, San Diego, pp. 182–199.

Paul, W. and Domsch, K.H. (1972) Ein mathematisches Modell für den Nitrifikationsprozeß im Boden. *Archives of Microbiology* 87, 77–92.

Powlson, D.S., Smith, P. and Smith, J.U. (eds) (1996) *Evaluation of Soil Organic Matter Using Existing Long-term Datasets*. Springer, Berlin and Heidelberg.

Richter, J. (1986) *Der Boden als Reaktor Modelle für Prozesse im Boden*. Ferdinand Enke Verlag, Stuttgart.

Schloesing, M.T. and Muntz, A. (1879) Recherches sur la nitrification. *Comptes Rendus hebdomadaires des Séances de l'Académie des Sciences* (Paris) 89, 1074–1077.

SPSS (1997) *SigmaPlot*. SPSS, Chicago.

Tate, R.L., III (1995) *Soil Microbiology*, 1st edn. John Wiley & Sons, New York.

Thornley, J. H. M. (1996) Modelling water in crops and plant ecosystems. *Annals of Botany* 77, 261–275.

Thornley, J.H.M. (1998) *Grassland Dynamics – an Ecosystem Simulation Model.* CAB International, Wallingford, Oxon.

Thornley, J.H.M. and Cannell, M.G.R. (1996) Temperate forest response to carbon dioxide, temperature and nitrogen: a model analysis. *Plant, Cell and Environment* 19, 1331–1348.

Thornley, J.H.M. and Johnson, I.R. (1990) *Plant and Crop Modelling: a Mathematical Approach to Plant and Crop Modelling.* Clarendon Press, Oxford.

Thornley, J.H.M. and Verberne, E.L.J. (1989) A model of nitrogen flows in grassland. *Plant, Cell and Environment* 12, 863–886.

Tiedje, J.M. (1982) Denitrification. In: Page, A.L. (ed.) *Methods of Soil Analysis: Part 2, Chemical and Microbiological Properties*, Vol. 9. Soil Science Society of America and American Society of Agronomy, Madison, Wisconsin, pp. 1011–1026.

van Veen, J.A. and Paul, E.A. (1981) Organic carbon dynamics in grassland soils. 1. Background information and computer simulation. *Canadian Journal of Soil Science* 61, 185–201.

van Veen, J.A., Ladd, J.N. and Frissel, M.J. (1984) Modelling C and N turnover through the microbial biomass in soil. *Plant and Soil* 76, 257–274.

Walker, A. (1997) *ModelMaker.* Cherwell Scientific Publishing, Oxford.

Wardle, D.A. (1998) Controls of temporal variability of the soil microbial biomass: a global-scale synthesis. *Soil Biology and Biochemistry* 30, 1627–1637.

Warrington, R. (1879) On nitrification (part 2). *Journal of the Chemical Society Transactions* 35, 429–456.

Warrington, R. (1884) On nitrification (part 3). *Journal of the Chemical Society Transactions* 45, 637–672.

Index

absolute humidity (air, HA) 211–212, 220
see also humidity
absorption 168, 203, 205
see also energy transfer
adsorption 60
see also process
aerobic respiration 248
see also respiration
air properties
density of air 217
specific heat 217
thermal conductivity 233
volumetric specific heat 217, 219
air temperature 211, 217, 219, 227–228
see also temperature
air-entry potential 184
see also water potential
air-entry suction 184
see also water potential
albedo (α) 209
ammonium 51–53, 60–61, 69, 106, 113, 125, 315
see also nitrogen
analytical solution 21–23, 40, 42–43, 168–169, 256, 260
see also integration
anthropogenic influence 1, 315
antidifferentiation 15
see also integration
<array notation> 39
see also ModelMaker
Arrhenius 31, 79
see also kinetics
artefact (numerical) 92
see also numerical
assumptions 4, 6
see also conceptual model
ATP 251

Beer's law 254, 304
see also light relationships
biosphere, processes 1
boundary conditions
definition 25–27
heat flow 167, 170, 174
leaching 322–323, 327–328
water flow 190–191, 194–198, 245
boundary layer
buoyancy effect 215
conductance (Kh) 216–219, 225, 302, 304
definition 214
diffusion 214
fetch 214
friction velocity 217, 225
heat flux 214
inertial forces 214
laminar 214
momentum transfer 214–216
neutral stability state 215–216, 218
resistance (RAC) 216, 218–219, 225
roughness elements 214
roughness length 214, 217, 305
soil–atmosphere exchange processes 213–219
stability parameter 217, 225, 304
stable stability state 215–216, 218
turbulent 214
unstable stability state 215–216, 218
vertical forces 214
water vapour flux 214
zero plane displacement 214, 217, 305
Brunt formula 211
see also humidity
bulk density 80, 161, 284
see also soil
buoyancy effect 215
see also boundary layer

calculation speed 247
see also ModelMaker
calculus
see also differentiation *and* integration
canopy
definition 254
ground heat flux (G) 301
see also soil energy balance
height 305
light relationships 254
see also light relationships
photosynthesis 254, 256, 258, 260, 269
carbohydrates 251–252
carbon (C)
^{14}C 134
organic 114
total 118
carbon dioxide (CO_2) 105, 107, 119, 140, 248–249, 251, 254, 294
diffusion resistance 259
external 258–259
internal 251, 258
stomatal control 300–301
see also stomata
carbon transformation
concept 107–110, 115
efficiency 107
gross rate 108, 112

carboxylation resistance 253, 258
  see also photosynthesis
cell wall rigidity 297, 299
  see also plant water
change of saturation vapour pressure 232, 234, 243, 302
  see also saturation vapour pressure
checking results 48–50
  see also conceptual model
clay
  fraction 235
  organic matter stabilization 105
  thermal property 161, 165
climate data 75, 203
  see also data
C:N ratio 52, 105, 111, 124–125, 128
  see also organic matter *and* microbial biomass
C/N transformation 52
  see also microbial biomass
<compartment> 38, 170
  see also ModelMaker
competitive weighting factors 89
  see also denitrification
<component event> 39, 287
  see also ModelMaker
components 37–39
  see also ModelMaker
computer
  Pentium 247
concentration gradient 23–24
conceptual model 3–4
  assumptions 4, 6
  checking results 48–50
  C/N transformation 104–106
  definition 4–6, 40–41
  denitrification 86–87
  equation 3
  leaching 318–319
  mathematical description 3, 41–45
  nitrification 58–59
  plant growth 249–250
  research question 39
conductance (Kh) 216–219, 225, 302, 304
  see also boundary layer
conductance (stomata) 300
  see also stomata
conduction 203, 205, 208
  see also energy transfer
continuity equation 27, 157–159, 166, 188, 319
  see also equations
convection (heat) 205, 208
  see also energy transfer
convection (leaching) 316–317
  see also transport processes *and* leaching

damping depth 169, 171
  see also heat flow
Darcy's law 184, 188
  see also water flow
data, climate 75, 203
<define> 37
  see also ModelMaker
definite integral 18, 44
  see also integration
<delays> 38, 102–103
  see also ModelMaker
denitrification 53
  competitive weighting factors 89
  intensity 88–90, 95
  generation time 95
  master parameter 98
  moisture influence 100–102
  N loss (C/N model) 126
  N_2 production 88–91
  N_2 emission 96–97
  N_2O production 88–91
  N_2O emission 96–97
  NO production 88–91
  NO emission 96–97
  temperature influence 102
density of air 217
  see also air properties
dependent variables 6–7
  see also variables
derepression enzyme dynamics 100–103
  see also enzymes
derivative 11–15
  see also differentiation
dewpoint temperature (DP) 211–212, 227–228
  see also temperature
differential equations 10–11, 18, 159
  see also equations
differentiation 11–15
  chain rule 15
  common functions 15
  constant multiple rule 13
  derivative 11–15
  power rule 13
  product rule 14
  quotient rule 14
  solution of differential equations 18
  sum rule 13
diffusion (leaching) 316–317
  see also transport processes
diffusion coefficient (leaching) 317, 322
  see also leaching
diffusivity
  gas 71–72
    see also gas
  water vapour (DV) 232–233, 239, 243
dinitrogen (N_2) 53, 88–91
  see also nitrogen
dispersion 316, 318
  see also transport processes
dispersion coefficient 320, 322
  see also leaching
<do..while> loop 226
  see also ModelMaker
drainage 191, 196–197, 245
  see also water flow
dry matter 278
  see also root *and* shoot
drying power 221, 303
  see also humidity
dummy variable 226–227
  see also iteration

Eadie–Hofstee transformation 36
  see also kinetics
eddies 214–215

efficiency of carbon transformation 107
 see also carbon transformation
electromagnetic waves 205
 see also radiation
emissivity constant 205, 211, 213
 see also radiation
energy relationships 203
 see also soil energy balance
energy status (water) 180–183
 see also water potential
energy transfer
 absorption 168, 203, 205
 conduction 203, 205, 208
 convection 205, 208
environmental pollution 315
enzyme
 denitrification 53, 87, 93–96
 derepression dynamics 100–103
 initial activity 63
 nitrification 52, 63, 69
 nitrite reductase (nitrification) 69
 reductase 93–94
 relative activities 70, 93, 95
 substrate complex 33–34, 87
equations
 continuity 27, 157–159, 166, 188, 319
 differential 10–11, 18, 159
 partial differential 158
 Penmann–Monteith 301–305, 309
 see also evapotranspiration
 proportionality 5
 proportionality constant 5, 23–24, 184, 215–216
 Richard's 188
 see also water flow
 simultaneous 223–224
 transport 23–25
evaporation
 actual 191–193, 246
 cumulative 193, 246
 demand 191–192, 194
 diurnal change 194
 potential 191–192, 196, 246
 rainfall 306
 see also rainfall, rate
 rate 193, 200, 210, 220, 226, 301, 328
 soil surface 191
evapotranspiration
 definition 294
 Penmann–Monteith equation 301–305, 309
exponents 7
 see also mathematics

fertilizer application 284, 287, 289–290
fetch 214
 see also boundary layer
field capacity 306
 see also soil moisture
finite difference notation 23–25, 157–158, 189–191
first-order kinetics 20, 28–29, 53–55, 69, 106, 129, 131
 see also kinetics
first-stage drying 191, 193
 see also water flow
flow
 heat *see* heat flow
 water *see* water flow
<flows> 38
 see also ModelMaker
<for> loop 226
 see also ModelMaker
Fourier's law 156–157
 see also heat flow
free drainage 196
 see also drainage
freezing–thawing 155
friction velocity 217, 225
 see also boundary layer
functions
 air-filled porosity 71
 difference form 11
 gas diffusivity 71–72
 hydraulic conductivity 187, 306
 intercept 10
 linear 10
 logarithm 8, 184, 214–215, 254, 260
 moisture 79–82
 partitioning 279–281
 see also partitioning
 pH 64–65, 68
 piecewise continuous 81
 quadratic 9
 root density 308
 root water status 307
 saturation vapour pressure (SVP) 212, 221, 225, 233
 sinusoidal 167–169, 195
 slope 10–11
 soil moisture characteristic 183–185
 see also soil moisture
 stomatal control 300–301
 see also stomata
 Taylor polynomials 23, 334–335
 temperature 72, 82–85, 102, 292
 thermal conductivity 161
 volumetric heat capacity 160
 water-filled porosity 80, 100–101, 179
 water status 310
 see also shoot
 zeros of 8–10

gas
 air concentration 72
 constant (R) 73–74, 212, 221, 222, 232, 234, 239, 243, 296, 299
 diffusivity NO 71–72
 diffusivity N_2O 72
 dissolved in water 74
 emission (diffusion approach)72, 96, 98
 emission (first-order approach) 77–78, 97–98
 law 73
 N production (denitrifier) 88–91
 see also denitrification, N_2O, NO, N_2 production
 N production (nitrifier) 70, 73
 see also nitrification, N_2O, NO production
 thermal property 161
 see also soil air
generation time
 denitrifier 95
 see also denitrification
 nitrifier 63–64
 see also nitrification

<GetChoice> 133, 149, 151, 163, 228, 237, 240, 244, 293, 311
see also ModelMaker
<GetValue> 84, 97, 149, 152, 170, 197, 328
see also ModelMaker
graph 49, 275
see also ModelMaker
grass *see* pasture
gravimetric water content 179
see also soil moisture
gravitational potential 182
see also water potential
gross rate
carbon transformation 108, 112
see also carbon transformation
photosynthesis 257, 259, 269
see also photosynthesis
ground heat flux (G) 210, 222, 226, 232, 301
see also soil energy balance
groundwater 315
growth
respiration 268–270, 276, 282
see also respiration
root 275–277, 281–282, 286
see also root
shoot 269, 281, 286
see also shoot
growth rate (nitrifier) 63
see also nitrification

Hanes–Wolf transformations 36
see also kinetics
harmonic function
see functions, sinusoidal
heat flow
absorption 168, 203, 205
see also energy transfer
analytical solution 168–170
boundary conditions 214
see also boundary conditions
conduction 203, 205, 208
see also energy transfer
constant thermal properties 165–169, 177
convection 205, 208
see also energy transfer
damping depth 169, 171
description 4–6, 23, 156
dynamic thermal properties 172–178
Fourier's law 156–157
graphical analysis 159–160
moisture relationship 172–179, 204
reflection 203
soil surface 203
specific heat 161
steady-state 156, 166
thermal conductivity 156,158, 161–167, 174–176, 203, 223,
thermal conductivity (water vapour) 229, 232–233, 237–238
thermal diffusivity 159, 163–168
volumetric heat capacity 156, 160, 161–167, 172–175, 203
humidity
absolute (air, HA) 211–212, 220
Brunt formula 211
drying power 221, 303
psychrometer constant 303
relative (air) 191
relative (soil) 191, 221, 232–234, 243
saturated 221, 225
soil (Hso) 234, 239
soil surface (HS) 220–221, 222, 225
soil vapour concentration 234
vapour pressure deficit (VPD) 302–303
units 212
Hurely pasture model 250
see also pasture
hydraulic conductivity
average 192
function 187, 306
see also functions
saturated 187
units 186
water vapour 188, 238–239
hydraulic gradients 198
see also water flow
hydrodynamic dispersion 316, 318
see also transport processes
hydroxylamine 52
see also nitrogen

ice 161, 208
ideal gas law 73
see also gas
<if...else> statement 135, 163–164
see also ModelMaker
immobilization 51–52, 105–106
ammonium 112, 149
inorganic nitrogen 141, 148
nitrate 112, 148
potential (nitrogen) 112, 141, 152
impermeable layer 191, 197
see also water flow
incremental specific leaf area 272, 286
see also leaf area index (LAI)
<independent event> 39, 133, 150, 163, 197, 204, 226
see also ModelMaker
independent variables 6–7
see also variables
<index> notation 39, 170
see also ModelMaker
index of physiological state 136–151
see also microorganisms
inertial forces 214
see also boundary layer
infiltration
capacity 199–200
cumulative 200
infiltrability 198–200
rate 200
warm water 208
<influences> 38
see also ModelMaker
inhibition notation 85, 131–132, 143
see also kinetics
initial guess value 224
see also iteration
inputs 7
see also variables
integral tables 18, 260
see also integration

ial biomass *continued*
iversity 108
rowth 105, 118, 129–133, 139–140
ndex of physiological state 136–151
see also microorganisms
nechanism 105
noisture influence 151, 153, 179
itrogen 113, 126, 139
hysiological state 136–151
see also microorganisms
rotection capacity 105, 134–136, 146
ubgroups 106
emperature influence 151
urrents of air *see* eddies
utrients 248
e also nutrients
ganisms
ctivity 136–138, 155
ormancy 137
dex of physiological state 136–151
components 138
components138
zation
rbon 108, 118, 122–123, 141
trogen 111, 119, 121, 123–125, 130, 141, 284
ocess 51–52, 105, 250
n law 84, 143, 248
e also Liebig
Exit> 92
e also ModelMaker
optimization> 98
e also ModelMaker
aker
ray notation 39
lculation speed 247
mpartments 38, 170
mponent event 39, 287
mponents 37–39
fine 37
lays 38, 102–103
..while loop 226
ws 38
loop 226
tChoice 133, 149, 151, 163, 228, 237, 240, 244, 293, 311
tValue 84, 97, 149, 152, 170, 197, 328
aph 49, 275
.else statement 135, 163–164
ependent event 39, 133, 150, 163, 197, 204, 226
ex notation 39, 170
uences 38
roduction 37–39
kup files 38, 75, 103, 229
kup tables 38, 75, 103
delExit 92
del optimization 98
options 91, 176, 204–205, 275
sitivity analysis 68–69, 77, 152–153
le 49
ger 227–228
iables 38
le 226
puts *see* simulation results
ults *see* simulation results

models
Arrhenius (*Mod2–1b*) 55
chain reactions (*Mod2–1c*) 56
C/N transformation (C concept, *Mod2–4a*) 110
C/N transformation (N concept, *Mod2–4b*) 115
C/N transformation (basic, *Mod2–4c*) 127
C/N transformation (substrate relationships, *Mod2–4d*) 133
combined temperature–moisture model (*Mod5–1a*) 206–207
denitrification (basic, *Mod2–3a, Mod2–3b*) 92
denitrification (enzyme dynamics, *Mod2–3c*) 95
denitrification (gas, *Mod2–3d*) 99
denitrification (moisture, temperature, *Mod2–3e*) 103
first-order kinetics (*Mod2–1a*) 53–55
heat flow (constant parameters, *Mod3–1b, Mod3–1c*) 171
heat flow (dynamic parameters, *Mod3–1d*) 177
hydraulic conductivity (*Mod4–1b*) 186
leaching (constant parameters, *Mod7–1a*) 324
leaching (dynamic parameters, *Mod7–1b*) 329
Michaelis–Menten (*Mod2–1d*) 57–58
Newton–Raphson algorithm (*ModA-1*) 336
nitrification (basic, *Mod2–2a*) 66
nitrification (gas, *Mod2–2b* and *Mod2–2d*) 76
nitrification (moisture, temperature, *Mod2–2c*) 83
one-compartment mixing (*Mod1–1a*) 41, 46
photosynthesis (rectangular hyperbola, *Mod6–1a*) 257
photosynthesis (advanced model, *Mod6–1b, Mod6–1c*) 265
plant growth (basic, *Mod6–1d*) 274
plant growth (root growth, *Mod6–1e*) 288
plant growth (temperature, *Mod6–1f*) 293
plant growth (water status, *Mod6–1g*) 312
second-order kinetics (*Mod2–1a*) 53–55
soil energy balance (basic, *Mod5–1b*) 230–231
soil energy balance (non-isothermal, *Mod5–1e*) 245
soil energy balance (vapour movement, *Mod5–1d*) 241
soil energy balance (water vapour, *Mod5–1c*) 236
soil moisture characteristics (*Mod4–1a*) 186
soil temperature–moisture (*Mod5–1a*) 206–207
thermal properties (*Mod3–1a*) 164
two-compartment mixing (*Mod1–1b*) 47–48
water flow (basic, *Mod4–1c*) 194
water flow (additional boundary conditions, *Mod4–1d*) 198
water flow (infiltrability, *Mod4–1e*) 201
zero-order kinetics (*Mod2–1a*) 53–55
momentum transfer 214–216
see also boundary layer
Monod 129, 133
see also kinetics
Monsi-Saeki 255

integration 15–18
- analytical solution 21–23, 40, 42–43, 168–169, 256, 260
- Burlisch-Stoer 23
- definite integral 18, 44
- Euler 23
- integral tables 18, 260
- mid-point rule 22
- numerical solution 21–23, 40, 169, 257–258
- parts, by 17
- rules 16–18, 262
- Runge–Kutta 23, 68
- substitution, by 16

introduction to ModelMaker 37–39
- *see also* ModelMaker

iteration
- dummy variable 226–227
- initial guess value 224
- loop calculation 224, 226, 228
- soil surface temperature 223–229

kinetics 27–37
- Arrhenius 31, 79
- chain reactions 32–33, 56–57
- competitive Michaelis–Menten 86
- Eadie–Hofstee transformation 36
- first-order 20, 28–29, 53–55, 69, 106, 129, 131
- Hanes–Wolf transformations 36
- inhibition notation 85, 131–132, 143
- Lineweaver–Burk transformation 35
- Michaelis–Menten 33–35, 57–58, 62–63, 106, 112, 114, 129, 251
- modelling 53–58
- Monod 129, 133
- rate constant 31, 55–56
- second-order 30–31, 53–55
- units 36–37
- zero-order 27–28, 53–55, 63

LAI *see* leaf area index

laminar boundary layer 214
- *see also* boundary layer

latent heat (LE) 208, 210, 213, 215–217, 219, 225, 229, 232, 302
- *see also* soil energy balance

latent heat of vaporization (L) 210, 220, 226, 232, 234–235, 302–303
- *see also* soil energy balance

laws 7–8
- *see also* mathematics

layered soil system 25–26, 158, 321
- *see also* soil

leaching
- diffusion coefficient 317, 322
- dispersion coefficient 320, 322
- dynamic parameters 324–328
- factor 284
- function 284
- simple model 284
- translocation process 316–322
  - *see also* transport process
- washing out effect 331

leaf
- first fully expanded 268, 270–271
- growth 248–250, 268, 270–271
- second fully expanded 268, 270–271
- senescing 268,
- tissue density 2

leaf area index (LAI) 272–273, 28
- incremental spe
- maximum spec
- plant–water dy
- specific leaf are
- temperature ad

Liebig, minimum law

light relationships
- absorption 25
- Beer's law 25
- canopy 254
- exponential ex
- Monsi–Saeki
- photoperiod 2
- reflection 254
- transmittance

lignin 106, 117
- *see also* organ

Lineweaver–Burk tr
- *see also* kineti

logarithm 8, 44
- *see also* math

long-wave radiation 232
- *see also* radia

<lookup files> 38, 7
- *see also* Mod

<lookup tables> 38,
- *see also* Mod

loop calculation 22
- *see also* iterat

macronutrients 248
- *see also* nutri

maintenance respirat
- *see also* respi

master parameter 9
- *see also* deni

matric potential 18 296, 306
- *see also* wate

mathematics
- exponent 7
- iteration 22
- laws 7–8,
- logarithm 8,
- roots 7
- solving prob
- transformati
- universal lar

maximum specific
- *see also* leaf

metabolites, organ

Michaelis–Menten 106, 112
- *see also* kin

microbial biomass
- biosynthesi
- carbon 126
- C:N ratio 5
- C/N transfo
- degradation 1

see also light relationships
multilayered root system 313, 331
see also root

^{15}N 134
see also nitrogen
N_2 53, 88–91, 96–97
see also nitrogen
N_2O 51, 52–53, 88–91
see also nitrogen
$N_2O:N_2$ ratio 53
see also nitrogen
NADH 251
net
flow (water) 193
see also water flow
plant growth 269
see also plant growth
radiation (Rn) 209–210, 227–228, 232, 302, 304
see also radiation
neutral stability state 215–216, 218
see also boundary layer
Newton–Raphson algorithm 23, 335–337
NH_4^+-N 51–53, 60–61, 69, 106, 113, 125, 315
see also nitrogen
nitrate 51, 62, 88–91, 106, 113, 315
see also nitrogen
nitric oxide (NO) 52–53, 88–91, 96–97
see also nitrogen
nitrification 52–53
ammonium oxidation 61
biosynthesis 59, 61
generation time 63–64
growth rate 63
N loss (C/N model) 126
inhibition notation 85
moisture influence 84–85
nitrifier-denitrification 52–53
nitrite oxidation 61
Nitrobacter 52
*Nitro*bacteria 62–64
Nitrosomonas 52
*Nitroso*bacteria 61–63
N_2O production 73
N_2O emission 70
NO production 70
NO emission 73
pH influence 62–65
temperature influence 84–85
nitrifier-denitrification 52–53
see also nitrification
nitrite 52–53, 61–62, 88–91
see also nitrogen
Nitrobacter 52
see also nitrification
nitrogen
^{15}N 134
ammonium (NH_4^+-N) 51–53, 60–61, 69, 106, 113, 125, 315
assimilation of 248
decomposable material 123
dinitrogen (N_2) 53, 88–91, 96–97
hydroxylamine 52
microbial biomass 113, 126, 139
see also microbial biomass
$N_2O:N_2$ ratio 53
nitrate (NO_3^-–N) 51, 62, 88–91, 106, 113, 315
nitric oxide (NO) 52–53, 88–91, 96–97
nitrite (NO_2^-–N) 52–53, 61–62, 88–91
nitrous acid 86
nitrous oxide (N_2O) 51, 52–53, 88–91
organic 1, 110, 114, 141
recycling 282
see also root
uptake 282–283, 289, 309
see also root
nitrogen transformations 51–155
concept (C/N model) 110–116, 141
denitrification *see* denitrification
immobilization *see* immobilization
mineralization *see* mineralization
nitrification *see* nitrification
Nitrosomonas 52
see also nitrification
nitrous acid 86 *see* nitrogen
nitrous oxide (N_2O) 51, 52–53, 88–91
see also nitrogen
NO 52–53, 88–91, 96–97
see also nitrogen
NO_2^-–N 52–53, 61–62, 88–91
see also nitrogen
NO_3^-–N 51, 62, 88–91, 106, 113, 315
see also nitrogen
non-isothermal
water flow 241–242
see also water flow
water vapour flow 242–243
see also water flow
non-rectangular hyperbola 250, 256, 260
see also photosynthesis
numerical
artefact 92
solution 21–23, 40, 169, 257–258
see also integration
stability 84–85
nutrients
macro 248
micro 248
phosphorus 248
potassium 248

one-compartment mixing process 41–46
see also process
organic carbon 114
see also carbon (C)
organic inputs (soil) 104, 116
organic matter
C:N ratios 105, 111, 124–125, 128
decomposable material 106, 121
degradation characteristics 104–105
fractions 104
lignin 106, 117
nitrogen 1, 110, 114, 141
see also nitrogen, organic
recalcitrant non-protected 106, 117
old or inert 106, 117
protected 106, 117, 142
stabilization 106, 120–121
thermal properties 161
turnover 250
organic metabolites 248
see also metabolites

organic nitrogen 1, 110, 114, 141
see also nitrogen
osmotically active substrate 296
see also plant water
osmotic potential 181, 296
see also water potential
outputs 7
see also variables
overburden potential 181
see also water potential
oxidation of ammonium 61
see also nitrification
oxidation of nitrite 61
see also nitrification

P components 138
see also microorganisms
partial differential equation 158
see also equations
particle surface 61
partitioning
empirical function 280
function 279–281
mechanistic function 280
root 281
shoot 281
teleonomic 280–281, 289
pasture 254, 266, 301
model 250
Penmann–Monteith equation 301–305, 309
see also evapotranspiration
Pentium computer 247
pH 53, 62, 64
see also soil
pH function 64–65, 68
see also functions
phosphorus 248
see also nutrients
photochemical efficiency 253, 258
see also photosynthesis (efficiency)
photons 205
see also radiation
photoperiod 257
see also light relationships
photosynthesis
activity 248
advanced model 258–264
analytical solution 256, 260
see also integration
biochemical equation 251
canopy *see* canopy
carboxylation resistance 253, 258
cumulative 257
daily rate 257
dark reaction 251
diffusion resistance (CO_2) 259
see also carbon dioxide
efficiency 253, 258
gross 257, 259, 269
instantaneous 257, 259, 286
leaf 259
light reaction 251
maximum rate 252–253, 256, 258–259, 310
model 257, 265
see also models
numerical solution 257
see also integration
non-rectangular hyperbola 250, 256, 260
process 1, 249
rectangular hyperbola 250–259, 267
units 253–254
water status function 310
photosynthetic active radiation (PAR) 300, 304
Planck's law 208
see also radiation
plant debris *see* plant residue
plant dry matter 277, 279
see also plant growth
plant growth
carbon 266
conceptual model 249–250
see also conceptual model
dry matter 277, 279
efficiency factor 269–270, 281
growth coefficient 269
leaf 248–250, 268, 270–271
see also leaf, growth
net 269
nitrogen 266
partitioning 279–281, 289
see also partitioning
recycling of substrate 282
root 275–277, 281–282, 286
see also root, growth
shoot 269, 281, 286
see also shoot, growth
stem *see* shoot, growth
storage weight 278
substrate relationships 266–291
temperature relationships 292–294
plant residue 250, 268, 282
plant storage weight 278
see also plant growth
plant–substrate relationships 266–291
see also plant growth
plant water
adjustment functions 310
cell wall rigidity 297, 299
chemical activities 310
osmotically active substrate 296
plant conductivity 309
potential 296–300, 313
relative water content 298–301, 310
see also shoot *or* root
resistance to water flow 295, 307
root 297
shoot 297
stomata 300–301
see also stomata, control
stress 298, 310
uptake 307
pollution 315
porosity
air-filled 72, 161, 233
minimum air-filled 77
total 72, 80,161, 187–188, 204
water-filled 80
potassium 248
see also nutrients
precipitation *see* rainfall
pressure potential 296–297
see also water potential

process
adsorption 60
one-compartment mixing 41–46
two-compartment mixing 47–48
programming languages 37, 40, 333
proportionality 5
see also equations
protection capacity 105, 134–136, 146
see also microbial biomass
psychrometer constant 303
see also humidity

quartz 161, 164

radiation
balance 205–211,
diffuse 209
direct 209
electromagnetic waves 205
emissivity constant 205, 211, 213
long-wave (down, Ld) 208–209, 211, 213, 232
long-wave (up, Lu) 209, 213, 223, 225, 232
net (Rn) 209–210, 227–228, 232, 302, 304
photons 205
photosynthetic active radiation (PAR) 300, 304
Planck's law 208
short-wave (St) 208–209, 211, 213, 227–229
soil emissivity 205, 211, 213
solar 203
Stefan–Boltzmann law 205, 211
wavelength 205, 208
Wien's law 205
rainfall 193, 197–198
cumulative 198, 200
diurnal change 195
rate 200, 211, 227–228, 245, 306, 328
steady 195
rate constant 31, 55–56
see also kinetics
rectangular hyperbola 250–259, 267
see also photosynthesis
reflectivity coefficient 209
see also albedo
relative humidity 191, 221, 232–234, 243
see also humidity
relative water content 298–301, 310
see also root *and* shoot
research question 39
see also conceptual model
resistance
boundary layer resistance (RAC) 216, 218–219, 225
see also boundary layer
carboxylation 253, 258
see also photosynthesis
diffusion (CO_2) 259
see also carbon dioxide
water flow in plants 295, 307
see also plant water
respiration
aerobic 248
growth (root) 276, 282
growth (shoot) 268–270, 282
maintenance (root) 277, 286, 293
maintenance (shoot) 268, 271–272, 286, 293
nitrogen uptake 285
see also root
Richard's equation 188
see also equations *and* water flow
root 248–250
activity 283
carbon 279
density 308
degradation 285, 287
dry matter fraction 278
growth 275–277, 281–282, 286
multilayered root system 313, 331
nitrogen recycling 282
nitrogen uptake 282–283, 285, 289, 309
osmotic potential 296
see also water potential
partitioning 281
see also partitioning
pressure potential 297
see also water potential
relative water content 298–301
solute uptake 331
structure 276–277, 282, 291
water content 297, 309
water status function 310
water potential (total) 298–299
water uptake 307
see also plant water
roots ($\sqrt{8}$) 7
see also mathematics
roughness elements 214
see also boundary layer
roughness length 214, 217, 305
see also boundary layer
<run options> 91, 176, 204–205, 275
see also ModelMaker
runoff 195, 197
cumulative 198, 200
see also water flow

sand *see* quartz
saturation moisture content 195
see also soil moisture
saturation vapour pressure (SVP)
change of 232, 234, 243, 302
empirical function 212, 221, 225, 233
see also functions
soil 234
second-order kinetics 30–31, 53–55
see also kinetics
second-stage drying 191, 193
see also water flow
sensible heat (C) 210, 213, 215, 217, 219, 225, 232
see also soil energy balance
<sensitivity analysis> 68–69, 77, 152–153
see also ModelMaker
separation of variable 19
see also variables
shoot 248–250
carbon 279
degradation 285, 287
dry matter fraction 278
growth 269, 281, 286
osmotic potential 297
see also water potential
partitioning 281
see also partitioning

shoot *continued*
pressure potential 297
see also water potential
relative water content 298–301, 310
structure 269, 273, 282, 291
water content 297, 299, 309
water potential (total) 298–299
water status function 310
see also function
short wave radiation (St) 208–209, 211, 213, 227–229
see also radiation
simulation results
actual evaporation 246
analytical solution 46
C:N ratio (microbial biomass) 154
cumulative evaporation 246
decomposable material 137
denitrification 103
evaporation 199
first-order kinetics 55, 57, 137
ground heat flux (G) 232
hydraulic conductivity 185, 187
immobilization 154
leaching 325, 330
leaf area index 266–267, 290, 294, 313
long-wave radiation (Lu and Ld) 232
matric potential 185
maximum photosynthesis (Pm) 291
Michaelis–Menten kinetics 57
microbial biomass carbon 137
net radiation (Rn) 232
nitrification 84–85
nitrous oxide (N_2O nitrification) emission 78
numerical solution 46
organic input 154
pH function 65
photosynthesis non-rectangular hyperbola 266–267
photosynthesis rectangular hyperbola 266–267
plant water status 313
potential evaporation 246
protection capacity 137
rainfall 199
root structural live weight 291
root water potential 313
second-order kinetics 55, 57
sensible heat flux (C) 232
shoot structural live weight 291
shoot water potential 313
short-wave radiation (St) 232
soil drying 199
soil energy balance 232
soil surface temperature 178
soil temperature 173, 178
soil water potential 313
soil wetting 199
solute concentration 330
temperature adjustment 294
temperature function 83
thermal conductivity 165, 246
two-compartment mixing process 48
volumetric heat capacity 165
volumetric water content 246
water-filled porosity function 81
water flow 330
water vapour flow 238
zero-order kinetics 55, 57
simulation scenarios
boundary conditions 197
C/N model 150, 153
denitrification 103
evaporation 202
leaching 325, 329
maximum photosynthesis 290–291
nitrification 84
non-isothermal conditions 244
photosynthesis 266
plant–water relationships 311
rainfall 202
soil temperature 163, 170
soil water 197
temperature adjustment 293–294
water vapour 237, 266, 289
simultaneous equations 223–224
see also equations
slope of a function 10–11
see also function
software, modelling packages 2
soil
bulk density 80, 161, 284
layered system 25–26, 158, 321
pH 53, 62, 64
profile 25–27, 172
solid density 80, 161
surface 156
temperature *see* soil temperature
texture 187
thermal properties 161, 179, 203
soil air, thermal property 161
soil atmosphere exchange 213–219
see also boundary layer
soil atmosphere interface 249
see also soil energy balance
soil emissivity 205, 211, 213
see also radiation
soil energy balance
bare soil 205–211
cropped soil 301
ground heat flux (G) 210, 222, 226, 232, 301
latent heat flux (LE) 208, 210, 213, 215–217, 219, 225, 229, 232, 302
latent heat of vaporization (L) 210, 220, 226, 232, 234–235, 302–303
overview 203
process 167
sensible heat flux (C) 210, 213, 215, 217, 219, 225, 232
soil–atmosphere interface 249
soil moisture
C/N model 151, 153, 179
see also microbial biomass
characteristics 183–184
denitrification 100–102
see also denitrification
field capacity 306
function 183, 185
see also functions
gravimetric water content 179
nitrification 84–85
see also nitrification
saturation 195

status 305–307, 309
tortuosity factor 317
volumetric water content (VWC) 72, 172, 177, 179, 183–184, 187–188, 195, 197, 235, 246, 305, 317, 320–322
water filled porosity 80
water vapour 162, 203, 229
soil–plant–atmosphere
continuum (SPAC) 249–250, 294–296, 298, 309, 311
relationship 247–250
soil porosity *see* porosity
soil surface evaporation 191
see also evaporation
soil surface temperature 167, 177, 203, 213, 219–229, 233
see also soil temperature
soil temperature
calculation procedure 223–229
see also iteration
C/N model 151
see also microbial biomass
denitrification 102
see also denitrification
diurnal changing 167–172
function 72, 82–85, 102, 292
see also functions
nitrification 84–85
see also nitrification
surface 167, 177, 203, 213, 219–229, 233
soil vapour concentration 234
see also humidity
soil water *see* soil moisture
soil water potential *see* water potential
soil water suction *see* water potential
solar radiation 203
see also radiation
solute 183, 316–317, 323–324, 328
transport 316–322
see also transport processes
uptake 331
see also root
solution of differential equation 18
see also differentiation
SPAC 249–250, 294–296, 298, 309, 311
see soil–plant–atmosphere, continuum
specific heat of air 217
see also air properties
specific heat 161
see also heat flow
stability parameter 217, 225, 304
see also boundary layer
stable stability state 215–216, 218
see also boundary layer
steady-state heat flow 156, 166
see also heat flow
Stefan–Boltzmann law 205, 211
see also radiation
stomata
carbon dioxide relationship 301
conductance of water vapour 300
control 300–301
function 300
structure of roots 276–277, 282, 291
see also roots
suction 183–184
see also water potential

<table> 49
see also ModelMaker
Taylor polynomials 23, 334–335
see also functions
temperature
air (TA) 211, 217, 219, 227–228
dewpoint (DP) 211–212, 227–228
influence on denitrification 102
see also denitrification
influence on microbial biomass 151
see also microbial biomass
influence on nitrification 84–85
see also nitrification
soil *see* soil temperature
thermal conductivity of air 233
see also air properties
thermal conductivity 156,158, 161–167, 174–176, 203, 223, 229, 232–233, 237–238
see also heat flow
thermal diffusivity 159, 163–168
see also heat flow
thermal properties 161, 179, 203
see also soil
thermally induced liquid flow 241–243
see also water flow, non-isothermal
thermodynamics 180
see also soil water potential
time conversion 66–68, 77
tortuosity factor 317
see also soil moisture
translocation *see* leaching
transpiration 249, 294–296, 301, 309
transport equations 23–25
see also equations
transport processes
convection 316–317
diffusion 316–317
dispersion 316, 318
hydrodynamic dispersion 316, 318
mathematical notation 316–322
<trigger> 227–228
see also ModelMaker
turbulent boundary layer 214
see also boundary layer
two-compartment mixing process 47–48
see also process

U components 138
see also microorganisms
units
conversion 6–7
evapotranspiration 302
gas flow 73
heat flow 166
humidity 212, 222
hydraulic conductivity 186
kinetics 36–37
latent heat (LE) 210, 220, 302
photosynthesis 253–254
sensible heat (C) 219
thermal properties 159–161
transpiration 302
water potential 180, 182
water vapour flow 239, 243–244
universal language 2–3
see also mathematics
unstable stability state 215–216, 218

unstable stability state *continued*
see also boundary layer
uptake
of nitrogen 282–283, 285, 289, 309
see also root, nitrogen uptake
of water 307
see also plant water

vapour pressure deficit (VPD) 302–303
see also humidity
variables
dependent 6–7
dummy 226–227
see also iteration
independent 6–7
input 7
ModelMaker notation 38
see also ModelMaker
output 7
separation of 19
vegetative plant growth *see* plant growth
vertical forces 214
see also boundary layer
volumetric heat capacity 156, 160, 161–167, 172–175, 203
see also heat flow
volumetric specific heat of air 217, 219
see also air properties
volumetric water content (VWC) 72, 172, 177, 179, 183–184, 187–188, 195, 197, 235, 246, 305, 317, 320–322
see also soil moisture

washing out effect 331
see also leaching
water content
plant 297–301, 310
see also plant water
root 297, 309
see also root
shoot 297, 299, 309
see also shoot
water filled porosity 80
see also porosity
water flow
average velocity 317–318, 320, 322
basic model 188–194, 307
boundary conditions 190–191, 194–198, 245
see also boundary conditions
Darcy's law 184, 188
drainage 191, 196–197, 245
see also drainage
first-stage drying 191, 193
heat relationship 167, 179
hydraulic gradients 198
impermeable layer 191, 197
net flow 193
non-isothermal (water) 241–242
non-isothermal (water vapour) 242–243
one-compartment model 305–307
see also models
Richard's equation 188
root to shoot 308–309
runoff 195, 197, 200
see also runoff
second-stage drying 191, 193
shoot to atmosphere 294, 301–305, 309
see also evapotranspiration
soil to root 307
water table 191, 196–197
water vapour 229–247
water vapour enhancement factor 235
zero-flux plane 189
water potential
air-entry 184
concept 180
conversion 182
definition 180–183
energy 183
function 80, 100–102, 179, 183–185
see also functions
gravitational 182
hydraulic 182, 186, 188–189, 192
matric 181, 183–185, 188, 221–222, 234, 296, 306
osmotic 181, 296–297
overburden 181
pressure 296–297
see also root *and* shoot
suction 183–184
thermodynamics 180
total plant 298–299
see also root *and* shoot
units 182
water table 191, 196–197
see also water flow
water vapour 162, 203, 229
see also soil moisture
concentration 191, 221, 232–234, 243
see also humidity
diffusivity 232–233, 239, 243
see also diffusivity
enhancement factor 235
see also water flow
flow 229–247
see also water flow
flux (boundary layer) 214
see also boundary layer
water-filled porosity
see also porosity
function 80, 100–101, 179
see also function *and* porosity
weather data 75, 203
see also data, climate
wetting–drying 155, 184
<while> loop 226
see also ModelMaker
Wien's law 205
see also radiation
wind
profile 214–215
speed (U) 211, 214, 227–228

zero flux plane 189
see also water flow
zero-order kinetics 27–28, 53–55, 63
see also kinetics
zero plane displacement 214, 217, 305
see also boundary layer